MATHEMATICS OF MODALITY

MATHEMATICS
OF
MODALITY

Robert Goldblatt

CSLI Publications
Center for the Study of Language and Information
Stanford, California

CSLI was founded early in 1983 by researchers from Stanford University, SRI International, and Xerox PARC to further research and development of integrated theories of language, information, and computation. CSLI headquarters and the publication offices are located at the Stanford site.

CSLI/SRI International   **CSLI/Stanford**   **CSLI/Xerox PARC**
333 Ravenswood Avenue   Ventura Hall   3333 Coyote Hill Road
Menlo Park, CA 94025   Stanford, CA 94305   Palo Alto, CA 94304

Library of Congress Cataloging-in-Publication Data
Goldblatt, Robert
    Mathematics of modality / Robert Goldblatt.
      p. cm. -- (CSLI lecture notes ; no. 43)
    Includes bibliography and index.
    ISBN 1-881526-24-0 (cloth) — ISBN 1-881526-23-2 (paper)
    1. Modality (Logic). I. Title. II. Series.
QA9.46.G66 1993
511.3--dc20   93-13522
                        CIP

"Metamathematics of Modal Logic" originally appeared in *Reports on Mathematical Logic*, vol. 6, 41–78 (Part I), and vol. 7, 21–52 (Part II). Copyright ©1976 by the Jagiellonian University of Cracow. Reprinted by permission.

"Semantic Analysis of Orthologic" originally appeared in the *Journal of Philosophical Logic*, vol. 3, 19–35. Copyright ©1974 by D. Reidel Publishing Company, Dordrecht-Holland. All Rights Reserved. Reprinted by permission of Kluwer Academic Publishers.

"Orthomodularity is Not Elementary" originally appeared in *The Journal of Symbolic Logic*, vol. 49, 401–404. Copyright ©1984 by The Association for Symbolic Logic. All Rights Reserved. This reproduction by special permission.

"Arithmetical Necessity, Provability and Intuitionistic Logic" originally appeared in *Theoria*, vol. 44, 38–46, 1978. Reprinted by permission.

"Diodorean Modality in Minkowski Spacetime" originally appeared in *Studia Logica*, vol. 39, 219–236. Copyright ©1980 by the Polish Academy of Sciences. Reprinted by permission.

"Grothendieck Topology as Geometric Modality" originally appeared in *Zeitschrift für Mathematische Logik und Grundlagen der Mathematik*, vol. 27, 495–529. Copyright ©1981 VEB Deutscher Verlag der Wissenschaften Berlin. Reprinted by permission.

"The Semantics of Hoare's Iteration Rule" originally appeared in *Studia Logica*, vol. 41, 141–158. Copyright ©1982 by the Polish Academy of Sciences. Reprinted by permission.

"An Abstract Setting for Henkin Proofs" originally appeared in *Topoi*, vol. 3, 37–41. Copyright ©1984 by D. Reidel Publishing Company, Dordrecht-Holland. Reprinted by permission of Kluwer Academic Publishers.

"The McKinsey Axiom is Not Canonical" originally appeared in *The Journal of Symbolic Logic*, vol. 56, 554–562. Copyright ©1991 by The Association for Symbolic Logic. All Rights Reserved. This reproduction by special permission.

Contents

Introduction

Modal logic is the study of *modalities*—logical operations that qualify assertions about the truth of statements. For example, we may say that a particular statement is *necessarily* true, or *possibly* true, *ought to be* true, *is known to be* true, *is believed to be* true, *has always been* true, *will eventually be* true, *is demonstrably* true, and so on.

The study of modalities is an ancient one, dating at least from Aristotle, but its most substantial progress has occurred in the last three decades, since the introduction by Saul Kripke [52] of the use of relational structures to provide a formal semantic analysis of languages containing modalities. The rich diversity of form supplied by relational structures has resulted in the method having a significant impact on a wide range of disciplines, including the philosophy of language ("possible worlds" semantics), constructive mathematics (intuitionistic logic), theoretical computer science (dynamic logic, temporal and other logics for concurrency), and category theory (sheaf semantics). It has led to the study of more "mathematically" motivated modalities, such as assertions that a statement is *provable in Peano arithmetic*, or is true *locally*, *at the next state*, *along some branch of a tree*, or *after the computation terminates*.

This volume collects together a number of my papers on modal logic, concerned with the general nature and capacity of Kripke semantics, its relationship with the use of algebraic models and methods, and its application to various mathematical modalities. The collection begins with my doctoral thesis, and includes two completely new papers, one on infinitary rules of inference (Chapter 9), and the other (Chapter 11) about recent results on the relationship between modal logic and first-order logic. Another paper (Chapter 8) on the "Henkin method" in completeness proofs has been substantially extended to include discussion of the Barcan formula in quantificational modal logic, and infinitary

1

propositional logic. Other articles are concerned with quantum logic, provability logic, the temporal logic of relativistic spacetime, modalities in topos theory, and the logic of programs.

The papers have been reproduced as originally published, with corrections, and in the original style, modified only by LaTeX's automatic conventions regarding layout and numbering of chapters, sections, theorems etc. A small amount of editing for uniformity has been undertaken, but the notation and terminology is by no means systematic throughout. The reader should bear in mind that the different chapters were produced at various times over a period of almost twenty years, and are written to be read independently.

Here now is an abstract of each of the chapters, with notes on any changes that have been made for this edition.

1. Metamathematics of Modal Logic

The first 18 sections comprise the content of my doctoral thesis, with the others being added for the published version [36, 37]. The work is concerned to develop the general structure theory of set-theoretic models (Kripke frames), analysing validity preserving operations—homomorphisms, substructures, disjoint unions, ultraproducts etc.—and determining their relationship with algebraic models (modal algebras). This theory is then applied to a range of questions about definability of classes of models, the use of "canonical" models, and the correspondence with first-order logic. Problems considered include: characterisations of classes of frames that are modal axiomatic, i.e. the class of models of a set of modal formulae; syntactic criteria for a logic to be determined by its canonical model; first-order definability of modal formulae; conditions under which a first-order definable class of frames is modal axiomatic.

I have added a paragraph at the beginning of Section 1.10, pointing out the priority of the work of Jónsson and Tarski [48] on representation of Boolean algebras with operators. This provided, a decade before Kripke's work, all that is needed to prove the completeness with respect to set-theoretic semantics of several of the more well-known modal systems.

Apart from correction of misprints, the one significant change concerns the description of the ultraproduct of KM-frames in Section 1.17 (cf. [37, p. 39]).

2. Semantic Analysis of Orthologic
3. Orthomodularity is Not Elementary

These two articles deal with quantum logic—the propositional logic of orthomodular lattices. The first develops a Kripke-style semantics for the logic of ortholattices, using *orthogonality* (irreflexive symmetric) relations, establishing completeness, decidability via the finite model property, and the existence of a translation into the Brouwerian modal system. A limited extension of the modelling is then given for the orthomodular law.

The second article shows that this programme cannot lead to a tractable modelling of orthomodularity: there is no elementary condition on orthogonality relations that characterises the orthomodular law. This is demonstrated by proving that a pre-Hilbert space is an elementary substructure, with respect to orthogonality, of its Hilbert space completion. The paper concludes with a list of open problems about orthomodular logic, including most of the important questions one would ask of a logical system. As far as I know, these are still unresolved.

4. Arithmetical Necessity, Provability and Intuitionistic Logic

The *provability interpretation* reads the modality $\Box$ as "it is provable in Peano arithmetic that". Here we modify this to "true and provable", which is not the same thing in view of Gödels Incompleteness Theorem on the existence of true but unprovable statements.

Building on the fundamental work of Solovay [91], it is shown that the modal logic corresponding to this interpretation is the system S4Grz determined by finite partially-ordered Kripke models. Then by means of the translation of intuitionistic logic into S4, an interpretation of non-modal propositional logic into formal arithmetic is obtained in which precisely the intuitionistic theorems turn out to be *arithmetically necessary* in the sense of being true in all models of Peano arithmetic.

5. Diodorean Modality in Minkowski Spacetime

Temporal logic studies such modalities as *it will eventually be, it has always been, it will be at the next moment*, etc. Most research has concerned linear time, focusing on the identification of the logics that result when the temporal ordering is regarded alternatively as being discrete, dense, or continuous.

Here the context is the non-linear ordering of four dimensional Minkowski spacetime $\mathbb{T}^4$. It is shown that under the Diodorean reading of "necessarily" as "now and forever", the resulting logic is the system S4.2, and that the same applies to spacetime $\mathbb{T}^n$ of any dimension $n \geq 2$. This is achieved by constructing an elaborate sequence of validity-preserving transformations leading from $\mathbb{T}^n$ to any finite S4.2-model.

Some discussion is given of other temporal orderings of spacetime, including the possibility of distinguishing different dimensions by the truth of certain formulae when the ordering is irreflexive.

The full temporal logic of $\mathbb{T}^4$, with past and future operators, has still not been investigated, and there remain some challenging open questions, as indicated at the end of the article.

6. Grothendieck Topology as Geometric Modality

In the axiomatic approach to sheaf theory due to Lawvere and Tierney, a Grothendieck topology on a category becomes a unary operator on the "object of truth values" of a topos, hence a suitable entity for interpreting a modality, which Lawvere suggested should be read "it is locally the case that".

Here the propositional modal logic defined by this interpretation is axiomatised and proven to be decidable. This is done by developing a modelling that combines the Kripke semantics for intuitionistic logic with that for modal logic, and then using it to construct a suitable characteristic topology on a topos.

The article includes an intuitive discussion of local truth as meaning "truth at all nearby points" or "truth throughout some neighbourhood", and on this basis formulates a variety of *relational* and *neighbourhood* models, as well as considering related algebraic models (operators on Heyting algebras).

Some associated logics are also discussed, including one arising from the interpretation of double negation as meaning "it is cofinally the case that".

7. The Semantics of Hoare's Iteration Rule

The modal logic of computer programs associates with each command α a modality $[\alpha]$ that is read "after α terminates, it will be the case that". This article examines the resulting logic for commands of the form (**while** e **do** α), and focuses on the Iteration Rule due to to Hoare for reasoning about the correctness of such programs. The exact semantic content of Hoare's Rule is determined, and a completeness theorem given (via the finite model property) which shows that additional principles are needed to axiomatise the logic of **while**-commands.

In an earlier monograph [23] I developed an axiomatisation of the program logic over a general first-order language, using an infinitary analogue of Hoare's Rule. The first stage of this was a completeness theorem for a propositional logic, using the same infinitary rule. However, whereas this rule is unavoidable in general in the presence of quantification, at the propositional level the set of valid formulae is decidable and can be given a finitary axiomatisation [23, p. 79]. This claim is verified in detail here.

8. An Abstract Setting for Henkin Proofs

There are many applications in model theory of the procedure of inductively constructing a maximally consistent theory satisfying certain prescribed closure conditions. An attempt is made to isolate the essence of this methodology in terms of a principle, stated in the language of abstract deducibility relations and inference rules, which specifies conditions under which a consistent set of sentences can be consistently enlarged to one that "decides" a given set of inferences.

In the original version of this article [24], the Abstract Henkin Principle was applied to give streamlined proof of completeness and omitting-types theorems for first-order logic, and for quantificational logic with infinitary conjunctions. For this edition further demonstrations of its utility are given, in the form of a discussion of completeness for the Barcan formula in quantificational modal logic, and for propositional modal logics with infinitary inference rules.

9. A Framework for Infinitary Modal Logic

There are natural modal logics that are complete (every consistent sentence is satisfiable) but not *strongly* complete because they have consistent *sets* of sentences that are not satisfiable. The problem arises of extending such a logic to a strongly complete one by the addition of infinitary inference rules. Similarly, we may ask for an axiomatisation of the smallest logic that contains a given one and is closed under some specified infinitary rules. Here a solution is provided in a general context, and is given in terms of the proof theory and model theory of an n-ary modality $\Box(A_1, \ldots, A_n)$, rather than just a unary connective.

For certain infinitary systems studied in Section 8.7, it was noted as a consequence of the completeness theorem that maximally *finitely* consistent sets having particular closure properties turn out to be fully consistent and deductively closed. The theory of Chapter 9 gives an account of this phenomenon that is purely proof-theoretic and prior to any model-theoretic analysis.

10. The McKinsey Axiom is Not Canonical

The McKinsey axiom $\Box\Diamond A \rightarrow \Diamond\Box A$ was shown in Section 1.17 not to be determined by any elementary (i.e. first-order definable) class of Kripke frames. Here it is shown to have a model on its canonical frame that falsifies it. This technical result, which was a long-standing open problem, has conceptual significance for the relationship between modal and first-order logic, as embodied in the question as to whether modal logics validated by their canonical frames are precisely those determined by an elementary class.

The McKinsey axiom is the simplest formula not belonging to a very general syntactically defined class, devised by Sahlqvist [79], whose members are known to be elementary and canonical. Thus the import of this article is that there is no natural way to extend Sahlqvist's scheme to obtain a larger class of canonical formulae.

In the original paper [30], the result that an elementary logic is canonical was attributed to van Benthem [102], whereas it was in fact first proven by Fine [15, Theorem 3], with van Benthem's contribution being to extend the result to show that such a logic is preserved by "ultra-filter extensions" of frames, which are the "completions" of Definition 1.20.2(ii).

11. Elementary Logics are Canonical and Pseudo-Equational

If a logic Λ is determined by some elementary class of Kripke frames, then it is valid in its canonical frame $\mathcal{F}^\Lambda$ [15, Theorem 3]. The conclusion of this result is strengthened here in several ways. First, it is shown that Λ is valid in any member of the class $\mathcal{K}_\Lambda$ of all models of the first-order theory of $\mathcal{F}^\Lambda$, i.e. in any frame elementarily equivalent to $\mathcal{F}^\Lambda$. Then it is shown that Λ is valid in any member of the class $Mod\,\Psi_\Lambda$ of models of the *pseudo-equational* theory of $\mathcal{F}^\Lambda$, a pseudo-equational sentence being one of the form $\forall x \phi$ with ϕ constructed from atomic formulae using only positive connectives and bounded quantifiers. On the way it is shown that if $\mathcal{K}$ is any elementary class determining Λ, then Λ is valid in all members of the class $Mod\,\Psi_\mathcal{K}$ of models of the pseudo-equational theory of $\mathcal{K}$, and indeed that

$$\mathcal{F}^\Lambda \in \mathcal{K}_\Lambda \subseteq Mod\,\Psi_\Lambda \subseteq Mod\,\Psi_\mathcal{K} \subseteq \{\mathcal{F} : \mathcal{F} \models \Lambda\},$$

with none of these set inclusions being an equality in general.

It remains unresolved at the point of writing as to whether every canonical logic is determined by some elementary class of frames, and so the final section explores a number of equivalent formulations of elementarity in terms of stability properties of classes of general frames with natural definability properties. The unresolved question is shown to reduce to the problem of proving Λ is valid in all ultrapowers of $\mathcal{F}^\Lambda$.

Acknowledgements

My work has benefited from many interactions over the years with a number of modal logicians who have contributed remarkably to the subject: Johan van Benthem, Robert Bull, Max Cresswell, Kit Fine, George Hughes, Krister Segerberg, and Steve Thomason. I am grateful also to

Wilf Malcolm for the inspiration of his teaching and for providing my introduction to the world of models and ultraproducts.

The new material was written during tenure of a Visiting Fellowship at the Centre for Information Science Research of the Australian National University, supported also by a sabbatical grant from Victoria University. I thank Professor Michael McRobbie for the conducive facilities that were made available to me at the Centre.

I am indebted to Jason Christopher for carrying out the initial reformatting of the previously published papers. Finally, I want to record my gratitude to Dikran Karagueuzian, friend and "faithful editor", for making the project possible, and for furnishing a marvelous typesetting and publishing environment within which to carry it out.

Rob.Goldblatt@vuw.ac.nz
Waitangi Day, 1993

1

Metamathematics of Modal Logic

Contents

Introduction and Summary

"The formal study of symbols of systems either in their relation to one another (syntax) or in their relation to assigned meanings (semantics) is called metamathematics...". (R. Feys and F. B. Fitch, *Dictionary of Symbols of Mathematical Logic*, North-Holland, 1969.)

This study is concerned with the mathematical objects that provide interpretations, or models, of formal propositional languages. Historically there have been two kinds of approach in this area. The first of these, *algebraic semantics,* employs algebras, typically lattices with operators, as models. Propositional variables range over elements of the lattice, and formal connectives correspond to its operators. In other words each formula induces a polynomial function on any of its algebraic models. Truth and validity of formulae are then defined in terms of designated polynomial values.

The other approach is *set-theoretic semantics.* Here the models, known as *frames,* carry structural features other than finitary operations, such as neighborhood systems and finitary relations. Formulae are then interpreted as subsets of the frame in a manner constrained by its particular structure.

These two kinds of model are closely related. Algebras may be constructed as subset lattices of frames. Frames may be obtained from algebras through various lattice representations. Furthermore the syntactical frame constructions in the Henkin style that are now widely employed in set-theoretic semantics may be mirrored on the algebraic level to recover the lattice representations.

The guiding theme of the present work is the relationship between frame and algebra, and the relative strengths and limitations of the semantical frameworks that these notions determine. The vehicle chosen for this study is normal modal logic. It should however be stressed from the outset that many of the concepts and results developed may be parallelled in other areas, or even stated for an abstract formal language. On the other hand modal logic provides a natural context for the discussion given. It is the most widely investigated and best understood branch of non-classical propositional logic. Indeed it was here that set-theoretic semantics began with the work of Saul Kripke [51, 52]. The significance of Kripke's method was quickly recognised, particularly since he showed that different logics could be characterised by imposing simple conditions on models. The 1960's saw these ideas being rapidly applied to tense, deontic, epistemic, and intuitionist logics, and to others besides. Currently they are proving relevant to such diverse areas as the foundations of physics (quantum logic) and the study of natural languages.

In 1966 E. J. Lemmon [59] conjectured that the method was completely general in its application, and that all modal logics possessed a characteristic class of Kripke models. Recently this has been shown to be false by Fine [13] and Thomason [97] who independently constructed logics for which no such class exists. These results gave the first real evidence that the Kripke modelling had its limitations, and have given impetus to a new line of research that is concerned rather more with the general nature of the discipline than with its particular applications.

It now seems that, in as much as formulae are interpreted as subsets of models, modal logic is fundamentally of the same species as second order quantificational logic. The Kripke models are then analogous to the *principal* models of second order logic, in which all possible sets and relations are present. As is well known, the exclusive use of principal models results in non-compact and incomplete logics. It is only by the introduction of *secondary models*, with restricted interpretations, that a uniform completeness theorem can be obtained.

The first part of the present study (Sections 1–11) develops the general structure theory of "first-order" frames. These are the secondary models for modal logic. The initial emphasis is on validity preserving constructions — subframes, homomorphisms, disjoint unions, ultraproducts — each of which corresponds to a polynomial-identity preserving construction on modal algebras. This is followed by the examination of a new kind of model — the *descriptive* frame. These structures are designed to provide an exact set-theoretic analogue to the modal algebra, and indeed they give rise to a category that is dual to the category of modal algebras. Because of this relationship, descriptive frames have proven invaluable in solving a wide range of problems. The later form the major pre-occupation of Sections 12-20. The problems considered there include: characterisations of classes of frames that are modal axiomatic, i.e. the class of models of a set of modal formulae; syntactic criteria for a logic to be determined by its principal models; elementary definability of modal formulae; conditions under which an elementary class of frames is modal axiomatic.

Sections 1–18 comprise the content of my doctoral thesis, written at the Victoria University of Wellington in late 1973 (Section 16 was rewritten in the light of subsequent developments). I would like to express my gratitude to my supervisors, Professors M. J. Cresswell, G. E. Hughes, and C. J. Seelye. To the various participants in the logic seminars at

VUW, and in particular Professor W. G. Malcolm, I am indebted for many opportunities to discuss problems and results.

The development and presentation of much of what appears below has benefited from some stimulating discussions and correspondence with Professor S. K. Thomason.

1.1   Syntax

Modal logic is designed to formalise philosophical discourse about the nature of necessity, possibility, and strict implication. A typical object language for such an inquiry (and one that will remain fixed throughout this chapter) has the following primitive symbols:

(i) a denumerable collection of *propositional variables* (p, q, p_1, q_1, etc.),

(ii) the Boolean connectives $\neg$ (negation), and $\wedge$ (conjunction),

(iii) the modal connective $\square$ (necessity),

(iv) brackets (and) .

The class Φ of all (*well-formed*) *formulae* (*wffs*) of this language is defined by the three formation rules:

(1) each variable is a wff,

(2) if α is a wff, so are $\neg\alpha$ and $\square\alpha$,

(3) if α and β are wffs, so is $(\alpha \wedge \beta)$.

The Boolean connectives $\vee$ (disjunction), $\rightarrow$ (material implication), $\leftrightarrow$ (material equivalence) and the modal $\Diamond$ (possibility) are introduced as the abbreviations

$$
\begin{array}{lll}
\alpha \vee \beta & \text{for} & \neg(\neg\alpha \wedge \neg\beta) \\
\alpha \rightarrow \beta & \text{for} & \neg\alpha \vee \beta \\
\alpha \leftrightarrow \beta & \text{for} & (\alpha \rightarrow \beta) \wedge (\beta \rightarrow \alpha) \\
\Diamond a & \text{for} & \neg\square\neg\alpha
\end{array}
$$

(Brackets may be omitted where convenient, the convention for reading formulae being that $\neg$, $\square$ and $\Diamond$ bind more strongly than $\vee$ and $\wedge$, the latter binding more strongly than $\rightarrow$ and $\leftrightarrow$.)

The *syntactical* study of formulae is concerned with formal relationships between wffs, and focuses on the notion of *derivability*. In this context a useful distinction can be made between *axiom systems* and *logics*. An axiom system S has two basic components—a set of wffs, called *axioms*, and a set of *rules of inference* that govern operations allowing certain formulae to be derived from others. A wff α is said to be a *theorem* of S, written $\vdash_S \alpha$, if there exists in S a *proof* of α, i.e. a finite sequence of wffs whose last member is α, and such that each member of the sequence is either an axiom, or derivable from earlier members by one of the rules of inference of S.

A logic, on the other hand, can be thought of as a set Λ of wffs closed under the application of certain inferential rules to its members. The members of Λ are called Λ-theorems, and in this case the symbolism $\vdash_\Lambda \alpha$ indicates merely that $\alpha \in \Lambda$.

For example, if S is an axiom system, than an S-logic can be defined as any set of wffs that includes all the axioms of S and is closed under the rules of S. In general the intersection Λ_S of all S-logics will be an S-logic whose members are precisely those wffs for which there are proofs in S. This is often described by saying that S is an *axiomatisation* of Λ_S, or that Λ_S is *generated* by S.

Thus each axiom system has a corresponding logic (the set of its theorems) and in some formal treatments little or no distinction is made between the two. The converse however is not true. Not every logic is axiomatisable. In any semantical framework, the set of wffs true in a particular model will be a logic of some kind, for which, in some cases, there may be no effectively specifiable generating procedure. A classic example is the first-order theory of the standard model of arithmetic.

Definition 1.1.1 *A **modal logic** is a set $\Lambda \subseteq \Phi$ satisfying*

(i) *Λ contains all tautologies of the classical propositional calculus PC,*
(ii) *if $\alpha, (\alpha \rightarrow \beta) \in \Lambda$, then $\beta \in \Lambda$ (Modus Ponens),*
(iii) *if $\alpha \in \Lambda$ and β is obtained from α by uniformly replacing some variable by some other wff, then $\beta \in \Lambda$ (Uniform Substitution).*

The symbol K (for Kripke) denotes the logic axiomatised by the system that has a standard basis for PC (including Modus Ponens and Uniform Substitution as rules of inference) together with the axiom

$$\Box(p \rightarrow q) \rightarrow (\Box p \rightarrow \Box q)$$

and the rule of Necessitation:

$$\text{from } \alpha \text{ to infer } \Box\alpha.$$

A logic is *normal* iff it contains K and is closed under Necessitation. If Γ is a set of wffs, $K\Gamma$ denotes the normal logic generated by adding the members of Γ as extra axioms to the system that generates K.

Definition 1.1.2 *Let Λ be a modal logic, $\Gamma \subseteq \Phi$, and $\alpha \in \Phi$. Then α is Λ-**derivable** from Γ, $\Gamma \vdash_\Lambda \alpha$ iff there exist $\alpha_1, \ldots, \alpha_n \in \Gamma$ such that $(\alpha_1 \wedge \ldots \wedge \alpha_n \rightarrow \alpha) \in \Lambda$. α is an Λ-**theorem**, $\vdash_\Lambda \alpha$, iff $\alpha \in \Lambda$. Γ is Λ-**consistent** iff there is at least one wff not Λ-derivable from Γ, and Λ-**inconsistent** otherwise. Γ is Λ-**maximal** iff Γ is Λ-consistent and* *for each wff α, either $\alpha \in \Gamma$ or $\neg\alpha \in \Gamma$. We denote by W_Λ the class of Λ-maximal subsets of Φ. For $\alpha \in \Phi$, $|\alpha|_\Lambda = \{x \in W_\Lambda : \alpha \in x\}$ and for $\Gamma \subseteq \Phi$, $|\Gamma|_\Lambda = \bigcap\{|\alpha|_\Lambda : \alpha \in \Gamma\} = \{x \in W_\Lambda : \Gamma \subseteq x\}$.*

A proof of the following two (equivalent) results may be found in [59, Section 0].

Theorem 1.1.3

(1) *Every Λ-consistent set of wffs has an Λ-maximal extension.*

(2) *$\Gamma \vdash_\Lambda \alpha$ iff $|\Gamma|_\Lambda \subseteq |\alpha|_\Lambda$ (i.e. iff every Λ-maximal extension of Γ contain α).*

1.2 Modal Algebras and Kripke Frames

Having set up the basic syntactical machinery for modal logic, we turn to the *semantical* study of formulae. This branch of metamathematics is concerned with the assignment of meanings or interpretations to wffs, and the setting out of conditions under which a wff is to be regarded as being true or false. Modal algebraic semantics has its origins in the work of McKinsey and Tarski [67, 68] and has been further developed by Lemmon [57, 58].

Definition 1.2.1 *A **normal modal algebra** (MA) is a structure*

$$\mathfrak{A} = \langle A, \cap, ', l \rangle,$$

where

(i) *$\langle A, \cap, ' \rangle$ is a Boolean algebra (BA),   and*

(ii) *l is a unary operator on A satisfying $l(a \cap b) = la \cap lb$, and $l1 = 1$, where 1 is the unit element of $\mathfrak{A}$.*

Now each wff $\alpha(p_1, \ldots, p_n)$ with n variables induces an n-ary polynomial function [39, p. 37] $h_\alpha^{\mathfrak{A}}$ on $\mathfrak{A}$ which may be defined inductively as follows:

$$\begin{aligned}
h_{p_i}^{\mathfrak{A}}(a_1, \ldots, a_n) &= a_i \\
h_{\neg\alpha}^{\mathfrak{A}}(a_1, \ldots, a_n) &= (h_\alpha^{\mathfrak{A}}(a_1, \ldots, a_n))' \\
h_{\alpha \wedge \beta}^{\mathfrak{A}}(a_1, \ldots, a_n) &= h_\alpha^{\mathfrak{A}}(a_1, \ldots, a_n) \cap h_\beta^{\mathfrak{A}}(a_1, \ldots, a_n) \\
h_{\Box\alpha}^{\mathfrak{A}}(a_1, \ldots, a_n) &= l(h_\alpha^{\mathfrak{A}}(a_1, \ldots, a_n)).
\end{aligned}$$

α is ***valid*** on $\mathfrak{A}$ ($\mathfrak{A} \models \alpha$) iff $h_\alpha^{\mathfrak{A}} = 1$ identically on $\mathfrak{A}$ (i.e. $h_\alpha^{\mathfrak{A}}$ takes the value 1 for all arguments in its domain). It is easy to see that every polynomial on $\mathfrak{A}$ is of the form $h_\alpha^{\mathfrak{A}}$, for some $\alpha \in \Phi$, and that $h_\alpha^{\mathfrak{A}} = h_\beta^{\mathfrak{A}}$ identically iff $\mathfrak{A} \models (\alpha \leftrightarrow \beta)$ iff $h_{\alpha \leftrightarrow \beta}^{\mathfrak{A}} = 1$ identically.

A logic Λ is said to be **determined** or **characterised** by a class $\mathfrak{C}$ of MA's iff for any $\alpha \in \Phi$, $\vdash_\Lambda \alpha$ iff $\mathfrak{A} \models \alpha$ for all $\mathfrak{A} \in \mathfrak{C}$.

That every logic is characterised by a single algebra is shown by the following classic construction.

Definition 1.2.2 *If Λ is a modal logic, then the **Lindenbaum algebra** for Λ is the structure $\mathfrak{A}_\Lambda = \langle A_\Lambda, \cap, ', l \rangle$ defined as follows:*

$$
\begin{aligned}
A_\Lambda &= \{\|\alpha\|_\Lambda : \alpha \in \Phi\}, \quad \text{where} \\
\|\alpha\|_\Lambda &= \{\beta : \alpha =_\Lambda \beta\}, \text{ and } \alpha =_\Lambda \beta \text{ iff } \vdash_\Lambda \alpha \leftrightarrow \beta \\
\|\alpha\|'_\Lambda &= \|\neg\alpha\|_\Lambda \\
l(\|\alpha\|_\Lambda) &= \|\Box\alpha\|_\Lambda \\
\|\alpha\|_\Lambda \cap \|\beta\|_\Lambda &= \|\alpha\wedge\beta\|_\Lambda.
\end{aligned}
$$

A proof that $\mathfrak{A}_\Lambda$ is well defined, characterises Λ, and is an MA if Λ is a normal logic, may be found in Lemmon [57, Section II] .

In an MA $\mathfrak{A}$, a unary operator m corresponding to the modal connective $\Diamond$ may be defined by the equation $ma = (l(a'))'$

Theorem 1.2.3

(1) $m(a \cup b) = ma \cup mb$, *where $\cup$ is the join operation of $\mathfrak{A}$, i.e. $a \cup b = (a' \cap b')'$.*

(2) $m0 = 0$, *where $0 = 1'$ is the least element of $\mathfrak{A}$.*

(3) l *and m are monotonic, i.e. $a \leq b$ only if $la \leq lb$ and $ma \leq mb$, where $\leq$ is the lattice ordering of $\mathfrak{A}$.*

(4) $h^{\mathfrak{A}}_{\Diamond\alpha} = m \circ h^{\mathfrak{A}}_\alpha$, *where $\circ$ denotes functional composition.*

(5) $la = (m(a'))'$.

Set-theoretic modelling of modal formulae seems to have been first explicitly described by Saul Kripke [51, 52], and has subsequently been developed by a number of authors (cf. [47, 59, 86]).

Definition 1.2.4 *A **Kripke frame** (K-frame) is a structure $\mathcal{F} = \langle W, R \rangle$, where W is a non-empty set, the **base-set** or **carrier** of $\mathcal{F}$, and R is a binary relation on W.*

*A **valuation** V on $\mathcal{F}$ is a function that associates with each variable p a subset $V(p)$ of W. The domain of V is extended to all of Φ by the stipulations*

(i) $x \in V(\alpha \wedge \beta)$ *iff $x \in V(\alpha)$ and $x \in V(\beta)$*

(ii) $x \in V(\neg\alpha)$ *iff $x \notin V(\alpha)$*

(iii) $x \in V(\Box\alpha)$ *iff for all y, if xRy then $y \in V(\alpha)$;   and hence*

(iv) $x \in V(\Diamond\alpha)$ *iff for some* y, xRy *and* $y \in V(\alpha)$.

α *is **valid** on* $\mathcal{F}$ *($\mathcal{F} \models \alpha$) iff* $V(\alpha) = W$ *for all valuations* V *on* $\mathcal{F}$. α *is valid on a class* $\mathfrak{C}$ *of frames ($\mathfrak{C} \models \alpha$) iff* $\mathcal{F} \models \alpha$ *for all* $\mathcal{F} \in \mathfrak{C}$. $\mathfrak{C}$ ***determines** or **characterises** a logic* Λ *iff for all* $\alpha \in \Phi$, $\vdash_\Lambda \alpha$ *iff* $\mathfrak{C} \models \alpha$.

The intuitive content of 1.2.4 is that W is a set of "possible worlds" or "states of affairs" and that worlds x and y are related (xRy) iff y is a world accessible to x, a conceivable alternative state of affairs to x. A proposition is identified with the set $V(\alpha)$ of worlds in which it is *true*. Clause (iii) then formalises the Leibnizian notion that a necessary truth is one that holds in all conceivable states of affairs.

A proof that the logic K is determined by the class of all K-frames is given in [59, Section 2] and employs the following construction:

Definition 1.2.5 *If* Λ *is a normal modal logic, the **canonical** K-**frame** for* Λ *is the structure* $\mathcal{F}_\Lambda^K = \langle W_\Lambda, R_\Lambda \rangle$, *where*

(i) W_Λ *is the class of* Λ-*maximal subsets of* Φ,   *and*

(ii) $xR_\Lambda y$ *iff* $\{A \in \Phi : \Box A \in x\} \subseteq y$ *iff* $\{\Diamond A : A \in y\} \subseteq x$.

*The **canonical valuation*** V_Λ *is defined by* $V_\Lambda(p) = |p|_\Lambda$ *(cf. 1.1.2).*

Theorem 1.2.6 *For all* $\alpha \in \Phi$, $V_\Lambda(\alpha) = |\alpha|_\Lambda$.

Proof. See Section 2, p. 7, of [59].   □

It follows from 1.2.6 and 1.1.3 that if $\nvdash_\Lambda \alpha$ then $V_\Lambda(\alpha) \neq W_\Lambda$ and hence that any non-theorem of Λ is not valid on $\mathcal{F}_\Lambda^K$. However, as we shall see in Section 1.18, the converse is not always true, and so not all normal logics are determined by their canonical K-frames.

Definition 1.2.7 *Two models (algebraic or set-theoretic) are **semantically equivalent** iff they validate precisely the same modal wffs.*

1.3   First-Order Frames

Connections between modal algebras and Kripke frames were studied by E. J. Lemmon in [57], where it is shown that each K-frame has a semantically equivalent MA.

Definition 1.3.1 *If* $\mathcal{F} = \langle W, R \rangle$ *is a* K-*frame then* $\mathcal{F}^+$ *is the* MA $\langle 2^W, \cap, -, l_R \rangle$, *where* 2^W *is the power set of* W *(the set of all subsets of* W*),* $\cap$ *and* $-$ *are set intersection and complementation respectively, and for* $S \subseteq W$,

$$l_R(S) = \{x \in W : \forall y (xRy \Rightarrow y \in S)\}.$$

That $\mathcal{F}^+$ satisfies 1.2.1 is easily checked. Note that

$$m_R(S) = -l_R(-S) = \{x \in W : \exists y(xRy \text{ and } y \in S)\}.$$

The operators l_R and m_R are of course interdefinable using $-$, and in subsequent computations we will use whichever of them is most convenient (in fact it usually turns out that m_R is easier to handle).

Theorem 1.3.2 *Let $\alpha(p_1, \ldots p_n) \in \Phi$. Then for any valuation V on $\mathcal{F}$,*

$$h_\alpha^{\mathcal{F}^+}(V(p_1), \ldots, V(p_n)) = V(\alpha).$$

Proof. By induction on the length of α, using 1.2.1, 1.2.4, and 1.3.1 as in the Lemma of [57, p. 61]. $\qquad\square$

Corollary 1.3.3 $\mathcal{F} \models \alpha$ *iff* $\mathcal{F}^+ \models \alpha$.

Proof. As in [57, Theorem 21], using 1.3.2. $\qquad\square$

We see from 1.3.3 that any set-theoretic falsification of a wff gives rise to an algebraic falsification of it, and so a wff is valid on all MA's only if it is valid on all K-frames. To prove the converse, Lemmon showed, by a refinement of Stone's Representation Theorem for BA's, that any MA can be isomorphically embedded in $\mathcal{F}^+$ for some K-frame $\mathcal{F}$. In this way an algebraic falsification of a wff gives rise to a set-theoretic one.

Thus as far as the condition of validity over a class of structures is concerned, the concepts of K-frame and modal algebra produce the same class of wffs. However, there are limitations on the correspondences described above. Kripke [53], in a review of Lemmon's work, pointed out that there are wffs that are valid on a particular algebra, but not valid on the associated frame of that algebra. The reason why this is possible is that the algebra is isomorphic in general not to $\mathcal{F}^+$, but only to a subalgebra of it. In other words, the elements of the original algebra correspond to some, but not all, subsets of the frame.

In an attempt to improve on Lemmon's work, D. C. Makinson [65] devised a new kind of set-theoretic model, and showed that to each MA $\mathcal{F}$ there was an equivalent model $\mathfrak{A}_+$ of his kind, and to each model $\mathcal{F}$ there was an equivalent MA $\mathcal{F}_+$. Furthermore, he proved

(1) $\mathfrak{A}$ is isomorphic to $(\mathfrak{A}_+)^+$,

but that we do not in general have

(2) $\mathcal{F}$ is isomorphic to $(\mathcal{F}^+)_+$.

Thus, while MA's and Makinson's models are semantically equivalent, they are not equivalent as mathematical objects. For the latter, at least according to the definition provided by Category Theory, we would require both (1) and (2) to hold.

Now Makinson's models involve a restriction on those subsets of a frame that can serve as the interpretation of a wff. That a constraint of this nature is needed would seem to be indicated by the above comments on Lemmon's construction of the frame corresponding to an MA. However, the models of [65] include reference to a set of valuations and so their very definition is dependent on a particular object language. In order to avoid this limitation, we adopt the following approach, due to S. K. Thomason [96].

Definition 1.3.4 *A (normal modal)* **frame** *is a structure*

$$\mathcal{F} = \langle W, R, P \rangle,$$

where

(i) $\langle W, R \rangle$ *is a* K*-frame,*   *and*

(ii) P *is a non-empty collection of subsets of* W *that is closed under* $\cap$, $-$, *and* l_R.

A **valuation** V *on a frame* $\mathcal{F}$ *is a function as in 1.2.4 with the added requirement*

(iii) $V(p) \in P$, *for all variables* p.

Condition (ii) then guarantees that

(iv) $V(\alpha) \in P$, *all* $\alpha \in \Phi$.

The definition of validity etc. remains as in 1.2.4.

Definition 1.3.5 *If* $\mathcal{F} = \langle W, R, P \rangle$ *is a frame, then* $\mathcal{F}^+$ *is the structure* $\langle P, \cap, -, l_R \rangle$, *an* MA *by 1.3.4(ii).*

The statement of Theorem 1.3.2 continues to hold for the new definition of frame, valuation, and associated MA. Thus for the structures of 1.3.5 we again have $\mathcal{F}$ semantically equivalent to $\mathcal{F}^+$.

A frame $\mathcal{F}$ is said to be **full** iff $P = 2^W$. It is clear from 1.3.3 and the above observations that the K-frame $\langle W, R \rangle$ is equivalent to the frame $\langle W, R, 2^W \rangle$, although formally they are not the same thing.

Example 1.3.6 If V is a valution on a K-frame $\mathcal{F} = \langle W, R \rangle$, then by 1.2.4, 1.3.1 and 1.3.4, $\mathcal{F}_V = \langle W, R, P_V \rangle$ is frame, where $P_V = \{V(\alpha) : \alpha \in \Phi\}$. Since P_V is countable (Φ being denumerable), $\mathcal{F}_V$ will not be full if W is not finite.

In order to distinguish the modelling of 1.2.4 from that of 1.3.4, Thomason [96] used the names *second-order* semantics for the former, and *first-order* semantics for the latter (the motivation for this terminology may be found in [96, p. 152]).

In Section 1.9 we will use first-order frames to develop a set-theoretic structure that in relation to MA's satisfies all the requirements of equivalence that have been discussed in the present section. Before doing that, however, we examine a number of frame constructions that preserve the validity of modal formulae.

1.4   Subframes

Our first observation about subobjects is effectively a reformulation of the general algebraic fact that polynomial identities are preserved under subalgebras.

Theorem 1.4.1 *Let $\mathcal{F} = \langle W, R, P \rangle$ and $\mathcal{F}_1 = \langle W, R, P_1 \rangle$ be frames with $P_1 \subseteq P$. Then $\mathcal{F} \models \alpha$ only if $\mathcal{F}_1 \models \alpha$.*

Proof. By hypothesis, and by 1.3.4(ii), $\mathcal{F}_1^+$ is a sub-MA of $\mathcal{F}^+$, so $h_\alpha^{\mathcal{F}^+} = 1$ identically only if $h_\alpha^{\mathcal{F}_1^+} = 1$. The result then follows by the equivalence of $\mathcal{F}$ and $\mathcal{F}_1$ to $\mathcal{F}^+$ and $\mathcal{F}_1^+$ respectively.   □

Our next concern is with frames whose base is a proper subset of the base of a given frame. For this we need some preliminary definitions.

Definition 1.4.2 *Let R be a binary relation on W. For each $k \in \mathbb{N}$ (the set of natural numbers) we define the relation $R^k \subseteq W^2$ by the inductive scheme*

$$\begin{aligned} xR^0y \quad & iff \quad x = y \\ xR^{k+1}y \quad & iff \quad \exists z(xRz \ and \ zR^ky). \end{aligned}$$

Definition 1.4.3 *If $R \subseteq W^2$, then $W' \subseteq W$ is R-hereditary iff*

$$if \ x \in W' \ and \ xRy, \ then \ y \in W'.$$

It is readily seen that the intersection of a class of R-hereditary sets is itself R-hereditary, and so for any $W' \subseteq W$ there is a smallest R-hereditary set W'_R containing W'.

Theorem 1.4.4 $W'_R = \{y : \exists x \exists k (x \in W' \ and \ xR^ky)\}$.

Proof. The right-hand set is R-hereditary, contains W' (by 1.4.2 with $k = 0$), and is contained in any other R-hereditary set that contains W'.   □

Definition 1.4.5 *If $\mathcal{F} = \langle W, R, P \rangle$ and $\mathcal{F}' = \langle W', R', P' \rangle$ are frames, then $\mathcal{F}'$ is a subframe of $\mathcal{F}$ (written $\mathcal{F}' \subseteq \mathcal{F}$) iff*

(i) W' is an R-hereditary subset of W,

(ii) $R' = R \cap (W' \times W')$,

(iii) $P' = \{W' \cap S : S \in P\}$.

Theorem 1.4.6 *If W' is an R-hereditary subset of W, and $R' = R \cap (W' \times W')$, then*

(i) $W' - (W' \cap S) = W' \cap (W - S)$,

(ii) $(W' \cap S) \cap (W' \cap S_1) = W' \cap (S \cap S_1)$,

(iii) $m_{R'}(W' \cap S) = W' \cap m_R(S)$,

(iv) $l_{R'}(W' \cap S) = W' \cap l_R(S)$.

Proof. (i) and (ii) are straightforward. For (iii), let $x \in m_{R'}(W' \cap S)$. Then $x \in W'$ and $xR'y$ for some $y \in W' \cap S$. Then xRy, and since $y \in S$, $x \in m_R(S)$. Conversely, if $x \in W' \cap m_R(S)$ then $x \in W'$ and xRy for some $y \in S$. Since W' is R-hereditary, $y \in W'$ (hence $y \in W' \cap S$) and so $xR'y$. Thus $x \in m_{R'}(W' \cap S)$.

(iv) follows by (i), (iii) and 1.2.3(5).   □

Corollary 1.4.7 *If* $\mathcal{F} = \langle W, R, P \rangle$ *is a frame and* $W' \subseteq W$ *is R-hereditary, then* $\mathcal{F}_{W'} = \langle W', R', P_{W'} \rangle$ *is a subframe of* $\mathcal{F}$, *where* $R' = R \cap (W' \times W')$ *and* $P_{W'} = \{W' \cap S : S \in P\}$.

Proof. The only requirement of 1.4.5 that is not automatically satisfied is that $\mathcal{F}_{W'}$ is a frame, i.e. $P_{W'}$ is closed under $\cap$, $-$, and $l_{R'}$. But this follows from 1.4.6 and the closure of P under the corresponding operations.   □

Corollary 1.4.8 *Let* $\{\mathcal{F}_i : i \in I\}$ *be a collection of subframes of* $\mathcal{F}$, *and* $W' = \bigcap_{i \in I} W_i$. *Then if* $W' \neq \emptyset$, $\mathcal{F}_{W'}$ *is a subframe of* $\mathcal{F}$.

Proof. W' is R-hereditary (each W_i being so), hence the result by 1.4.7.   □

1.4.8 states in effect that the intersection of subframes is a subframe, so each $W' \subseteq W$ has a smallest subframe of $\mathcal{F}$ containing it. This subframe is of course $\mathcal{F}_{W'_R}$ (cf. 1.4.4, 1.4.7), and will be called the **subframe of** $\mathcal{F}$ **generated by** W'. If $W' = \{x\}$ then $\mathcal{F}_{W'_R}$ will be written simply as $\mathcal{F}_x$, the **subframe generated by** x.

If $\langle W, R \rangle$ is a K-frame and $W' \subseteq W$, then $\langle W', R \cap (W' \times W') \rangle$ is a substructure of $\langle W, R \rangle$, as that term is understood in the theory of relational structures, but will not be regarded as a subframe unless W' is R-hereditary. That 1.4.5 provides the appropriate subobjects for modal logic is shown by the next result and its corollary.

Theorem 1.4.9 *If* $\mathcal{F}' \subseteq \mathcal{F}$, $\alpha(p_1 \ldots, p_n) \in \Phi$, *and* $S_1, \ldots, S_n \in P$,

$$h_\alpha^{\mathcal{F}'^+}(W' \cap S_1, \ldots, W' \cap S_n) = W' \cap h_\alpha^{\mathcal{F}^+}(S_1, \ldots, S_n).$$

Proof. By induction on the length of α, using 1.4.6.   □

Corollary 1.4.10 *If* $\mathcal{F}' \subseteq \mathcal{F}$ *and* $\mathcal{F} \models \alpha$ *then* $\mathcal{F}' \models \alpha$.

Proof. It suffices to prove that $\mathcal{F}^+ \models \alpha$ only if $\mathcal{F}'^+ \models \alpha$. If $\mathcal{F}'^+ \not\models \alpha$, there exist $T_1, \ldots, T_n \in P'$ such that $h_\alpha^{\mathcal{F}'^+}(T_1, \ldots, T_n) \neq W'$ (W' being the unit element of $\mathcal{F}'^+$). By 1.4.5(iii), for $1 \leq i \leq n$ there is $S_i \in P$ such that $T_i = W' \cap S_i$. Thus by 1.4.9, $W' \cap h_\alpha^{\mathcal{F}^+}(S_1, \ldots, S_n) \neq W'$, so $W' \nsubseteq h_\alpha^{\mathcal{F}^+}(S_1, \ldots, S_n)$. Hence $h_\alpha^{\mathcal{F}^+}(S_1, \ldots, S_n) \neq W$, so $\mathcal{F}^+ \nvDash \alpha$. □

In the context of K-frames, we have the following special case of 1.4.9.

Theorem 1.4.11 *Let V be a valuation on $\langle W, R \rangle$, $W' \subseteq W$, and $\langle W'_R, R' \rangle$ the sub-K-frame generated by W' (W'_R is as in 1.4.4 and $R' = R \cap (W'_R \times W'_R)$). Define $V_{W'}$ by $V_{W'}(p) = W'_R \cap V(p)$, all variables p. Then $V_{W'}(\alpha) = W'_R \cap V(\alpha)$, all $\alpha \in \Phi$.*

The function $V_{W'}$ (above) will be called the *valuation derived from (generated by) V*.

From now on, if no confusion arises, we will use the same symbol to denote the binary relation on a frame and any of its subframes.

1.5 Homomorphisms

That subframes preserve validity (1.4.10) may in fact be established indirectly from the preservation of polynomial identities under homomorphisms of algebras. For if $\mathcal{F}' \subseteq \mathcal{F}$, then the map $S \mapsto W' \cap S$ is by 1.4.5 and 1.4.6 an MA-homomorphism of $\mathcal{F}^+$ onto $\mathcal{F}'^+$. There are a number of other algebraic constructions that preserve identities (subalgebras, direct products, direct limits, ultraproducts) and, as we shall see, each of these is associated with a particular frame construction. In this section, it is shown that structure preserving maps between frames are linked with sub-MA's.

Definition 1.5.1 *If $\mathcal{F}$ and $\mathcal{F}'$ are frames, a map $Q : W \to W'$ is a frame homomorphism of $\mathcal{F}$ into $\mathcal{F}'$ iff*

(1) *xRy only if $Q(x)R'Q(y)$,*

(2) *$Q(x)R'z$ only if $\exists y(xRy$ and $Q(y) = z)$,*

(3) *$S \in P'$ only if $Q^{-1}(S) \in P$, where $Q^{-1}(S) = \{x \in W : Q(x) \in S\}$.*

If Q is surjective (onto, i.e. $Q(W) = W'$), then $\mathcal{F}'$ is a *homomorphic image* of $\mathcal{F}$ (written $\mathcal{F}' \preccurlyeq \mathcal{F}$).

Q is an *embedding* iff it is injective (one-to-one) and satisfies

(4) *$S \in P \Rightarrow \exists T \in P'(Q(S) = Q(W) \cap T)$.*

If Q is bijective (injective and onto) and Q^{-1} is a homomorphism then Q is an *isomorphism*, in which case $\mathcal{F}$ and $\mathcal{F}'$ are *isomorphic* ($\mathcal{F} \cong \mathcal{F}'$). An isomorphism may alternatively be described as a surjective embedding. Note that a bijective homomorphism need not be an isomorphism, e.g.

iι $P \neq 2^W$ then the identity map from $\langle W, R, 2^W \rangle$ to $\langle W, R, P \rangle$ is a bijective homomorphism whose inverse does not satisfy 1.5.1(3).

Theorem 1.5.2 *If $Q : \mathcal{F} \to \mathcal{F}'$ is a frame homomorphism, then for $S, T \subseteq W'$*

(i) $Q^{-1}(W' - S) = W - Q^{-1}(S)$,
(ii) $Q^{-1}(S \cap T) = Q^{-1}(S) \cap Q^{-1}(T)$,
(iii) $Q^{-1}(m_{R'}(S)) = m_R(Q^{-1}(S))$,
(iv) $Q^{-1}(l_{R'}(S)) = l_R(Q^{-1}(S))$.

Proof. (i) and (ii) are basic properties of inverses of set maps. For (iii), let $x \in Q^{-1}(m_{R'}(S))$. Then $Q(x) \in m_{R'}(S)$, so $Q(x)R'z$, for some $z \in S$. By 1.5.1(2), xRy for some y such that $Q(y) = z$, hence $y \in Q^{-1}(S)$. Thus $x \in m_R(Q^{-1}(S))$. Conversely, if $x \in m_R(Q^{-1}(S))$, xRy for some $y \in Q^{-1}(S)$, i.e. $Q(y) \in S$. But by 1.5.1(1), $Q(x)R'Q(y)$, so $Q(x) \in m_{R'}(S)$ and therefore $x \in Q^{-1}(m_{R'}(S))$.

(iv) follows by (i), (iii) and 1.2.3(5).   □

Theorem 1.5.3 *Let $Q : \mathcal{F} \to \mathcal{F}'$ be a homomorphism and $\mathcal{F}_Q = \langle W_Q, R', P_Q \rangle$ be the subframe of $\mathcal{F}'$ generated by $Q(W)$. Then*

(1) *$\mathcal{F}_Q$ is a homomorphic image of $\mathcal{F}$ under Q,   and*
(2) *$\mathcal{F}_Q \cong \mathcal{F}$ if Q is an embedding.*

Proof.

(1) By 1.5.1(2), $Q(W)$ is an R'-hereditary set, so W_Q, the base of $\mathcal{F}_Q$, is just $Q(W)$ itself, so Q maps onto $\mathcal{F}_Q$. If $S \in P_Q$, $S = Q(W) \cap T$ for some $T \in P'$ (1.4.7). Then $Q^{-1}(S) = Q^{-1}(Q(W) \cap T) = Q^{-1}(Q(W)) \cap Q^{-1}(T) = W \cap Q^{-1}(T) = Q^{-1}(T) \in P$ by 1.5.1(3). Thus $Q : \mathcal{F} \to \mathcal{F}_Q$ is a homomorphism.

(2) If Q is injective then $(Q^{-1})^{-1} = Q$. The definition of embedding then gives $S \in P$ only if $(Q^{-1})^{-1}(S) = Q(S) \in P_Q$, so Q^{-1} is a homomorphism. This, together with (1), yields Q as an isomorphism between $\mathcal{F}$ and $\mathcal{F}_Q$.   □

Theorem 1.5.4 *If $Q : \mathcal{F} \to \mathcal{F}'$ is a homomorphism, $\alpha(p_1, \ldots, p_n) \in \Phi$, and $S_1, \ldots, S_n \in P'$, then*

$$Q^{-1}(h_\alpha^{\mathcal{F}'+}(S_1, \ldots, S_n)) = h_\alpha^{\mathcal{F}+}(Q^{-1}(S_1), \ldots, Q^{-1}(S_n)).$$

Proof. By induction on the length of α, using 1.5.2. Note that 1.5.1(3) is needed to ensure that the indicated argument appears in the domain of $h_\alpha^{\mathcal{F}+}$.   □

Corollary 1.5.5 *If $\mathcal{F}' \preccurlyeq \mathcal{F}$ and $\mathcal{F} \models \alpha$ then $\mathcal{F}' \models \alpha$.*

Proof. Let Q be a homomorphism from $\mathcal{F}$ onto $\mathcal{F}'$. Then

$$
\begin{aligned}
h_\alpha^{\mathcal{F}'+}(S_1,\ldots,S_n) &\supseteq Q(Q^{-1}(h_\alpha^{\mathcal{F}'+}(S_1,\ldots,S_n))) \\
&= Q(h_\alpha^{\mathcal{F}+}(Q^{-1}(S_1),\ldots,Q^{-1}(S_n))) \quad (1.5.4) \\
&= Q(W) \quad (\mathcal{F}^+ \models \alpha) \\
&= W' \quad (Q \text{ onto}).
\end{aligned}
$$

Thus $\mathcal{F}'^+ \models \alpha$, so $\mathcal{F}' \models \alpha$.   □

Corollary 1.5.6 *Isomorphic frames are semantically equivalent.*

Theorem 1.5.7 *If $\mathcal{F}$ is embeddable in $\mathcal{F}'$, then $\mathcal{F}' \models \alpha$ only if $\mathcal{F} \models \alpha$.*

Proof. Let $Q : \mathcal{F} \to \mathcal{F}'$ be an embedding. Then $\mathcal{F}' \supseteq \mathcal{F}_Q$ and $\mathcal{F}_Q \cong \mathcal{F}$ (1.5.3). The result follows by 1.4.10 and 1.5.6.   □

Theorem 1.4.1 may be regarded as a special case of 1.5.5, for in that theorem the identity map is a homomorphism from $\mathcal{F}$ onto $\mathcal{F}_1$.

We now show how frame homomorphisms give rise to structure preserving maps of the associated MA's.

Definition 1.5.8 *If $Q : \mathcal{F} \to \mathcal{F}'$ is a frame homomorphism, then $Q^+ : \mathcal{F}'^+ \to \mathcal{F}^+$ is defined by $Q^+(S) = Q^{-1}(S)$, all $S \in P'$.*

By 1.5.1(3), Q^+ is indeed a mapping into P.

Theorem 1.5.9

(1) Q^+ *is an MA-homomorphism.*
(2) Q^+ *is injective if Q is onto.*
(3) Q^+ *is onto if Q is an embedding.*
(4) Q^+ *is an MA-isomorphism if Q is a frame isomorphism.*

Proof.

(1) An MA-homomorphism is a mapping of MA's that preserves the MA operators, so we have to show that $Q^+(S \cap T) = Q^+(S) \cap Q^+(T)$, $Q^+(-S) = -Q^+(S)$, and $Q^+(m_{R'}(S)) = m_R(Q^+(S))$. But this is immediate from 1.5.8 and 1.5.2.
(2) Suppose $Q^+(S) = Q^+(T)$, i.e. $Q^{-1}(S) = Q^{-1}(T)$. Then $Q(Q^{-1}(S)) = Q(Q^{-1}(T))$. But as Q is onto, $Q(Q^{-1}(S)) = S$, $Q(Q^{-1}(T)) = T$, whence $S = T$ as required.
(3) Let $S \in P$. Then as Q is an embedding, $Q(S) = Q(W) \cap T$, for some $T \in P'$. Clearly $S \subseteq Q^{-1}(T)$. But if $x \in Q^{-1}(T)$, $Q(x) \in T \cap Q(W) = Q(S)$, so $Q(x) = Q(y)$ for some $y \in S$. But Q is injective, so $x = y \in S$. Hence $Q^{-1}(T) = S$, i.e. $Q^+(T) = S$, and Q^+ is onto.

(4) Follows from (1) - (3).   □

Theorem 1.5.5 may now be obtained indirectly from 1.5.9, for if $\mathcal{F}'$ is a homomorphic image of $\mathcal{F}$, by 1.5.9 there is an injective homomorphism from $\mathcal{F}'^+$ to $\mathcal{F}^+$. Thus $\mathcal{F}'^+$ is isomorphic to a sub-MA of $\mathcal{F}^+$, and polynomial identities are preserved under subalgebras and isomorphisms.

Similarly for 1.5.7, if $\mathcal{F}$ is embeddable in $\mathcal{F}'$, by 1.5.9 $\mathcal{F}^+$ is a homomorphic image of $\mathcal{F}'^+$, so validity is preserved in passing from the latter to the former.

1.6   Disjoint Unions

In this section we examine the frame construction that corresponds to direct products of MA's.

Definition 1.6.1 *Let $\{\mathcal{F}_i : i \in I\}$ be a family of pairwise disjoint frames, i.e. $W_i \cap W_j = \emptyset$, for $i \neq j \in I$. The **disjoint union** of the $\mathcal{F}_i$'s is the frame*

$$\sum_{i \in I} \mathcal{F}_i = \langle W, R, P \rangle,$$

where

(i) $W = \bigcup_{i \in I} W_i$,

(ii) $R = \bigcup_{i \in I} R_i$,

(iii) $P = \{S \subseteq W : S \cap W_i \in P_i, \text{ all } i \in I\}$.

Since the W_i's are disjoint, (iii) is equivalent to

(iv) $S \in P$ iff there exists $S_i \in P_i$, all $i \in I$, such that $S = \bigcup_{i \in I} S_i$.

In fact $S_i = S \cap W_i$, so the expression of S as $\bigcup S_i$ is uniquely determined.

That $\sum \mathcal{F}_i$ is a frame, i.e. P is closed under $\cap$, $-$, and m_R, is shown by the next theorem, whose proof follows easily from the disjointness of the W_i's and the R_i's.

Theorem 1.6.2 *Let $S = \bigcup S_i$, $T = \bigcup T_i$, where S_i, $T_i \in P_i$, all $i \in I$. Then*

(1) $W - S = \bigcup_{i \in I}(W_i - S_i)$,

(2) $S \cap T = \bigcup_{i \in I}(S_i \cap T_i)$,

(3) $m_R(S) = \bigcup_{i \in I}(m_{R_i}(S_i))$.   □

Theorem 1.6.3 *If $\alpha(p_1 \ldots, p_n) \in \Phi$, and $S_j = \bigcup_{i \in I} S_{ji} \in P$ for $1 \leq j \leq n$, then*

$$h_\alpha^{(\Sigma \mathcal{F}_i)^+}(S_1, \ldots, S_n) = \bigcup_{i \in I}(h_\alpha^{\mathcal{F}_i^+}(S_{1i}, \ldots, S_{ni})).$$

Proof. By induction on α using 1.6.2. $\qquad\square$

Corollary 1.6.4 $\sum \mathcal{F}_i \models \alpha$ *iff* $\mathcal{F}_i \models \alpha$ *for all* $i \in I$.

Proof. It is readily seen that each $\mathcal{F}_i$ is a subframe of $\sum \mathcal{F}_i$, so the "only if" part follows from 1.4.10. Conversely, suppose $\mathcal{F}_i \models \alpha$, all $i \in I$. Then for $S_j = \bigcup_{i \in I} S_{ji} \in P$,

$$
\begin{aligned}
h_\alpha^{(\Sigma \mathcal{F}_i)^+}(S_1, \ldots, S_n) &= \bigcup_{i \in I} h_\alpha^{\mathcal{F}_i^+}(S_{1i}, \ldots, S_{ni}) && (1.6.3) \\
&= \bigcup_{i \in I} W_i && (\text{as } \mathcal{F}_i^+ \models \alpha) \\
&= W,
\end{aligned}
$$

so $\sum \mathcal{F}_i \models \alpha$. $\qquad\square$

Theorem 1.6.5 $(\sum \mathcal{F}_i)^+$ *is isomorphic to the direct product of the family* $\{\mathcal{F}_i^+ : i \in I\}$ *of* MA's.

Proof. Define $Q : P \to \prod_{i \in I} P_i$, the Cartesian product of the P_i's, by

$$Q(S)(i) = S_i = S \cap W_i, \quad \text{all } i \in I.$$

By 1.6.1(iii), we indeed have $Q(S) \in \prod P_i$. By the uniqueness of the expression $S = \bigcup S_i$, Q is readily shown to be a bijection. The MA-homomorphism properties of Q, with respect to the usual "point-wise" definitions of operations on $\prod P_i$, follow from 1.6.2, e.g.

$$Q(m_R(S))(i) = m_R(S) \cap W_i = m_{R_i}(S_i) = m_{R_i}(Q(S)(i)) = m(Q(S))(i),$$

so $Q(m_R(S)) = m(Q(S))$. $\qquad\square$

Theorem 1.6.5, and the fact that an identity is satisfied by a direct product of algebras iff it is satisfied by each of them, yield an alternative proof of 1.6.4.

If $\{\mathcal{F}_i : i \in I\}$ is a family of frames that are not pairwise disjoint, we replace each $\mathcal{F}_i$ by its isomorphic copy $\mathcal{F}_i'$ where

$$
\begin{aligned}
W_i' &= W_i \times \{i\} \\
\langle x, i \rangle R_i' \langle y, i \rangle &\text{ iff } x R_i y \\
P_i' &= \{\{\langle x, i \rangle : x \in S\} : S \in P_i\}.
\end{aligned}
$$

The *disjoint union* of the $\mathcal{F}_i$'s is then defined to be the disjoint union of the $\mathcal{F}_i'$'s. Clearly 1.6.4 continues to hold for this construction.

1.7 Ultraproducts

The theory of ultraproducts plays a central role in the model theory of first-order quantificational logic. It forms the basis of many compactness results and characterisations of semantic concepts, such as equivalence of models, and elementary classes. The frame constructions associated with ultraproducts of MA's are in fact ultraproducts themselves, and prove, as we shall see, to be just as useful in the modal context as they are for first-order logic.

Definition 1.7.1 *Let* $\mathfrak{A} = \langle A, \cap, {}' \rangle$ *be a Boolean algebra. Then* $G \subseteq A$ *is an* **ultrafilter** *in* $\mathfrak{A}$ *iff*

(1) $1 \in G$,

(2) *if* $a, a' \cup b \in G$, *then* $b \in G$,

(3) *exactly one of* $a, a' \in G$, *for all* $a \in A$.

A set satisfying (1) and (2) is known as a **filter**. A filter is **proper** iff it does not contain the least element 0. Ultrafilters may alternatively be defined as maximal proper filters, i.e. those not properly included in any other proper filter. All filters G satisfy

(4) $a, b \in G$ *iff* $a \cap b \in G$,

(5) *if* $a \in G$ *and* $a \leq b$, *then* $b \in G$.

Condition (4), or the "only if" part of (4) together with (5), give alternative definitions of a filter. In the presence of (1) and (2), (3) is equivalent to

(6) $0 \notin G$, *and for* $a, b \in A$, $a \cup b \in G$ *only if* $a \in G$ *or* $b \in G$.

An ultrafilter on a set I is, by definition, an ultrafilter in the BA $\langle 2^I, \cap, - \rangle$.

A set $G \subseteq A$ has the **finite intersection property** (fip) iff the lattice meet ($\cap$) of every finite subset of G is not equal to 0. The following standard result will be used repeatedly to establish the existence of ultrafilters.

Theorem 1.7.2 *Any subset of a BA that has the fip is included in an ultrafilter.*

We begin our discussion of ultraproducts with the definition for K-frames.

Definition 1.7.3 *Let* W_i *be a set, for all* $i \in I$, *and* G *an ultrafilter on* I. *An equivalence relation is defined on* $\prod_{i \in I} W_i$, *the Cartesian product of the* W_i, *by*

$$f \sim g \text{ iff } \{i : f(i) = g(i)\} \in G.$$

If $\hat{f}$ *is the* $\sim$-*equivalence class of* $f \in \prod W_i$, *then the* **ultraproduct of the** W_i **over** G *is the set* $W_G = \prod W_i / G = \{\hat{f} : f \in \prod_{i \in I} W_i\}$. *If* $\mathcal{F}_i = \langle W_i, R_i \rangle$ *is a K-frame, all* $i \in I$, *then the* **ultraproduct of the** $\mathcal{F}_i$ **over** G *is the frame* $\mathcal{F}_G = \prod \mathcal{F}_i / G = \langle W_G, R_G \rangle$, *where*

$$\hat{f} R_G \hat{g} \text{ iff } \{i : f(i) R_i g(i)\} \in G.$$

Intuitively, the elements of the ultrafilter G may be thought of as sets that contain "almost all" members of I. Properties are then defined to hold of $\mathcal{F}_G$ iff they hold of almost all of the $\mathcal{F}_i$'s. A proof that $\sim$ is an equivalence relation and that R_G is well-defined may be found in Bell and Slomson [4, p. 88].

Our definition of ultraproduct for first-order frames $\langle W_i, R_i, P_i \rangle$ is an adaptation of the higher order ultraproduct construction presented by Malcolm [66]. The basic idea is to define second-order individuals as $\sim$-equivalence classes of the Cartesian product of the P_i, and then by "normalising", to turn these into genuine subsets. The formal development is facilitated by a preliminary description of ultraproducts of modal algebras.

Definition 1.7.4 *Let* $\mathfrak{A}_i = \langle A_i, \cap, {}', m_i \rangle$ *be an MA, for each* $i \in I$, *and* G *an ultrafilter on* I. *Define operations on* $\prod_{i \in I} A_i$ *in the usual point-wise manner:*

$$
\begin{aligned}
(\sigma \cap \theta)(i) &= \sigma(i) \cap \theta(i) \\
\sigma'(i) &= (\sigma(i))' \\
m\sigma(i) &= m_i(\sigma(i)), \quad \text{for } \sigma, \theta \in \textstyle\prod A_i.
\end{aligned}
$$

Then the **ultraproduct of the** $\mathfrak{A}$**'s over** G *is the MA*

$$
\prod_{i \in I} \mathfrak{A}_i / G = \langle \prod A_i / G, \cap, {}', m \rangle
$$

where $\prod A_i / G$ *is defined as in 1.7.3, and*

$$
\hat{\sigma} \cap \hat{\theta} = \widehat{\sigma \cap \theta}, \quad (\hat{\sigma})' = \widehat{\sigma'}, \quad m(\hat{\sigma}) = \widehat{m\sigma}
$$

Thus $\prod_{i \in I} \mathfrak{A}_i / G$ is the quotient algebra of $\prod_{i \in I} \mathfrak{A}_i$, the direct product of the $\mathfrak{A}_i$'s, with respect to the congruence $\sim$ defined as in 1.7.3.

Theorem 1.7.5 *Suppose* $P_i \subseteq 2^{W_i}$, *all* $i \in I$. *If* $f \in \prod W_i$, $\sigma \in \prod P_i$, *let* $[f, \sigma] = \{i : f(i) \in \sigma(i)\}$. *Then if* G *is an ultrafilter on* I,

(1) *if* $f, g \in \prod W_i$, $\sigma \in \prod P_i$ *and* $f \sim g$, *then* $[f, \sigma] \in G$ *iff* $[g, \sigma] \in G$;

(2) *if* $\sigma, \theta \in \prod P_i$, *then* $\sigma \sim \theta$ *iff for all* $f \in \prod W_i$, $[f, \sigma] \in G$ *iff* $[f, \theta] \in G$.

Proof.

(1) Since $f \sim g$, $A = \{i : f(i) = g(i)\} \in G$. But $[f, \sigma] \cap A \subseteq [g, \sigma]$, and $[g, \sigma] \cap A \subseteq [f, \sigma]$, so the result follows by 1.7.1(4),(5).

(2) Let $\sigma \sim \theta$. Then $A = \{i : \sigma(i) = \theta(i)\} \in G$. Then if $f \in \prod W_i$, $[f, \sigma] \cap A \subseteq [f, \theta]$, and we proceed as in (1). For the converse, we observe that I is partitioned into the disjoint sets $A = \{i : \sigma(i) = \theta(i)\}$, $B = \{i : \sigma(i) \subset \theta(i)\}$, $C = \{i : \theta(i) \subset \sigma(i)\}$, so by 1.7.1, exactly one of A, B, C is in G.

Now suppose not $\sigma \sim \theta$. Then $A \notin G$. Suppose $B \in G$. Then for $i \in B$, there exists $f_i \in \theta(i) - \sigma(i)$. Choose $f \in \prod W_i$ such that $f(i) = f_i$, all $i \in B$. Then $B \subseteq [f, \theta]$, so $[f, \theta] \in G$ by 1.7.1(5). But $[f, \sigma] \subseteq -B$ and $-B \notin G$ (1.7.1(3)) so again by 1.7.1(5), $[f, \sigma] \notin G$. Similarly if $C \in G$ we may construct an f such that $[f, \sigma] \in G$ and $[f, \theta] \notin G$.

$\square$

Definition 1.7.6 Let $\mathcal{F}_i = \langle W_i, R_i, P_i \rangle$ be a first-order frame, for all $i \in I$, and G be ultrafilter on I. For $\hat{\sigma} \in \prod P_i / G$, define $S_{\hat{\sigma}} \subseteq W_G$ (1.7.3) by

$$S_{\hat{\sigma}} = \{\hat{f} : [f, \sigma] \in G\}$$

(by 1.7.5 this definition is unambiguous). The **ultraproduct of the** $\mathcal{F}_i$'s **over** G is defined as the frame $\mathcal{F}_G = \prod \mathcal{F}_i / G = \langle W_G, R_G, P_G \rangle$, where $\langle W_G, R_G \rangle$ is as in 1.7.3 and $P_G = \{S_{\hat{\sigma}} : \sigma \in \prod P_i\}$.

That P_G is closed under $\cap, -, m_{R_G}$ is shown by

Theorem 1.7.7

(1) $\hat{\sigma} = \hat{\theta}$ iff $S_{\hat{\sigma}} = S_{\hat{\theta}}$

(2) $S_{\hat{\sigma}} \cap S_{\hat{\theta}} = S_{\widehat{\sigma \cap \theta}}$

(3) $W_G - S_{\hat{\sigma}} = S_{\widehat{\sigma'}}$

(4) $m_{R_G}(S_{\hat{\sigma}}) = S_{\widehat{m\sigma}}$

where $\sigma \cap \theta$, σ', and $m\sigma$ are defined in 1.7.4.

Proof. (1) is the content of 1.7.5(2). (2) and (3) are established via 1.7.1(4) and 1.7.1(3). We shall give a detailed proof only for (4). Let $\hat{f} \in m_{R_G}(S_{\hat{\sigma}})$. Then $\hat{f} R_G \hat{g}$ for some $\hat{g} \in S_{\hat{\sigma}}$. Then $A = \{i : f(i) R_i g(i)\} \in G$ (1.7.3) and $[g, \sigma] \in G$ (1.7.6). But $A \cap [g, \sigma] \subseteq \{i : f(i) \in m_{R_i}(\sigma(i))\} = [f, m\sigma]$ (1.7.4, 1.7.5). Thus $[f, m\sigma] \in G$, and so $\hat{f} \in S_{\widehat{m\sigma}}$.

Conversely, let $\hat{f} \in S_{\widehat{m\sigma}}$. Then $[f, m\sigma] = \{i : f(i) \in m_{R_i}(\sigma(i))\} \in G$. For each $i \in [f, m\sigma]$, $f(i) R_i g_i$ for some $g_i \in \sigma(i)$. Choose $g \in \prod W_i$ such that $g(i) = g_i$ all $i \in [f, m\sigma]$. Then $[f, m\sigma] \subseteq \{i : f(i) R_i g(i)\}$ and $[f, m\sigma] \subseteq [g, \sigma]$, so $\hat{f} R_G \hat{g}$ and $\hat{g} \in S_{\hat{\sigma}}$, whence $\hat{f} \in m_{R_G}(S_{\hat{\sigma}})$. $\square$

Corollary 1.7.8 $(\prod \mathcal{F}_i / G)^+ \cong (\prod \mathcal{F}_i^+)/G$.

Proof. The map $\hat{\sigma} \to S_{\hat{\sigma}}$ is, by 1.7.7, an MA isomorphism from $\prod P_i / G$ onto P_G. $\square$

We shall see in Section 1.17 that, even if $P_i = 2^{W_i}$, all $i \in I$, we may still have $P_G \neq 2^{W_G}$. Thus, whereas the ultraproduct of a class of K-frames is *by definition* (1.7.3) a K-frame, the ultraproduct of a class of full first-order frames is a first-order frame that may not be full. So, for ultraproducts at least, the distinction between K-frames and full frames becomes significant.

Our next two results hold for either kind of ultraproduct.

Theorem 1.7.9 *Let $Q_i : \mathcal{F}_i \to \mathcal{F}'_i$ be a frame homomorphism, for each $i \in I$, and G an ultrafilter on I. Then there is a homomorphism from $\mathcal{F}_G$ to $\mathcal{F}'_G$ which is onto if (almost all) of the Q_i's are onto.*

Proof. The construction is a standard one in the theory of ultraproducts (cf. [42, p. 107]). Let $\hat{f} \in W_G$. Define $f^* \in \prod W'_i$ by $f^*(i) = Q_i(f(i))$, all $i \in I$. Then $Q : W_G \to W'_G$, defined by $Q(\hat{f}) = \widehat{f^*}$, is the required homomorphism. The only new feature, in the event that the frames are of the first-order kind, is to check that Q satisfies 1.5.1(3). So let $S_{\hat{\sigma}} \in P'_G$, where $\sigma \in \prod P'_i$. Let $\theta(i) = Q_i^{-1}(\sigma(i))$, all $i \in I$. Then, as each Q_i satisfies 1.5.1(3), $\theta \in \prod P_i$, so $S_{\hat{\theta}} \in P_G$. But a straightforward argument shows $S_{\hat{\theta}} = Q^{-1}(S_{\hat{\sigma}})$. $\qquad\qquad\square$

Corollary 1.7.10 *If each $\mathcal{F}_i$ is embeddable in $\mathcal{F}'_i$, $\mathcal{F}_G$ is embeddable in $\mathcal{F}'_G$.*

Proof. Suppose each Q_i (above) is an embedding. Then $Q : \mathcal{F}_G \to \mathcal{F}'_G$ is easily proved to be injective.

Now let $S_{\hat{\sigma}} \in P_G$. Then $\sigma(i) \in P_i$, all $i \in I$, so there exists $\theta(i) \in P'_G$ such that $Q_i(\sigma(i)) = Q_i(W_i) \cap \theta(i)$ (as Q is an embedding). But from this one may show $Q(S_{\hat{\sigma}}) = Q(W_G) \cap S_{\hat{\theta}}$. Hence Q is an embedding. $\square$

Corollary 1.7.11 *If each $\mathcal{F}_i$ is (isomorphic to) a subframe of $\mathcal{F}'_i$, then $\mathcal{F}_G$ is isomorphic to a subframe of $\mathcal{F}'_G$.*

Proof. Since each subframe is embedded in any of its parent frames by the identity map, the result follows by 1.7.10 and 1.5.3 $\qquad\qquad\square$

Theorem 1.7.12 *Let $\mathcal{F}_i$ be a first-order frame, all $i \in I$, and G an ultrafilter on I. If $\alpha(p_1, \ldots, p_n) \in \Phi$, $\sigma_1, \ldots, \sigma_n \in \prod P_i$, and $f \in \prod W_i$, then*

$$\hat{f} \in h_\alpha^{\mathcal{F}_G^+}(S_{\widehat{\sigma_1}}, \ldots, S_{\widehat{\sigma_n}}) \text{ iff } \{i : f(i) \in h_\alpha^{\mathcal{F}_i^+}(\sigma_1(i), \ldots, \sigma_n(i))\} \in G.$$

Proof. By induction on α, the basis of which is given by the definition of $S_{\hat{\sigma}}$. We shall discuss only the case for the modal connective $\Diamond$, assuming for convenience (as we often do) that $n = 1$, i.e. α has only one variable.

If $\hat{f} \in h_{\Diamond\alpha}^{\mathcal{F}_G^+}(S_{\hat{\sigma}})$ then $\hat{f} R_G \hat{g}$, for some $\hat{g} \in h_\alpha^{\mathcal{F}_G^+}(S_{\hat{\sigma}})$. Then by 1.7.3 and IH (induction hypothesis) we have $A = \{i : f(i)R_i g(i)\} \in G$ and $B = \{i : g(i) \in h_\alpha^{\mathcal{F}_i^+}(\sigma(i))\} \in G$. But, as in 1.7.7(4), $A \cap B \subseteq C =$

$\{i : f(i) \in h_{\diamond\alpha}^{\mathcal{F}_i^+}(\sigma(i))\}$, so $C \in G$. The converse follows also by a modification of the proof of 1.7.7(4).   □

The key fact about ultraproducts for first-order logic, known as Łoś's Theorem, is that a sentence is valid on an ultraproduct of structures iff it is valid on almost all of the structures themselves [4, pp. 90-91]. The analogous result for modal logic follows from 1.7.12.

Corollary 1.7.13 *If $\mathcal{F}_i$ is a first-order frame, all $i \in I$, and G is an ultrafilter on I, then for any $\alpha \in \Phi$, $\mathcal{F}_G \models \alpha$ iff $\{i : \mathcal{F}_i \models \alpha\} \in G$.*

Proof. Let $A = \{i : \mathcal{F}_i \models \alpha\} = \{i : \mathcal{F}_i^+ \models \alpha\}$. Suppose $A \in G$. Then clearly for any $S_{\hat{\sigma}} \in P_G$, $\hat{f} \in W_G$ we have $A \subseteq \{i : f(i) \in h_{\alpha}^{\mathcal{F}_i^+}(\sigma(i))\}$, so by 1.7.1(5) and 1.7.12, $\hat{f} \in h_{\alpha}^{\mathcal{F}_G^+}(S_{\hat{\sigma}})$. Thus $\mathcal{F}_G \models \alpha$.

Conversely, suppose $A \notin G$. Now for $i \notin A$, $\mathcal{F}_i \not\models \alpha$, so there exists $f_i \in W_i$, $\sigma_i \in P_i$, such that $f_i \notin h_{\alpha}^{\mathcal{F}_i^+}(\sigma_i)$. If f, σ are chosen so that $f(i) = f_i$, $\sigma(i) = \sigma_i$, all $i \notin A$, then $\{i : f(i) \in h_{\alpha}^{\mathcal{F}_i^+}(\sigma(i))\} \subseteq A$, so by 1.7.1(5) and 1.7.12, $\hat{f} \notin h_{\alpha}^{\mathcal{F}_G^+}(S_{\hat{\sigma}})$, whence $\mathcal{F}_G \not\models \alpha$.   □

Corollary 1.7.13 may be indirectly derived from 1.7.8 and Łoś's Theorem for modal algebras. The 'if' part of 1.7.13 is not true in general for ultraproducts of K-frames (cf. Section 1.17 for an example). However we do have

Corollary 1.7.14 *If $\mathcal{F}_i = \langle W_i, R_i \rangle$ is a K-frame, all $i \in I$, G an ultrafilter on I, and $\alpha \in \Phi$, $\langle W_G, R_G \rangle \models \alpha$ only if $\{i : \mathcal{F}_i \models \alpha\} \in G$.*

Proof. If $\{i : \mathcal{F}_i \models \alpha\} \notin G$, then $\{i : \langle W_i, R_i, P_i \rangle \models \alpha\} \notin G$, where $P_i = 2^{W_i}$, all $i \in I$. By 1.7.13, $\langle W_G, R_G, P_G \rangle \not\models \alpha$, hence by 1.4.1, $\langle W_G, R_G, 2^{W_G} \rangle \not\models \alpha$, so $\langle W_G, R_G \rangle \not\models \alpha$.   □

We end this section with a reformulation, in terms of valuations on frames, of Theorem 1.7.12. This result holds also for ultraproducts of K-frames.

Theorem 1.7.15 *Let V_i be a valuation on $\mathcal{F}_i$, all $i \in I$. If V is any valuation on $\mathcal{F}_G$ such that $\hat{f} \in V(p)$ iff $\{i : f(i) \in V_i(p)\} \in G$, for all variables p, then for any $\alpha \in \Phi$, $\hat{f} \in V(\alpha)$ iff $\{i : f(i) \in V_i(\alpha)\} \in G$.*

1.8   Compactness and Semantic Consequence

Ultraproducts may be used in first-order logic to prove the Compactness Theorem. In this section we use them to discuss the compactness properties of two modal semantic consequence relations.

Definition 1.8.1 *Let $\mathfrak{C}$ be a class of frames, $\Gamma \subseteq \Phi$, and $\alpha \in \Phi$. Then*

(1) $\Gamma \models_0 \alpha(\mathfrak{C})$ *iff for all $\mathcal{F} \in \mathfrak{C}$, $\mathcal{F} \models \Gamma$ only if $\mathcal{F} \models \alpha$*
   (where $\mathcal{F} \models \Gamma$ iff $\mathcal{F} \models \beta$, all $\beta \in \Gamma$).
(2) $\Gamma \models \alpha(\mathfrak{C})$ *iff for all $\mathcal{F} \in \mathfrak{C}$, and all valuations V on $\mathcal{F}$, $V(\Gamma) \subseteq V(\alpha)$*
   (where $V(\Gamma) = \bigcap \{V(\beta) : \beta \in \Gamma\}$).

Theorem 1.8.2 $\Gamma \models \alpha(\mathfrak{C})$ *only if $\Gamma \models_0 \alpha(\mathfrak{C})$.*

Proof. Take V on $\mathcal{F} \in \mathfrak{C}$, where $\mathcal{F} \models \Gamma$. Then $V(\Gamma) = W$. But as $\Gamma \models \alpha(\mathfrak{C})$, $V(\Gamma) \subseteq V(\alpha)$, so $V(\alpha) = W$. Since this holds for arbitrary V, $\mathcal{F} \models \alpha$. Hence $\Gamma \models_0 \alpha(\mathfrak{C})$. □

The converse of 8.2 does not always hold. For example, let $\Gamma = \{\Diamond(p \vee \neg p)\}$, $\alpha = \Box\Diamond(p \vee \neg p)$. Then as Necessitation is validity-preserving, $\Gamma \models_0 \alpha(\mathfrak{C})$ for any $\mathfrak{C}$. But let $\mathfrak{C} = \{\mathcal{F}\}$, where $\mathcal{F} = \langle W, R, 2^W \rangle$, with $W = \{0,1\}$ and $R = \{\langle 0,1 \rangle\}$. Then for any V on $\mathcal{F}$, $V(\Gamma) = \{0\}$, $V(\alpha) = \{1\}$, so $\Gamma \not\models \alpha(\mathfrak{C})$.

Definition 1.8.3 $\Gamma \subseteq \Phi$ *is **satisfiable** on $\mathcal{F}$ iff for some valuation V on $\mathcal{F}$, $V(\Gamma) \neq \emptyset$. Γ is satisfiable in a class $\mathfrak{C}$ of frames iff it is satisfiable on some frame in $\mathfrak{C}$.*

A logic Λ is **strongly determined** by $\mathfrak{C}$ iff for all $\Gamma \subseteq \Phi$, $\alpha \in \Phi$,

(1) $\Gamma \vdash_\Lambda \alpha$ iff $\Gamma \models \alpha(\mathfrak{C})$.

To make the distinction between this concept and that of 1.2.4 quite explicit, we may sometimes say that $\mathfrak{C}$ *simply determines* Λ if for $\alpha \in \Phi$,

(2) $\vdash_\Lambda \alpha$ iff $\mathcal{F} \models \alpha$, all $\mathcal{F} \in \mathfrak{C}$.

(1) is equivalent to

(1') Γ is Λ-consistent iff satisfiable in $\mathfrak{C}$, for all $\alpha \in \Phi$.

(2) is equivalent to

(2') α is Λ-consistent iff satisfiable in $\mathfrak{C}$, for all $\alpha \in \Phi$

(where "α is Λ-consistent" means $\{\alpha\}$ is Λ-consistent).

Clearly Λ is simply determined by $\mathfrak{C}$ if it is strongly determined by $\mathfrak{C}$. We shall shortly present a sufficient condition for the converse to hold. To do that however we need the following result, which may be regarded as a modal version of the Compactness Theorem.

Theorem 1.8.4 *Let $\mathfrak{C}$ be a class of first-order frames that is closed under the formation of ultraproducts. Then*

(1) $\Gamma \models \alpha(\mathfrak{C})$ *iff $\Gamma' \models \alpha(\mathfrak{C})$ for some finite $\Gamma' \subseteq \Gamma$,*
(2) $\Gamma \models_0 \alpha(\mathfrak{C})$ *iff $\Gamma' \models \alpha(\mathfrak{C})$ for some finite $\Gamma' \subseteq \Gamma$.*

Proof. (1) Sufficiency is trivial. For the converse, let $I = \{i : i \text{ is a finite subset of } \Gamma\}$, and suppose that $i \nVDash \alpha(\mathfrak{C})$, all $i \in I$. Then for each i there exist $\mathcal{F}_i \in \mathfrak{C}$, V_i on $\mathcal{F}_i$, and some $f_i \in W_i$ such that

(i) $f_i \in V_i(i)$ and

(ii) $f_i \notin V_i(\alpha)$.

For $\beta \in \Gamma$, let $F_\beta = \{i : f_i \in V(\beta)\}$ and $G_0 = \{F_\beta : \beta \in \Gamma\}$. Then clearly by (i), $\{\beta, \gamma\} \in F_\beta \cap F_\gamma$, so G_0 has the fip and by 1.7.2 is included in an ultrafilter G on I. Let $\mathcal{F}_G$ be the ultraproduct of the $\mathcal{F}_i$'s over G, and define V on $\mathcal{F}_G$ by $V(p) = S_{\hat{\sigma}}$, where $\sigma(i) = V_i(p)$, all $i \in I$. Let $f(i) = f_i$, all $i \in I$.

Now for $\beta \in \Gamma$, $\{i : f(i) \in V_i(\beta)\} = F_\beta \in G$, so by 1.7.15, $\hat{f} \in V(\beta)$. Hence $\hat{f} \in V(\Gamma)$. But by (ii), $\{i : f(i) \in V_i(\alpha)\} = \emptyset \notin G$, so by 1.7.15 again $\hat{f} \notin V(\alpha)$. Thus $\Gamma \nVDash \alpha(\mathcal{F}_G)$. But $\mathfrak{C}$ is closed under ultraproducts, so $\mathcal{F}_G \in \mathfrak{C}$ and therefore $\Gamma \nVDash \alpha(\mathfrak{C})$.

The proof of (2) follows similar lines, replacing (i) by (i)': $\mathcal{F}_i \models i$, and putting $F_\beta = \{i : \mathcal{F}_i \models \beta\}$. We then use Theorem 1.7.13 to obtain $\mathcal{F}_G \models \Gamma$. $\qquad\square$

It is clear from the construction just given, and from 1.7.15, that 1.8.4(1) continues to hold for a class $\mathfrak{C}$ of K-frames closed under the ultraproducts of 1.7.3. That 1.8.4(2) is not in general true for such classes was shown by Thomason [95].

Corollary 1.8.5 *Let $\mathfrak{C}$ be a class of frames closed under ultraproducts ($\mathfrak{C}$ may be a class of first-order frames or a class of K-frames). Then $\Gamma \subseteq \Phi$ is satisfiable in $\mathfrak{C}$ iff every finite subset of Γ is satisfiable in $\mathfrak{C}$.*

Proof. Let $\bot = p \wedge \neg p$. Then Γ is satisfiable in $\mathfrak{C}$ iff $\Gamma \nVDash \bot (\mathfrak{C})$, and so the result follows from 1.8.4(1). $\qquad\square$

Corollary 1.8.6 *If Λ is a modal logic and $\mathfrak{C}$ is closed under ultraproducts, then $\mathfrak{C}$ simply determines Λ only if $\mathfrak{C}$ strongly determines Λ.*

Proof. Suppose that 1.8.3(2') holds. For each finite $\Delta \subseteq \Gamma$, let α_Δ be the conjunction of the members of Δ. Then Δ is Λ-consistent iff α_Δ is Λ-consistent (1.1.2) iff α_Δ is satisfiable in $\mathfrak{C}$ (1.8.3(2')) iff Δ is satisfiable in $\mathfrak{C}$ (1.2.4(i)). Thus we have $\Gamma \subseteq \Phi$ is Λ-consistent iff each finite subset of Γ is Λ-consistent iff each finite subset of Γ is satisfiable in $\mathfrak{C}$ iff (by 1.8.5) Γ is satisfiable in $\mathfrak{C}$. Thus 1.8.3(1') holds and $\mathfrak{C}$ strongly determines Λ. $\qquad\square$

1.9   Descriptive Frames

We return now to the problem discussed in Section 1.3 of finding a concept of "possible-worlds" model structure that is equivalent semantically and mathematically to that of a modal algebra. The structures we define are obtained by imposing constraints on frames $\langle W, R, P \rangle$ that we hope to show have some philosophical significance.

We recall that the members of P are called the *propositions* of the frame, and that if $x \in S$, then proposition S is to be understood as being true of the world x. Our first condition is a kind of converse to the thesis that a proposition is determined by the set of worlds in which it is true. It asserts that each world is uniquely determined by the propositions true in that world. For, if two states of affairs are to be different there must be something about one that does not hold of the other, and hence a statement that is true of one and not the other. We formalise this as follows:

For $x \in W$, let $Px = \{S \in P : x \in S\}$. Then our principle is

Axiom I. *For all $x, y \in W$, $Px = Py$ only if $x = y$.*

The interpretation of xRy is that y is a world accessible to x, in which case anything necessarily true of x is true of y. We now wish to assert the converse of this, viz. if all propositions necessarily true of x are true of y, then y is a conceivable alternative state of affairs to x. For, an individual living in world x could understand what y is like. He could comprehend at least all those facts about y that correspond to necessary truths of his own world. Thus we have

Axiom II. *If $x \in l_R(S)$ only if $y \in S$, for all $S \in P$, then xRy.*

Our third and final constraint is concerned with the set Px, which might be called the "truth-description" of x. Each Px is an ultrafilter in $\mathcal{F}^+$, i.e.

(1) $W \in Px$,
(2) $S, -S \cup T \in Px$ only if $T \in Px$,
(3) exactly one of $S, -S \in Px$, all $S \in P$.

Since the set $-S \cup T$ corresponds to the proposition "S implies T", these conditions may be interpreted as stating that in any particular world x: the universal proposition (W) is true; any proposition implied by a true proposition is true; and exactly one of each proposition and its negation (denial) is true. Our contention is that any collection of propositions meeting these conditions must be the truth description of some world. For, if a world is to be uniquely specified by the propositions true of it, then to make a consistent and exhaustive selection from amongst all possible propositions is to describe a state of affairs—that state in which all the selected propositions obtain, and all the rejected ones, the negations of selected ones, are false. We therefore have

Axiom III. *Every ultrafilter in $\mathcal{F}^+$ is of the form Px for some $x \in W$.*

Definition 1.9.1

(1) *A frame $\mathcal{F} = \langle W, R, P \rangle$ is **descriptive** iff it satisfies Axioms **I**, **II** and **III**.*

(2) *$\mathcal{F}$ is **refined** iff it satisfies **I** and **II**.*

Refined frames were first defined and studied by Thomason [96].

Theorem 1.9.2 *Any full frame is refined. Any finite full frame is descriptive, but no infinite full frame can be descriptive.*

Proof. Consider $\mathcal{F} = \langle W, R, P \rangle$, where $P = 2^W$. If $x \neq y$, then $x \in Px - Py$, whence $Px \neq Py$, and **I** holds. For **II**, suppose that not xRy. Let $S = \{z : xRz\}$, then $x \in l_R(S)$ and $y \notin S$. Thus $\mathcal{F}$ is refined. It is a standard result that if W is finite then every ultrafilter on W is principal and has singleton intersection, whence **III** holds. On the other hand every infinite set has non-principal ultrafilters [4, p. 108] with empty intersection, in which case **III** fails. $\square$

Descriptive frames are extremely rich in structure. Axiom **III** is reminiscent of the notion of convergence of ultrafilters, which in topology is equivalent to compactness. In fact the set P, being closed under $\cap$ and containing W, is a base for a topology on W which is compact in the presence of **III**. Axioms **I** and **III** make $\langle W, P \rangle$ a *perfect reduced set field* in the sense of Sikorski [90, p. 20].

Theorem 1.9.3 *Axiom **III** is equivalent, given P closed under $\cap$ and $-$, to each of*

IV *Every subset of P with the fip has non-empty intersection;*
    and

V *If $W = \bigcup_{i \in I} S_i$, with $S_i \in P$ all $i \in I$, then $W = \bigcup_{i \in I_0} S_i$ for some finite $I_0 \subseteq I$.*

Proof. **III** $\Rightarrow$ **IV**. If $P_0 \subseteq P$ has the fip, by 1.7.2 $P_0 \subseteq G$ for some ultrafilter G on $\mathcal{F}^+$. By **III**, $G = Px$ for some $x \in W$. Then clearly $x \in \bigcap P_0$.

IV $\Rightarrow$ **V**. If $W \neq \bigcup_{i \in I_0} S_i$ for all finite $I_0 \subseteq I$, then $\bigcap_{i \in I_0} -S_i \neq \emptyset$, hence $P_0 = \{-S_i : i \in I\}$ has the fip. But as P is closed under $-$, $P_0 \subseteq P$, so by **IV** there exists $x \in \bigcap_{i \in I} P_0 = \bigcap_{i \in I} -S_i$. Thus $\bigcup_{i \in I} S_i \neq W$.

$V \Rightarrow III$. Let G be an ultrafilter on $\mathcal{F}^+$. If $G \neq Px$, all $x \in W$, then by 1.7.1(3) and closure of P under $-$, for each x there exists $Sx \in P$ such that $Sx \in Px$ and $-Sx \in G$. Then $\bigcup_{x \in W} S_x = W$, so by V there exist $x_1, \ldots, x_n \in W$ such that $\bigcup_{i \leqslant n} Sx_i = W$, hence $\bigcap_{i \leqslant n} -Sx_i = \emptyset$. But each $-Sx_i \in G$, so by 1.7.1(4) $\emptyset \in G$, which is in contradiction with 1.7.1(6). $\square$

Theorem 1.9.4 *Let* $\mathcal{F} = \langle W, R, P \rangle$ *be a descriptive frame. Define*

$$\zeta = \{\bigcap Q : Q \subseteq P\}, \quad \tau = \{\bigcup Q : Q \subseteq P\}.$$

Then

(1) $\langle W, \tau \rangle$ *is a compact, Hausdorff topological space,*

(2) $S \in \tau$ *iff* $-S \in \zeta$,

(3) τ *is closed under finite intersections and* ζ *under finite unions,*

(4) *Every finite set is in* ζ *and every cofinite set in* τ,

(5) *If* $S \in \zeta$ *and* $S \subseteq \bigcup_{i \in I} S_i$, *where* $S_i \in \tau$ *all* $i \in I$, *then* $S \subseteq \bigcup_{i \in I_0} S_i$ *for some finite* $I_0 \subseteq I$,

(6) *If* $S \in \tau$ *and* $\bigcap_{i \in I} S_i \subseteq S$, *where* $S_i \in \zeta$ *all* $i \in I$, *then* $\bigcap_{i \in I_0} S_i \subseteq S$ *for some finite* $I_0 \subseteq I$.

Proof.

(1) τ is of course the class of open sets in the topology with P as base. By Axiom **I** and the closure of P under $-$, any two distinct points of W are separated by disjoint τ-open sets. Thus $\langle W, \tau \rangle$ is Hausdorff. Compactness is given by V (cf. 1.9.3).

(2) $-(\bigcup_{i \in I} S_i) = \bigcap_{i \in I} -S_i$, and P is closed under $-$.

(3)
$$(\bigcup_{i \in I} S_i) \cap (\bigcup_{j \in J} T_j) = \bigcup_{\substack{i \in I \\ j \in J}} (S_i \cap T_j),$$

and P is closed under finite intersections. Similarly $(\bigcap S_i) \cup (\bigcap T_j) = \bigcap (S_i \cup T_j)$, and P is closed under finite unions.

(4) By Axiom **I** and the closure of P under $-$, if $x \neq y$, there exists $S \in P$ such that $x \in S$ and $y \notin S$. Thus $\{x\} = (\bigcap Px) \in \zeta$. Since any finite set is a finite union of singletons, it follows from (3) that every finite set is in ζ. Hence by (2) every confinite set is in τ.

(5) By (2), ζ is the class of τ-closed subsets of W. Then (5) follows as the standard result that a closed subset of a compact space is compact.

(6) May be derived from (5), (2) and De Morgan's Laws for set algebra.
                                                               $\square$

Definition 1.9.5 *If $R \subseteq W^2$, then for each $n \in \mathbb{N}$ we define $l_R^n : 2^W \to 2^W$ inductively by*

$$l_R^0(S) = S \text{ and } l_R^{n+1}(S) = l_R(l_R^n(S)).$$

$m_R(S)$ is defined as $-(l_R^n((-S)))$.
  For $\alpha \in \Phi$ and $n \in \mathbb{N}$,

$$\square^n \alpha = \underbrace{\square \ldots \square}_{n \ times} \alpha.$$

Similarly $\Diamond^n \alpha = \underbrace{\Diamond \ldots \Diamond}_{n \ times} \alpha$, which is equivalent in K to $\neg\square^n\neg\alpha$.

The following properties are easily established.

Theorem 1.9.6

(1) $\bigcap_{i \in I} l_R^n(S_i) = l_R^n(\bigcap_{i \in I} S_i)$.

(2) $\bigcup_{i \in I} m_R^n(S_i) = m_R^n(\bigcup_{i \in I} S_i)$.

(3) $S \subseteq T$ only if $l_R^n(S) \subseteq l_R^n(T)$ and $m_R^n(S) \subseteq m_R^n(T)$.

(4) $l_R^n(S) = \{x : xR^ny \text{ only if } y \in S\}$.

(5) $m_R^n(S) = \{x : xR^ny \text{ for some } y \in S\}$.

(6) $\{S : x \in l_R^n(S)\}$ is closed under arbitrary intersections.

$\square$

Theorem 1.9.7 *If $\langle W, R, P \rangle$ is descriptive, then*

(1) *if $x \in l_R^n(S)$ only if $y \in S$, all $S \in P$, then xR^ny; or equivalently*

(2) *if $y \in S$ only if $x \in m_R^n(S)$, all $S \in P$, then xR^ny.*

Proof. By induction on n.

For $n = 0$, suppose $x \in S$ only if $y \in S$, all $S \in P$, i.e. $Px \subseteq Py$. But Px is a maximal filter, so $Px = Py$, whence by Axiom **I** $x = y$, i.e. xR^0y.

Now suppose the result holds for n, and that

(i) $x \in l_R^{n+1}(S)$ only if $y \in S$.

Let $P_0 = \{S \in P : x \in l_R(S)\} \cup \{m_R^n(S) : y \in S \in P\}$. Then if P_0 does not have the fip, by 1.9.6(6) there is $S \in P$ such that $x \in l_R(S)$, and $m_R^n(S_i)$ such that $y \in S_i$ for $i \leq m$, some m, and

$$S \cap m_R^n(S_1) \cap \ldots \cap m_R^n(S_m) = \emptyset.$$

Hence by 1.9.6(3), $S \cap m_R^n(S') = \emptyset$, where $S' = \bigcap_{i \leq m} S_i$. But then $S \subseteq -m_R^n(S')$, so using 1.9.6(3),

$$l_R(S) \subseteq l_R(-m_R^n(S')) = l_R^{n+1}(-S').$$

But $x \in l_R(S)$, so $x \in l_R^{n+1}(-S')$, thus by (i) $y \in -S'$, which is impossible as $y \in S'$.

Therefore we conclude that P_0 has the fip, whence by **IV** (1.9.3) there exists $z \in \bigcap P_0$. Then by Axiom **II**, IH, and the definition of P_0, we have xRz and $zR^n y$, hence $xR^{n+1}y$ as required. □

Corollary 1.9.8 *For all* $x \in W$ *and* $n \in \mathbb{N}$, $R_x^n = \{t : xR^n t\} \in \zeta$.

Proof. If $t \notin R_x^n$, by 1.9.7 there is $S_t \in P$ such that $x \in l_R^n(S_t)$ and $t \notin S_t$. Let $B = \bigcap \{S_t : t \notin R_x^n\} \in \zeta$. Then clearly $B \subseteq R_x^n$. But by 1.9.6(1), $x \in l_R^n(B)$, so by 1.9.6(4), $R_x^n \subseteq B$. □

Corollary 1.9.9 *For any* $S \subseteq W$,

(1) $l_R^n(S) = \bigcup_{S \supseteq T \in \zeta} l_R^n(T) = \bigcap_{S \subseteq T \in \tau} l_R^n(T)$;

(2) $m_R^n(S) = \bigcap_{S \subseteq T \in \tau} m_R^n(T) = \bigcup_{S \supseteq T \in \zeta} m_R^n(T)$.

Proof. We begin with the first part of (1). By 1.9.6(3)

$$\bigcup\nolimits_{S \supseteq T \in \zeta} l_R^n(T) \subseteq l_R^n(S).$$

Conversely if $x \in l_R^n(S)$, $R_x^n \subseteq S$ by 1.9.6(4). But $R_x^n \in \zeta$ (1.9.8) and clearly $x \in l_R^n(R_x^n)$. Hence $x \in \bigcup_{S \supseteq T \in \zeta} l_R^n(T)$.

For the second part of (2), we have

$$m_R^n(S) \supseteq \bigcup\nolimits_{S \supseteq T \in \zeta} m_R^n(T)$$

by 1.9.6(3). Conversely if $x \in m_R^n(S)$, $xR^n y$ for some $y \in S$. Then $x \in m_R^n\{y\}$, and $S \supseteq \{y\} \in \zeta$ by 1.9.4(4).

Now for the second part of (1),

$$l_R^n(S) = -m_R^n(-S) = -(\bigcup\nolimits_{-S \supseteq T \in \zeta} m_R^n(T)) \quad \text{(above)}$$

$$= \bigcap\nolimits_{-S \supseteq T \in \zeta} -m_R^n(T) = \bigcap\nolimits_{-S \supseteq T \in \zeta} l_R^n(-T) = \bigcap\nolimits_{S \subseteq T \in \tau} l_R^n(T) \quad (1.9.4(2)).$$

The first part of (2) follows similarly from the first part of (1). □

We end this section with some observations about the constructibility of descriptive frames from given ones.

Theorem 1.9.10 *The disjoint union of a finite number of descriptive frames is descriptive, but a disjoint union of infinitely many frames is never descriptive.*

Proof. Let $\mathcal{F}$ be the disjoint union of disjoint descriptive frames $\mathcal{F}_1$ and $\mathcal{F}_2$. Then $W = W_1 \cup W_2$, $R = R_1 \cup R_2$ and $P = \{S \cup T : S \in P_1 \text{ and } T \in P_2\}$, so clearly $P_1, P_2 \subseteq P$ (since $\emptyset \in P_1 \cap P_2$). Now for Axiom **I**, suppose $x \neq y$. If $x \in W_1$, $y \in W_2$, then $W_1 \in Px - Py$, so $Px \neq Py$. If $x, y \in W_1$, then as $\mathcal{F}_1$ satisfies **I**, there exists $S \in P_1$ such that $x \in S$, $y \notin S$. Then $S \in Px - Py$. A similar result holds if $x, y \in W_2$.

For **II**, suppose not xRy. If $x, y \in W_1$, for some $S \in P_1$ we have $x \in l_{R_1}(S)$ and $y \notin S$. Then clearly $x \in l_R(S)$ and $S \in P$. If $x \in W_1$, $y \in W_2$, then $x \in l_R(W_1)$ and $y \notin W_1 \in P$.

For **III**, suppose $W = \bigcup_{i \in I}(S_i \cup T_i)$ where for all $i \in I$, $S_i \in P_1$ and $T_i \in P_2$. Then $W = (\bigcup_{i \in I} S_i) \cup (\bigcup_{i \in I} T_i)$, and since W is the disjoint union of W_1 and W_2, $W_1 = \bigcup S_i$, $W_2 = \bigcup T_i$. But $\mathcal{F}_1$, $\mathcal{F}_2$ satisfy **V**, so $W_1 = \bigcup_{i \in I_0} S_i$, $W_2 = \bigcup_{i \in J} T_i$, for some finite $I_0, J \subseteq I$. Then $W = \bigcup_{i \in I_0 \cup J}(S_i \cup T_i)$. Hence $\mathcal{F}$ satisfies **V** and therefore **III** by 1.9.3.

To complete the theorem we observe that if $\mathcal{F}$ is the disjoint union of an infinite disjoint family $\{\mathcal{F}_i : i \in I\}$, then $W = \bigcup_{i \in I} W_i$. But each $W_i \in P$, and since the W_i's are pairwise disjoint and non-empty we cannot have $W = \bigcup_{i \in I_0} W_i$ for any finite $I_0 \subseteq I$. Hence $\mathcal{F}$ does not satisfy **V** so cannot be descriptive. $\qquad\qquad\Box$

We note that the proof just given is easily adapted to show that the disjoint union of any collection of refined frames is refined.

Theorem 1.9.11 *An ultraproduct of refined frames is refined.*

Proof. Let $\mathcal{F}_G$ be the ultraproduct of the $\mathcal{F}_i$'s over an ultrafilter G on a set I.

For Axiom **I**, suppose $\hat{f} \neq \hat{g}$. Then $A = \{i : f(i) \neq g(i)\} \in G$. For $i \in A$, as $\mathcal{F}_i$ is refined there is $\sigma_i \in P_i$ such that $f(i) \in \sigma_i$, $g(i) \notin \sigma_i$. If σ is such that $\sigma(i) = \sigma_i$, all $i \in A$, then $A \subseteq [f, \sigma]$, so $\hat{f} \in S_{\hat{\sigma}}$. But $[g, \sigma] \subseteq -A \notin G$, so $\hat{g} \notin S_{\hat{\sigma}}$. For **II**, suppose that not $\hat{f} R_G \hat{g}$. Then $B = \{i : \text{not } f(i) R_i g(i)\} \in G$. For each $i \in B$ there exists $\sigma(i) \in P_i$ such that $f(i) \in l_{R_i}(\sigma(i))$ and $g(i) \notin \sigma(i)$. Then, reasoning as above, we obtain $\hat{f} \in l_{R_G}(S_{\hat{\sigma}}) \in P_G$ and $\hat{g} \notin S_{\hat{\sigma}}$. $\qquad\qquad\Box$

Theorem 1.9.11 still goes through under the weaker hypothesis that almost all $\mathcal{F}_i$'s are refined. However $\mathcal{F}_G$ may not be descriptive even if all $\mathcal{F}_i$'s are. For example, suppose that each $\mathcal{F}_i$ is finite and full, hence descriptive by 1.9.2, but $\mathcal{F}_G$ is infinite (such ultraproducts exist—one will be displayed in Section 1.17). Let $\hat{f} \in W_G$. Define $\sigma(i) = \{f(i)\}$, all $i \in I$. Then as each $\mathcal{F}_i$ is full, $\sigma \in \prod P_i$. But it is easily seen that $S_{\hat{\sigma}} = \{\hat{f}\}$, so each singleton subset of W_G is in P_G. Since W_G is infinite, it follows that $\mathcal{F}_G$ does not satisfy **V**, so is not descriptive.

1.10   The Categories of Descriptive Frames and Modal Algebras

It was shown in Section 1.3 that each frame has a semantically equivalent MA. Conversely, as will now be shown, each MA has a corresponding frame that is descriptive. The construction is an adaptation of Lemmon [58, Section III], which itself is an extension of Stone's Representation Theorem for BA's.

It should however be recognised that the fact that an MA is isomorphic to an algebra of a descriptive frame is due originally to Jónsson and Tarski [48, Theorem 3.10], who developed a general theory of representation of additive n-ary operators on Boolean algebras in terms of $n+1$-ary relations on sets. Modal algebras are essentially the case $n = 1$ of this theory. Moreover, Theorem 3.5 of [48] showed that various elementary properties of a binary relation (reflexivity, symmetry, transitivity, functionality) are equivalent to satisfaction of certain MA-equations by the algebra of the associated frame. Thus Jónsson and Tarski provided the mathematical tools and results to give the completeness theorem with respect to set-theoretic semantics of a number of the more well-known modal logics. This development occurred more than a decade before the work of Kripke, Lemmon et. alia., but was not taken up by the latter authors.

Definition 1.10.1 *If* $\mathfrak{A} = \langle A, \cap, ', l \rangle$ *is an* MA, *then*

$$\mathfrak{A}_+ = \langle W_{\mathfrak{A}}, R_{\mathfrak{A}}, P^{\mathfrak{A}} \rangle,$$

where

$$W_{\mathfrak{A}} = \{x : x \text{ is an ultrafilter on } \mathfrak{A}\},$$

$$x R_{\mathfrak{A}} y \text{ iff } \{a : la \in x\} \subseteq y \text{ iff } \{ma : a \in y\} \subseteq x,$$

$$P^{\mathfrak{A}} = \{|a|^{\mathfrak{A}}\}, \text{ where } |a|^{\mathfrak{A}} = \{x \in W_{\mathfrak{A}} : a \in x\}.$$

That $P^{\mathfrak{A}}$ is closed under $\cap, -, l_{R_{\mathfrak{A}}}$ is shown by the next theorem, which is given by [58, Theorem 32].

Theorem 1.10.2

(1) $|a|^{\mathfrak{A}} = |b|^{\mathfrak{A}}$ *iff* $a = b$.
(2) $W_{\mathfrak{A}} - |a|^{\mathfrak{A}} = |a'|^{\mathfrak{A}}$.
(3) $|a|^{\mathfrak{A}} \cap |b|^{\mathfrak{A}} = |a \cap b|^{\mathfrak{A}}$.
(4) $l_{R_{\mathfrak{A}}}(|a|^{\mathfrak{A}}) = |la|^{\mathfrak{A}}$.

Corollary 1.10.3 $\mathfrak{A} \cong (\mathfrak{A}_+)^+$.

Proof. The map $a \mapsto |a|^{\mathfrak{A}}$ is by 1.10.1 and 1.10.2 an MA isomorphism of A onto $P^{\mathfrak{A}}$. $\qquad\square$

Corollary 1.10.4 $\mathfrak{A} \models \alpha$ *iff* $\mathfrak{A}_+ \models \alpha$.

Proof. Isomorphic MA's satisfy the same identities, so $\mathfrak{A} \models \alpha$ iff $(\mathfrak{A}_+)^+ \models \alpha$ (1.10.3) iff $\mathfrak{A}_+ \models \alpha$ (Section 1.3). $\qquad\square$

Theorem 1.10.5 $\mathfrak{A}^+$ *is a descriptive frame.*

Proof. For **I**, suppose $x \neq y \in W_{\mathfrak{A}}$. Then (using 1.7.1(3)) there is $a \in A$ such that $a \in x$, $a \notin y$. Clearly it follows that $|a|^{\mathfrak{A}} \in P^{\mathfrak{A}}x - P^{\mathfrak{A}}y$. For **II**, suppose not $xR_{\mathfrak{A}}y$. Then for some $a \in A$, $\boldsymbol{l}a \in x$ and $a \notin y$. Hence $x \in |\boldsymbol{l}a|^{\mathfrak{A}} = \boldsymbol{l}_{R_{\mathfrak{A}}}(|a|^{\mathfrak{A}})$ (1.10.2) and $y \notin |a|^{\mathfrak{A}} \in P^{\mathfrak{A}}$. Axiom **III** is established as in [90, p. 24], and we briefly repeat the argument. If G is an ultrafilter on $P^{\mathfrak{A}}$, then by 1.7.1 and 1.10.2, $g = \{a : |a|^{\mathfrak{A}} \in G\}$ is an utrafilter on $\mathfrak{A}$, i.e. $g \in W_{\mathfrak{A}}$. But $|a|^{\mathfrak{A}} \in G$ iff $a \in g$ iff $g \in |a|^{\mathfrak{A}}$, so $G = P^{\mathfrak{A}}g$.   □

Corollary 1.10.6 *Every frame has a semantically equivalent descriptive frame.*

Proof. $\mathcal{F}$ is equivalent to $\mathcal{F}^+$ (Section 1.3) which by 1.10.4 and 1.10.5 is equivalent to the descriptive frame $(\mathcal{F}^+)+$.   □

For descriptive frames themselves we have the following strengthening of 1.10.6:

Theorem 1.10.7 *If $\mathcal{F}$ is a descriptive frame, then $\mathcal{F} \cong (\mathcal{F}^+)_+$.*

Proof. Since each Px is an ultrafilter on $\mathcal{F}^+$, the correspondence $Q : x \mapsto Px$ is by 1.10.1 a map of W into $W_{\mathcal{F}^+}$. Q is injective by Axiom **I** and onto by Axiom **III**. Furthermore

$$\begin{aligned} xRy \quad &\text{iff} \quad \{S \in P : \boldsymbol{l}_R(S) \in Px\} \subseteq Py \quad \text{(Axiom \textbf{II}, def. } \boldsymbol{l}_R) \\ &\text{iff} \quad PxR_{\mathcal{F}^+}Py \quad\quad\quad\quad\quad\quad\quad (1.10.1) \\ &\text{iff} \quad Q(x)R_{\mathcal{F}^+}Q(y). \end{aligned}$$

To complete the proof that Q is an isomorphism, it suffices to show that $S \in P$ iff $Q(S) \in P^{\mathcal{F}^+}$. But for any $S \in P$

$$Q(S) = \{Q(x) : x \in S\} = \{Px : S \in Px\} = |S|^{\mathcal{F}^+},$$

so $S \in P$ only if $Q(S) \in P^{\mathcal{F}^+}$ by 1.10.1. Conversely, if $Q(S) \in P^{\mathcal{F}^+}$, by 1.10.1 $Q(S) = |T|^{\mathcal{F}^+} = Q(T)$ for some $T \in P$. But Q is bijective so $S = T \in P$.   □

We saw in Section 5 that a frame homomorphism induces an MA homomorphism of the corresponding MA's. A similar duality holds between mappings of MA's and their corresponding descriptive frames.

Definition 1.10.8 *Let $\psi : \mathfrak{A} \to \mathfrak{B}$ be an MA homomorphism of $\mathfrak{A}$ into $\mathfrak{B}$. Then $\psi_+ : \mathfrak{B}_+ \to \mathfrak{A}_+$ is defined, for $x \in W_{\mathfrak{B}}$, by*

$$\psi_+(x) = \{a \in A : \psi(a) \in x\}.$$

This construction is a standard one in the representation theory of BA's [90, Section 11]. That $\psi_+(x)$ is an ultrafilter in $\mathfrak{A}$ follows from the homomorphism properties of ψ, e.g. exactly one of $\psi(a)$ and $(\psi(a))' = \psi(a')$ is in x, so exactly one of a, a' is in $\psi_+(x)$. ψ_+ may alternately be described as follows. By the isomorphism of 1.10.3 ψ may be regarded as a map from $P^{\mathfrak{A}}$ to $P^{\mathfrak{B}}$. For $x \in W_{\mathfrak{B}}$, $\{S \in P^{\mathfrak{A}} : x \in \psi(S)\}$ is an ultrafilter on $P^{\mathfrak{A}}$ so, as $\mathfrak{A}_+$ is descriptive, is of the form $P^{\mathfrak{A}}y$ for some $y \in W_{\mathfrak{A}}$. We then put $\psi_+(x) = y$.

Theorem 1.10.9

(1) ψ_+ is a frame homomorphism
(2) ψ is injective only if ψ_+ is onto
(3) ψ is onto only if ψ_+ is an embedding
(4) ψ is an MA isomorphism only if ψ_+ is a frame isomorphism.

Proof.

(1) Suppose $xR_{\mathfrak{B}}y$. Then for any $a \in A$, $la \in \psi_+(x)$ only if $\psi(la) = l(\psi(a)) \in x$. Then as $xR_{\mathfrak{B}}y$, $\psi(a) \in y$ and so $a \in \psi_+(y)$. By 1.10.1 this implies $\psi_+(x)R_{\mathfrak{A}}\psi_+(y)$.

Now suppose $\psi_+(x)R_{\mathfrak{A}}z$. The sets $y_1 = \{a : la \in x\}$ and $y_2 = \{\psi(b) : b \in z\}$ are each closed under $\cap$ (the latter since ψ is an MA homomorphism and z is closed under $\cap$), so if $y_0 = y_1 \cup y_2$ does not have the fip there exist $a, \psi(b)$ such that $la \in x$, $b \in z$ and $a \leq (\psi(b))' = \psi(b')$. Then $la \leq l(\psi(b')) = \psi(l(b'))$. Then $\psi(l(b')) \in x$, so $l(b') \in \psi_+(x)$. Since $\psi_+(x)R_{\mathfrak{A}}z$, we get $b' \in z$, which is impossible as $b \in z$. Thus y_0 has the fip, so by 1.7.2 $y_0 \subseteq y$ for some $y \in W_{\mathfrak{B}}$. As $y_1 \subseteq y$, we have $xR_{\mathfrak{B}}y$ by 1.10.1. As $y_2 \subseteq y$, $z \subseteq \psi_+(y)$, whence by the maximality of z, $\psi_+(y) = z$.

Now if $S \in P^{\mathfrak{A}}$, $S = |a|^{\mathfrak{A}}$ for some $a \in A$. Then $\psi_+^{-1}(S) = \psi_+^{-1}(|a|^{\mathfrak{A}}) = \{x : \psi_+(x) \in |a|^{\mathfrak{A}}\} = \{x : a \in \psi_+(x)\} = \{x : \psi(a) \in x\} = |\psi(a)|^{\mathfrak{B}} \in P^{\mathfrak{B}}$. Thus ψ_+ satisfies 1.5.1 and is a frame homomorphism.

(2) Let $x \in W_{\mathfrak{A}}$. Put $y = \{\psi(a) : a \in x\}$. If ψ is injective then $\psi(a) \in y$ iff $a \in x$. From this it follows easily that $y \in W_{\mathfrak{B}}$ and $\psi_+(y) = x$.

(3) Suppose $\psi_+(x) = \psi_+(y)$. If ψ is onto, each element of $\mathfrak{B}$ is of the form $\psi(a)$, some $a \in A$. Then we have $\psi(a) \in x$ only if $a \in \psi_+(x)$, only if $a \in \psi_+(y)$, only if $\psi(a) \in y$. Thus $x \subseteq y$, whence by maximality $x = y$ and ψ_+ is injective.

Now suppose $|b|^{\mathfrak{B}} \in P^{\mathfrak{B}}$. As ψ is onto, $b = \psi(a)$, some $a \in A$. But then, from the proof of (1), $\psi_+(|\psi(a)|^{\mathfrak{B}}) = \psi_+(\psi_+^{-1}(|a|^{\mathfrak{A}})) = |a|^{\mathfrak{A}} \cap \psi_+(W_{\mathfrak{B}})$. Thus ψ_+ is an embedding.

(4) Follows from (1)–(3).

$\square$

Since ψ_+ is a frame homomorphism, $(\psi_+)^+$ is an MA homomorphism of $(\mathfrak{A}_+)^+$ into $(\mathfrak{B}_+)^+$ (1.5.9). But these two algebras are isomorphic to $\mathfrak{A}$ and $\mathfrak{B}$ respectively (1.10.3) so one might expect there to be some relationship between $(\psi_+)^+$ and the original map ψ.

Theorem 1.10.10 *For each MA $\mathfrak{A}$, let $\psi_{\mathfrak{A}} : \mathfrak{A} \rightarrow (\mathfrak{A}_+)^+$ be the iso-morphism of 1.10.3. Then for any MA homomorphism $\psi : \mathfrak{A} \rightarrow \mathfrak{B}$, $(\psi_+)^+ \circ \psi_{\mathfrak{A}} = \psi_{\mathfrak{B}} \circ \psi$.*

Proof. For any $a \in A$,

$$
\begin{aligned}
(\psi_+)^+(\psi_{\mathfrak{A}}(a)) &= (\psi_+)^+(|a|^{\mathfrak{A}}) & (1.10.3) \\
&= \psi_+^{-1}(|a|^{\mathfrak{A}}) & (1.5.8) \\
&= |\psi(a)|^{\mathfrak{B}} & (\text{proof of } 1.10.9(1)) \\
&= \psi_{\mathfrak{B}}(\psi(a)) & (1.10.3) \qquad \square
\end{aligned}
$$

The analogous result for frame homomorphisms is

Theorem 1.10.11 *For any descriptive frame $\mathcal{F}$, let $Q_{\mathcal{F}} : \mathcal{F} \rightarrow (\mathcal{F}^+)_+$ be the isomorphism of 1.10.7. Then for any frame homomorphism $Q : \mathcal{F} \rightarrow \mathcal{F}'$, $(Q^+)_+ \circ Q_{\mathcal{F}} = Q_{\mathcal{F}'} \circ Q$.*

Proof. For any $x \in W$,

$$
\begin{aligned}
(Q^+)_+(Q_{\mathcal{F}}(x)) &= (Q^+)_+(Px) & (1.10.7) \\
&= \{S \in P' : Q^+(S) \in Px\} & (1.10.8) \\
&= \{S \in P' : x \in Q^{-1}(S)\} & (1.5.8) \\
&= \{S \in P' : Q(x) \in S\} & \\
&= P'_{Q(x)} & \\
&= Q_{\mathcal{F}'}(Q(x)). & (1.10.7)
\end{aligned}
$$

$\square$

Now the identity map on any frame is a homomorphism, and the composition of frame homomorphisms is a frame homomorphism, so the collect $\mathfrak{D}$ of all descriptive frames and homomorphisms between descriptive frames forms a *category* in the sense of Pareigis [69, p. 1]. Similarly the class $\mathfrak{M}$ of MA's and MA homomorphisms is a category. The correspondence $\mathcal{F} \mapsto \mathcal{F}^+$, $Q \mapsto Q^+$ defines functor $(-)^+$ from $\mathfrak{D}$ to $\mathfrak{M}$ that is contravariant ([69, p. 7]; it is easily checked that $(Q_1 \circ Q_2)^+ = Q_2^+ \circ Q_1^+$ and $(id_{\mathcal{F}})^+ = id_{\mathcal{F}^+}$, where id denotes identity maps). Similarly, the constructions of 1.10.1 and 1.10.8 yield a contravariant functor $(-)_+$ from $\mathfrak{M}$ to $\mathfrak{D}$. 1.10.10 shows that the collection of isomorphisms $\psi_{\mathfrak{A}}$ constitutes a natural isomorphism between the composite functor $((-)_+)^+$ and the identity functor on $\mathfrak{M}$ ([69, pp. 9, 18]). Similarly, by 1.10.11 the $Q_{\mathcal{F}}$'s are a natural isomorphism between $((-)^+)_+$ and the identity functor on $\mathfrak{D}$. Hence the categories $\mathfrak{D}$ and $\mathfrak{M}$ are *dual* to each other ([69, p. 18]).

Subcategories $\mathfrak{D}'$ and $\mathfrak{M}'$ of $\mathfrak{D}$ and $\mathfrak{M}$ may be formed by keeping the same objects but retaining only isomorphisms as the maps. By inverting these maps appropriately, we obtain *covariant* functors between $\mathfrak{D}'$ and $\mathfrak{M}'$ whose composites are naturally isomorphic to the respective identity functors. This means that the category of descriptive frames and frame isomorphisms is *equivalent* ([69, p. 18]) to the category of MA's and MA homomorphisms. It is in this sense that we assert the two kinds of object are mathematically equivalent.

1.11   Inverse Limits of Descriptive Frames

It is known [39, p. 156] that polynomial identities are preserved under direct limits of algebras. It might therefore be expected that the analogous result holds for inverse limits of frames. However, it appears that this construction can only be effectively carried out for descriptive frames. The development we give depends heavily on the compactness properties of the latter.

Definition 1.11.1 *An **inverse family** of descriptive frames consists of:*

(1) *a directed partially ordered set $\langle I, \leq \rangle$;*
(2) *a descriptive frame $\mathcal{F}_i = \langle W_i, R_i, P_i \rangle$ for each $i \in I$;*
(3) *frame homomorphisms $Q_j^i : \mathcal{F}_i \to \mathcal{F}_j$ for all $i \geq j$, such that*

$$Q_k^j \circ Q_j^i = Q_k^i \text{ whenever } i \geq j \geq k, \text{ and } Q_i^i = id_{W_i}.$$

*The **inverse limit** of the $\mathcal{F}_i$'s is the structure $\mathcal{F}^\infty = \langle W^\infty, R^\infty, P^\infty \rangle$, where*

$$W^\infty = \{f \in \textstyle\prod_{i \in I} W_i : Q_j^i(f(i)) = f(j), \text{ all } i \geq j\},$$
$$fR^\infty g \text{ iff } f(i)R_i g(i) \text{ for all } i \in I,$$
$$P^\infty = \textstyle\bigcup_{i \in I}\{(Q_i^\infty)^{-1}(S) : S \in P_i\}, \quad \text{where}$$
$$Q_i^\infty : W^\infty \to W_i \text{ is the projection map given by } Q_i^\infty(f) = f(i).$$

We show that $\mathcal{F}^\infty$ is a descriptive frame by the following sequence of results:

Theorem 1.11.2

(1) *if $i \geq j$, $Q_j^i \circ Q_i^\infty = Q_j^\infty$, hence $Q_j^{\infty-1}(S) = Q_i^{\infty-1}(Q_j^{i-1}(S))$.*
(2) *if $k \geq i, j$, $Q_i^{\infty-1}(S) \cap Q_j^{\infty-1}(T) = Q_k^{\infty-1}(Q_i^{k-1}(S) \cap Q_j^{k-1}(T))$.*
(3) *P^∞ is a field of sets (i.e. closed under $\cap, -$).*
(4) *$\mathcal{F}^\infty$ satisfies Axioms **I**, **V** of descriptive frames.*
(5) *Q_i^∞ is a frame homomorphism.*
(6) *$m_{R^\infty}(Q_i^{\infty-1}(S)) = Q_i^{\infty-1}(m_{R_i}(S))$.*

(7) $\mathcal{F}^\infty$ is a frame.

(8) $\mathcal{F}^\infty$ satisfies Axiom **II** and is a descriptive frame.

Proof.

(1) [39, p. 131].

(2) By (1) and distribution of $Q_k^{\infty-1}$ over $\cap$.

(3) Since $-Q_i^{\infty-1}(S) = Q_i^{\infty-1}(-S)$ and P_i is closed under $-$, so is P^∞. If $Q_i^{\infty-1}(S), Q_i^{\infty-1}(T) \in P^\infty$, choose some $k \geq i, j$ (k exists by the directedness of $\leq$). As Q_i^k, Q_j^k are homomophisms, $Q_i^{k-1}(S), Q_j^{k-1}(T) \in P_k$, so $Q_i^{\infty-1}(S) \cap Q_j^{\infty-1}(T) \in P^\infty$ by (2), closure of P_k under $\cap$, and the definition of P^∞.

(4) If $f \neq g \in W^\infty$, then for some i, $f(i) \neq g(i)$. But $\mathcal{F}_i$ satisfies **I**, so there is $S_i \in P_i$ such that $f(i) \in S_i$, $g(i) \notin S_i$. Then clearly $Q_i^{\infty-1}(S) \in P_f^\infty - P_g^\infty$.

   To prove **V** we appeal to the topological properties of descriptive frames. By 1.9.4(1) each $\langle W_i, \tau_i \rangle$ is compact and Hausdorff, where τ_i is the topology with P_i as base. Hence by Theorem 3.6, p. 217, of Eilenberg and Steenrod [11], W^∞ is a non-empty subset of $\prod W_i$ that is compact in the subspace topology of the product topology on $\prod W_i$. But if $S \in P_i$, $Q_i^{\infty-1}(S) = W^\infty \cap Q_i^{-1}(S)$, where Q_i is the projection map on $\prod W_i$. Thus each member of P^∞ is open in the subspace topology, hence by compactness **V** holds for P^∞.

(5) That Q_i^∞ satisfies 1.5.1(1) and 1.5.1(3) follows easily from the definitions. For 1.5.1(2), let $f \in W^\infty$ and suppose $Q_i^\infty(f)R_i z$, i.e. $f(i)R_i z$. For each $j \in I$, let $A_j \subset \prod W_i$ contain those g such that

$$
\begin{aligned}
&\text{(i)} \quad g(i) = z, \\
&\text{(ii)} \quad Q_k^j(g(j)) = g(k) \quad \text{if} \quad j \geq k, \\
&\text{(iii)} \quad f(k)R_k g(k) \quad \text{if} \quad j \geq k.
\end{aligned}
$$

Then if $g \in \bigcap_{j \in I} A_j$, by (iii) $f R^\infty g$, by (ii) $g \in W^\infty$ and by (i) $Q_i^\infty(g) = z$. We must therefore show there exists such a g.

Lemma 1. $A_j \neq \emptyset$.

Proof. Take $l \geq i, j$. As Q_i^l is a homomorphism and $f(i) = Q_i^l(f(l))$, for some $t \in W_l$ we have $f(l)R_l t$ and $Q_i^l(t) = z$. Put $g(k) = Q_k^l(t)$ for all $k \leq l$. Then (i) holds with $i = k$, (iii) holds as Q_k is a homomorphism and $j \geq k$ only if $l \geq k$, and for (ii) we have $j \geq k$ only if $Q_k^j(g(j)) = Q_k^j(Q_j^l(t)) = Q_k^l(t) = g(k)$. Thus, if we choose g($k$) arbitrarily if not $k \leq l$, we have $g \in A_j$.   □

Lemma 2. A_j is closed in $\prod W_i$.

Proof. We show that $-A_j$ is open, i.e. for $g \notin A_j$ there is an open set A such that $g \in A \subseteq -A_j$. Now if $g \notin A_j$ then g fails to satisfy one of (i)-(iii) above.

(i) $g(i) \neq z$. As $\mathcal{F}_i$ satisfies Axiom **I** there is $S \in P_i$ such that $g(i) \in S$, $z \notin S$. Let $A = Q_i^{\infty-1}(S)$. Then $g \in A$, and if $h \in A$, $h(i) \in S$, so $h(i) \neq z$ and hence $h \notin A_j$.

(ii) $Q_k^j(g(j)) \neq g(k)$ for some $k \leq j$. Then there is $S \in P_k$ such that $g(k) \in S$, $Q_k^j(g(j)) \notin S$. Let $T = Q_k^{j-1}(-S) \in P_j$. Then $g(j) \in T$. Putting $A = Q_j^{\infty-1}(T) \cap Q_k^{\infty-1}(S)$, we have $g \in A$, and $h \in A$ only if $h(k) \in S$, $h(j) \in T$, so $Q_k^j(h(j)) \in Q_k^j(T) \subseteq -S$, whence $Q_k^j(h(j)) \neq h(k)$ and therefore $h \notin A_j$.

(iii) For some $k \leq j$, not $f(k)R_k g(k)$. As $\mathcal{F}_k$ is descriptive, for some $S \in P_k$, $g(k) \in S$ and $f(k) \notin m_{R_k}(S)$. Let $A = Q_k^{\infty-1}(S)$. Then $g \in A$, and $h \in A$ only if $h(k) \in S$, whence not $f(k)R_k h(k)$ and so $h \notin A_j$.   □

Now clearly $j \geq k$ only if $A_j \subseteq A_k$, so as $\leq$ is directed it follows from Lemmata 1 and 2 that $\{A_j : j \in I\}$ is a family of closed sets with the fip. By compactness this implies $\bigcap_{j \in I} A_j \neq \emptyset$, which, as explained, completes the proof that Q_i^∞ is a homomorphism.

(6) From (5) and 1.5.2(iii).

(7) By (3), P^∞ is closed under $\cap, -$. Since each P_i is closed under m_{R_i}, (6) yields P^∞ closed under m_{R^∞}.

(8) If not $fR^\infty g$, for some i not $f(i)R_i g(i)$. Then as $\mathcal{F}_i$ satisfies **II**, there is $S \in P_i$ such that $g(i) \in S$, $f(i) \notin m_{R_i}(S)$. Then $g \in Q_i^{\infty-1}(S) \in P^\infty$, and $f \notin Q_i^{\infty-1}(m_{R_i}(S)) = m_{R^\infty}(Q_i^\infty(S))$. Thus $\mathcal{F}^\infty$ satisfies **II** and this, together with (4) (and 1.9.3) completes the proof that $\mathcal{F}^\infty$ is a descriptive frame.   □

For convenience we state the next theorem for functions with two-placed arguments. It generalises easily to any polynomial.

Theorem 1.11.3 *If* $\alpha(p,q) \in \Phi$,

$$h_\alpha^{\mathcal{F}^{\infty+}}(Q_i^{\infty-1}(S), Q_j^{\infty-1}(T)) = Q_k^{\infty-1}(h_\alpha^{\mathcal{F}_k^+}(Q_i^{k-1}(S), Q_j^{k-1}(T)))$$

whenever $k \geq i, j$.

Proof. By induction on α, using 1.11.2(1), (2), (3), (6).   □

Corollary 1.11.4 $\mathcal{F} \models \alpha$ *if* $\mathcal{F}_i \models \alpha$ *for all* $i \in I$.

Proof. If $\mathcal{F}_i \models \alpha$ for all $i \in I$, then the argument of $Q_k^{\infty-1}$ in 1.11.3 is always W_k (a suitable k can always be found by the directedness of $\leq$). The result follows as $Q_k^{\infty-1}(W_k) = W^\infty$. $\qquad\square$

Definition 1.11.5 *A **direct family** of MA's consists of*

(i) *a set I direct by $\leq$;*

(ii) *an MA $\mathfrak{A}_i = \langle A_i, \cap, ', \mathbf{l} \rangle$ for each $i \in I$;*

(iii) *MA homomorphisms $\psi_j^i : \mathfrak{A}_i \to \mathfrak{A}_j$ whenever $i \leq j$, such that $\psi_k^j \circ \psi_j^i = \psi_k^i$ if $i \leq j \leq k$, and $\psi_i^i = id_{\mathfrak{A}_i}$.*

An equivalence relation is defined on $\bigcup A_i$, the (disjoint) union of the A_i's by:

$$\text{if } x \in A_i, y \in A_j, \text{ then } x \sim y \text{ iff for some } k \geq i,j, \ \psi_k^i(x) = \psi_k^j(y).$$

Let $A_\infty = \{[x] : x \in \bigcup A_i\}$ be the resulting set of equivalence classes, and define

$$
\begin{aligned}
\mathbf{l}[x] &= [\mathbf{l}x] \\
[x]' &= [x'] \\
[x] \cap [y] &= [\psi_k^i(x) \cap \psi_k^j(y)], \text{ for any } k \geq i,j.
\end{aligned}
$$

*Then $\mathfrak{A}_\infty = \langle A_\infty, \cap, ', \mathbf{l} \rangle$ is the **direct limit** of the $\mathfrak{A}_i$'s. (A proof that $\mathfrak{A}_\infty$ is well defined is given in [39, p. 129]).*

Now if $\{\mathcal{F}_i : i \in I\} \cup \{Q_j^i : i \geq j\}$ is an inverse family of descriptive frames, then it is easily seen that $\{\mathcal{F}_i^+ : i \in I\} \cup \{\psi_j^i : i \leq j\}$ is a direct family of MA's, where $\psi_j^i = Q_i^{j+}$ as in 1.5.8.

Theorem 1.11.6 *If $\mathcal{F}^\infty$ is the inverse limit of the $\mathcal{F}_i$'s, then $\mathcal{F}^{\infty+}$ is a homomorphic image of the direct limit of the $\mathcal{F}_i^+$'s. If each Q_j^i is onto, the homomorphism is an isomorphism.*

Proof. Define $\theta : P_\infty \to P^\infty$ by: for $S \in \bigcup P_i$

$$\theta([S]) = Q_i^{\infty-1}(S), \quad \text{where } S \in P_i.$$

To check that θ is well-defined, suppose $S \sim T$ with $T \in P_j$. By 1.11.5, for some $k \geq i,j$, $\psi_k^i(S) = \psi_k^j(T)$, i.e. $Q_i^{k-1}(S) = Q_j^{k-1}(T)$ (1.5.8).
Then by 1.11.2,

$$
\begin{aligned}
\theta([S]) &= Q_i^{\infty-1}(S) = Q_k^{\infty-1}(Q_i^{k-1}(S)) = Q_k^{\infty-1}(Q_j^{k-1}(T)) \\
&= Q_j^{\infty-1}(T) = \theta([T]).
\end{aligned}
$$

Clearly θ is onto. That θ is an MA homomorphism may be shown using 1.11.2 and 1.11.5, e.g.

$$
\begin{aligned}
\theta(m[S]) &= \theta[m_{R_i}(S)] &&(1.11.5) \\
&= Q_i^{\infty-1}(m_{R_i}(S)) = m_{R^\infty}(Q_i^{\infty-1}(S)) &&(1.11.2(6)) \\
&= m_{R^\infty}(\theta([S])).
\end{aligned}
$$

Now if each Q_j^i is onto, then each Q_j^∞ is onto [11, p. 218]. Then if $\theta([S]) = \theta([T])$, with $S \in P_i$, $T \in P_j$, we have $Q_i^{\infty-1}(S) = Q_j^{\infty-1}(T)$. Choosing $k \geq i,j$, by 1.11.2(1) $Q_k^{\infty-1}(Q_i^{k-1}(S)) = Q_k^{\infty-1}(Q_j^{k-1}(T))$. But Q_k^∞ is onto, so it follows that $Q_i^{k-1}(S) = Q_j^{k-1}(T)$, whence $S \sim T$ and $[S] = [T]$. Thus θ is injective.   □

We note that 1.11.6 and the preservation of identities under direct limits and homomorphisms, gives an alternative proof of 1.11.4.

1.12   Modal Axiomatic Classes

Definition 1.12.1 *If $\Gamma \subseteq \Phi$, $\Gamma^* = \{\mathcal{F} \in \mathfrak{D} : \mathcal{F} \models \alpha \text{ all } \alpha \in \Gamma\}$.*

*A class of descriptive frames is **modal axiomatic** iff it is Γ^* for some $\Gamma \subseteq \Phi$. A class is **modal elementary** iff it is α^* for some $\alpha \in \Phi$ (where $\alpha^* = \{\alpha\}^*$).*

If Δ is a set of MA polynomial identities,

$$
\Delta^* = \{\mathfrak{A} \in \mathfrak{M} : \mathfrak{A} \models \delta \text{ all } \delta \in \Delta\}.
$$

*A class of MA's is **equational** iff it is Δ^* for some Δ.*

If X is a class of descriptive frames,

$$
X^+ = \{\mathfrak{A} \in \mathfrak{M} : \mathfrak{A} \cong \mathcal{F}^+ \text{ for some } \mathcal{F} \in X\}.
$$

The purpose of this section is to characterise in terms of frame constructions the modal axiomatic and modal elementary classes of descriptive frames.

Theorem 1.12.2 *Let X be a class of descriptive frames closed under isomorphism. Then X is modal axiomatic iff X^+ is equational.*

Proof. Suppose $X = \Gamma^*$. Define $\Delta = \{(h_\alpha = 1) : \alpha \in \Gamma\}$. Since in general $\mathcal{F}$ is semantically equivalent to $\mathcal{F}^+$, clearly $X^+ \subseteq \Delta^*$. Now let $\mathfrak{A} \in \Delta^*$. But $\mathfrak{A} \cong (\mathfrak{A}_+)^+$ (1.10.3) and $\mathfrak{A}$ is equivalent to $\mathfrak{A}_+$ (1.10.4). Since $\mathfrak{A}_+$ is descriptive, $\mathfrak{A}_+ \in \Gamma^* = X$, and so $\mathfrak{A} \in X^+$. Thus $X^+ = \Delta^*$ is equational.

Conversely, suppose $X^+ = \Delta^*$ for some Δ. As explained in 1.2.1, each member of Δ may be presumed to be of the form $h_\alpha = 1$ for some $\alpha \in \Phi$. Let $\Gamma = \{\alpha : (h_\alpha = 1) \in \Delta\}$. If $\mathcal{F} \in X$, $\mathcal{F}^+ \in X^+ = \Delta^*$ so $X \subseteq \Gamma^*$. Now suppose $\mathcal{F} \in \Gamma^*$. Then $\mathcal{F}^+ \in \Delta^* = X^+$, so $\mathcal{F}^+ \cong \mathcal{G}^+$ for some $\mathcal{G} \in X$. But then $(\mathcal{F}^+)_+ \cong (\mathcal{G}^+)_+$ so by 1.10.7 $\mathcal{F} \cong \mathcal{G}$. Since X is closed under isomorphism, $\mathcal{F} \in X$. Hence $X = \Gamma^*$ is modal axiomatic. □

To obtain our characterisation, we need two new concepts. We saw in Section 1.9 that ultraproducts and infinite disjoint unions of descriptive frames need not be descriptive. We therefore introduce

Definition 1.12.3 *Let $\{\mathcal{F}_i : i \in I\}$ be a collection of frames. Then*

(1) *the **descriptive union** of the $\mathcal{F}_i$'s is the frame*

$$\sum \mathcal{F}_i^0 = \begin{cases} \sum \mathcal{F}_i & \text{if } I \text{ is finite} \\ ((\sum \mathcal{F}_i)^+)_+ & \text{if } I \text{ is infinite.} \end{cases}$$

(2) *if G is an ultrafilter on I, the **descriptive ultraproduct of the** $\mathcal{F}_i$'s **over** G is the frame*

$$\mathcal{F}_G^0 = \begin{cases} \mathcal{F}_G & \text{if } \mathcal{F}_G \in \mathfrak{D} \\ (\mathcal{F}_G^+)_+ & \text{if } \mathcal{F}_G \notin \mathfrak{D}. \end{cases}$$

By 1.10.5 we always have $\mathcal{F}_G^0, \sum \mathcal{F}_i^0 \in \mathfrak{D}$. In general $\mathcal{F}$ is equivalent semantically to $(\mathcal{F}^+)_+$ (1.10.6) so the new constructions are validity preserving, i.e. 1.6.4 and 1.7.13 hold with $\sum \mathcal{F}_i^0, \mathcal{F}_G^0$ in place of $\sum \mathcal{F}_i, \mathcal{F}_G$.

Theorem 1.12.4 *Let X be a class of descriptive frames closed under isomorphism. Then*

(1) *X is closed under subframes only if X^+ is closed under homomorphic images;*

(2) *X is closed under homomorphic images only if X^+ is closed under subalgebras;*

(3) *X is closed under finite disjoint unions only if X^+ is closed under finite direct products;*

(4) *X is closed under descriptive unions only if X^+ is closed under direct products;*

(5) *X is closed under onto inverse limits only if X^+ is closed under one-one direct limits.*

Proof. A subframe (homomorphic image) of a descriptive frame need not be descriptive, so by "X is closed under subframes (homomorphic images)" we mean that if $\mathcal{F} \in X$ and $\mathcal{F}_1 \subseteq \mathcal{F}$ ($\mathcal{F}_1 \preccurlyeq \mathcal{F}$) and $\mathcal{F}_1 \in \mathfrak{D}$, then $\mathcal{F}_1 \in X$.

Now in general if $\mathfrak{A} \in X^+$, $\mathfrak{A} \cong \mathcal{F}^+$ for some $\mathcal{F} \in X$, so $\mathfrak{A}_+ \cong (\mathcal{F}^+)_+ \cong \mathcal{F}$ (1.10.7) and since X is closed under $\cong$, $\mathfrak{A}_+ \in X$.

(1) Suppose $\mathfrak{A} \in X^+$ and $\mathfrak{B}$ is a homomorphic image of $\mathfrak{A}$. Then by 1.10.9(3) and 1.5.3(2) $\mathfrak{B}_+$ is isomorphic to a subframe of $\mathfrak{A}_+ \in X$.

But X is closed under subframes and isomorphism, so $\mathfrak{B}_+ \in X$. Since $(\mathfrak{B}_+)^+ \cong \mathfrak{B}$ (1.10.3), it follows that $\mathfrak{B} \in X^+$.

(2) Suppose $\mathfrak{B}$ is a sub-MA of $\mathfrak{A} \in X^+$. By 1.10.9(2) (with $\psi = id_{\mathfrak{B}}$) $\mathfrak{B}_+$ is a homomorphic image of $\mathfrak{A}_+ \in X$. Thus $\mathfrak{B}_+ \in X$, so as in (1), $\mathfrak{B} \in X^+$.

(3) Let $\{\mathfrak{A}_i : i \in I\} \subseteq X^+$. By 1.6.5 $(\sum \mathfrak{A}_{i+})^+ \cong \prod((\mathfrak{A}_{i+})^+) \cong \prod \mathfrak{A}_i$ (1.10.3) and each $\mathfrak{A}_{i+} \in X$. If I is finite, by hypothesis $\sum \mathfrak{A}_{i+} \in X$, so $\prod \mathfrak{A}_i \in X^+$.

(4) Suppose $\sum \mathfrak{A}_{i+}^0 \in X$. Then $(\sum \mathfrak{A}_{i+}^0)^+ = (((\sum \mathfrak{A}_{i+})^+)_+)^+ \cong ((\prod \mathfrak{A}_i)_+)^+ \cong \prod \mathfrak{A}_i$, so $\prod \mathfrak{A}_i \in X^+$.

(5) Let $\{\mathfrak{A}_i : i \in I\} \cup \{\psi_j^i : i \leq j\}$ be a one-one direct family in X^+ (i.e. all the ψ_j^i's are injective). By 1.10.9 $\{\mathfrak{A}_{i+} : i \in I\} \cup \{Q_j^i : i \geq j\}$ is an onto inverse family of descriptive frames (i.e. the Q_j^i's are onto), where $Q_j^i = (\psi_i^j)_+$. By hypotheses $\mathfrak{A}_+^\infty$, the inverse limit of the $\mathfrak{A}_{i+}$'s, is in X. But by 1.11.6 $(\mathfrak{A}_+^\infty)^+$ is isomorphic to the direct limit of the $(\mathfrak{A}_{i+})^+$'s, which by 1.10.3 is clearly isomorphic to the direct limit of the $\mathfrak{A}_i$'s, and so the latter is in X^+.

$\square$

We note that the converses of Theorem 1.12.4 are all true. The results as proven however suffice for the applications we have in mind.

Theorem 1.12.5 *A class of descriptive frames is modal axiomatic iff it is closed under subframes, homomorphic images, and descriptive unions.*

Proof. We have seen that these three constructions preserve the validity of modal wffs, and so necessity follows. For the converse, suppose X is closed under the three given constructions and so by 1.12.4 X^+ is closed under subalgebras, homomorphic images and direct products. By Birkhoff's Theorem on equational classes (cf. [39, p. 171]), it follows that X^+ is equational and therefore by 1.12.2 that X is modal axiomatic. $\square$

To obtain a characterisation that does not involve the rather ad hoc notion of descriptive union, we appeal to a result of Tarski [93] that equational classes of algebras of finite type (which MA's are) are characterised by closure under subalgebras, homomorphic images, finite direct products, and unions of chains of algebras.

Theorem 1.12.6 *A class of descriptive frames is modal axiomatic iff it is closed under subframes, homomorphic images, finite disjoint unions, and inverse limits.*

Proof. We saw in 1.11.4 that inverse limits are validity preserving, so necessity follows as before. For the converse, if X is closed under isomorphism and inverse limits, then in particular X is closed under onto inverse limits, so by 1.12.4(5), X^+ is closed under one-one direct limits. But the union of a chain of algebras is isomorphic to a one-one direct limit of algebras isomorphic to the members of the chain [50, p. 116]. Since by definition X^+ is closed under isomorphism, it follows that X^+ is closed under chain unions. The rest of the proof goes through, via 1.12.4 and Tarski's result, as for 1.12.5. $\square$

We turn now to modal elementary classes.

Theorem 1.12.7 *Let $\mathfrak{N}$ be an equational class of MA's. Then the following are equivalent:*

(1) $\mathfrak{N} = \delta^*$ *for some identity δ.*

(2) $c\mathfrak{N} = \{\mathfrak{A} \in \mathfrak{M} : \mathfrak{A} \notin \mathfrak{N}\}$ *is closed under ultraproducts.*

Proof. (1) $\Rightarrow$ (2): Suppose $\{\mathfrak{A}_i : i \in I\} \subseteq c\mathfrak{N}$ and $\mathfrak{N} = \delta^*$. Then for any ultrafilter G on I, $\{i : \mathfrak{A}_i \models \delta\} = \emptyset \notin G$ so by Łoś's Theorem $\prod \mathfrak{A}_i/G \nvDash \delta$, hence $\prod \mathfrak{A}_i/G \in c\mathfrak{N}$.

(2) $\Rightarrow$ (1): We have $\mathfrak{N}$ equational, so $\mathfrak{N} = \Delta^*$ for some set of identities Δ. It follows that $\mathfrak{N}$ and $c\mathfrak{N}$ are closed under isomorphism, and by Łoś's Theorem that $\mathfrak{N}$ is closed under ultraproducts. If further $c\mathfrak{N}$ is closed under ultraproducts, then by [4, Theorem 7.3.11] $\mathfrak{N}$ is an elementary class in the first order language of MA's, i.e. $\mathfrak{N} = \delta^*$ for some sentence δ in this language. Thus $\Delta^* = \delta^*$, so by the Compactness Theorem for first order logic, $\Delta_0^* = \delta^* = \mathfrak{N}$ for some finite $\Delta_0 \subseteq \Delta$. But any finite set of MA identities is equivalent to a single identity ($\mathfrak{N} \models (h_\alpha = 1)$ & $(h_\beta = 1)$ iff $\mathfrak{N} \models (h_{\alpha \wedge \beta} = 1)$), and so (1) follows.

$\square$

Theorem 1.12.8 *Let X be a class of descriptive frames that is modal axiomatic. Then X is modal elementary iff X^+ is closed under ultraproducts.*

Proof. If X is modal axiomatic, then by 1.12.2 X^+ is equational. But by the proof of 1.12.2 we also have $X = \alpha^*$ iff $X^+ = (h_\alpha = 1)^*$ and so the result follows by 1.12.7. $\square$

Theorem 1.12.9 *Let X be a class of descriptive frames closed under isomorphism. Then $cX = \{\mathcal{F} \in \mathfrak{D} : \mathcal{F} \notin X\}$ is closed under descriptive ultraproducts only if cX^+ is closed under MA ultraproducts.*

Proof. Suppose $\{\mathfrak{A}_i : i \in I\} \subseteq cX^+$. By 1.10.3 there exists $\mathcal{F}_i \in \mathfrak{D}$ such that $\mathcal{F}_i^+ \cong \mathfrak{A}_i$, all $i \in I$. Then $\mathcal{F}_i \notin X$, or else $\mathfrak{A}_i \in X^+$. If G is an ultrafilter on I, then by 1.7.8, $\mathcal{F}_G^+ \cong \prod \mathcal{F}_i^+/G \cong \prod \mathfrak{A}_i/G$, whence by 1.12.3(2) we have in general $(\mathcal{F}_G^0)^+ \cong \prod \mathfrak{A}_i/G$. But by hypothesis $\mathcal{F}_G^0 \in cX$. Now if $\prod \mathfrak{A}_i/G \in X^+$, $\prod \mathfrak{A}_i/G \cong \mathcal{G}^+$ for some $\mathcal{G} \in X$. Thus

$(\mathcal{F}_G^0)^+ \cong \mathcal{G}^+$. But from this, as in the proof of 1.12.2, we get $\mathcal{F}_G^0 \in X$, a contradiction. Hence $\prod \mathfrak{A}_i / G \in cX^+$.   □

Theorem 1.12.10 *A class X of descriptive frames is modal elementary iff X is closed under subframes, homomorphic images, and descriptive unions (or finite disjoint unions and inverse limits) and cX is closed under descriptive ultraproducts.*

Proof. Necessity uses 1.7.13 in addition to previous cases. Sufficiency follows from 1.12.5, 1.12.6, 1.12.9 and 1.12.8.   □

In order to consider modal axiomatic classes of nondescriptive frames we recall that any frame $\mathcal{F}$ is semantically equivalent to $(\mathcal{F}^+)_+$, and so any axiomatic class will contain one iff it contains the other. This condition (which by 1.10.7 is automatically satisfied by a class of descriptive frames closed under isomorphism) turns out to be precisely what is needed to generalize our discussion. We leave it to the reader to modify the preceding results of this section to obtain proofs of the following.

Theorem 1.12.11 *A class X of (refined) frames is modal axiomatic iff*

(1) $\mathcal{F} \in X$ *iff* $(\mathcal{F}^+)_+ \in X$ *for any (refined) frame $\mathcal{F}$, and*

(2) X *is closed under disjoint unions, (refined) subframes, and (refined) homomorphic images.*

Furthermore X is modal elementary iff it satisfies (1), (2) and

(3) *the class of (refined) frames not in X is closed under ultraproducts.*

1.13   Characteristic Models Revisited

We observed in Section 1.2 that the Lindenbaum algebra for any normal modal logic determines that logic but that the canonical K-frame does not always enjoy this property. The construction of 1.2.5 can however be adapted to provide each logic with a characteristic model in the first-order semantics.

Definition 1.13.1 *If Λ is a normal logic, the **canonical frame** for Λ is*

$$\mathcal{F}_\Lambda = \langle W_\Lambda, R_\Lambda, P_\Lambda \rangle,$$

where

(i) $\langle W_\Lambda, R_\Lambda \rangle$ *is the canonical K-frame for Λ (cf. 1.2.5), and*

(ii) $P_\Lambda = \{|\alpha|_\Lambda : \alpha \in \Phi\}$ *(cf. 1.1.2).*

Note that the canonical valuation $V_\Lambda(p) = |p|_\Lambda$ of 1.2.5 is a valuation on $\mathcal{F}_\Lambda$.

There is known to be (cf. Rasiowa and Sikorski [74, VII §§9–10])
a bijective correspondence between Λ-maximal sets and ultrafilters in
$\mathfrak{A}_\Lambda$, the Lindenbaum algebra for Λ. This, together with the obvious
similarities between Definitions 1.13.1 and 1.10.1 leads to

Theorem 1.13.2 $\mathfrak{A}_{\Lambda+} \cong \mathcal{F}_\Lambda$

Proof. Define $Q : W_\Lambda \to W_{\mathfrak{A}_\Lambda}$ by $Q(x) = \{\|\alpha\|_\Lambda : \alpha \in x\}$ (cf. 1.2.2). Q is
the bijection given in [74]. Using 1.10.1, 1.2.2 and 1.2.5, it follows easily
that $xR_\Lambda y$ iff $Q(x)R_{\mathfrak{A}_\Lambda}Q(y)$, and that for $\alpha \in \Phi$, $Q(|\alpha|_\Lambda) = \|\|\alpha\|_\Lambda|^{\mathfrak{A}_\Lambda}$,
and hence that $S \in P_\Lambda$ iff $Q(S) \in P^{\mathfrak{A}_\Lambda}$. $\square$

Corollary 1.13.3 $\mathcal{F}_\Lambda$ *is descriptive.*

Proof. 1.10.5 and the preservation of "descriptiveness" under isomorphism. $\square$

Theorem 1.13.4 $\mathcal{F}_\Lambda$ *strongly determines* Λ.

Proof. If not $\Gamma \vdash_\Lambda \alpha$, by 1.1.3 and 1.2.6 $V_\Lambda(\Gamma) \not\subseteq V_\Lambda(\alpha)$, hence by
1.8.2(2) $\Gamma \not\models_\Lambda \alpha(\mathcal{F}_\Lambda)$. On the other hand if $\Gamma \vdash_\Lambda \alpha$, then $\vdash_\Lambda \beta \to \alpha$
where β is the conjunction of some finite subset of Γ. But $\mathfrak{A}_\Lambda$ determines
Λ, so $\mathfrak{A}_\Lambda \models \beta \to \alpha$ whence by 1.10.4 and 1.13.2, $\mathcal{F}_\Lambda \models \beta \to \alpha$. Then for
any V on $\mathcal{F}_\Lambda$, $V(\Gamma) \subseteq V(\beta) \subseteq V(\alpha)$, so $\Gamma \models \alpha(\mathcal{F}_\Lambda)$. $\square$

As a further illustration of the structural properties of descriptive
frames, we include the following result.

Theorem 1.13.5 *Let V be a valuation on a descriptive frame $\mathcal{F} = \langle W, R, P \rangle$. If $\Delta v = \{\alpha : V(\alpha) = W\}$, then there is a homomorphism
from $\langle W, R \rangle$ onto the canonical K-frame for $K\Delta v$.*

Proof. Let $\langle W', R' \rangle$ be the canonical frame for $K\Delta v$. Define $Q : W \to W'$ by $Q(x) = \{\alpha : x \in V(\alpha)\}$. Clearly $\Delta v \subseteq Q(x)$ and for any $\alpha \in \Phi$,
$\alpha \in Q(x)$ iff $\neg\alpha \notin Q(x)$, so $Q(x)$ is $K\Delta v$-maximal as required. To show
that Q is onto, let y be a $K\Delta v$-maximal set. Put $P_0 = \{V(\alpha) : \alpha \in y\} \subseteq P$. Then P_0 is closed under finite $\bigcap$'s, so if P_0 does not have the fip, for
some $\alpha \in y$, $V(\alpha) = \emptyset$, whence $V(\neg\alpha) = W$, so $\neg\alpha \in \Delta v$ and therefore
$\neg\alpha \in y$, contrary to the $K\Delta v$-consistency of y. Since $\mathcal{F}$ is descriptive,
it follows by 1.9.3IV that there exists $x \in \bigcap P_0$. Then clearly $y \subseteq Q(x)$,
so by properties of maximal sets, $Q(x) = y$.

Now if xRy, $\Box\alpha \in Q(x)$ only if $x \in V(\Box\alpha)$, only if $y \in V(\alpha)$, only
if $\alpha \in Q(y)$. Hence $Q(x)R'Q(y)$ (1.2.5(ii)).

Finally suppose $Q(x)R'z$, where $z \in W'$. Let $P_0 = \{S \in P : x \in l_R(S)\} \cup \{V(\alpha) : \alpha \in z\}$. If P_0 does not have the fip, there exist $S \in P$
such that $x \in l_R(S)$, and $\alpha \in z$ such that $S \subseteq -V(\alpha)$, so $l_R(S) \subseteq l_R(-V(\alpha)) = V(\Box\neg\alpha)$. Since $x \in l_R(S)$ we obtain $x \in V(\Box\neg\alpha)$, i.e.

$\Box \neg \alpha \in Q(x)$. But $Q(x)R'z$, so $\neg \alpha \in z$, contrary to the fact that $\alpha \in z$. Hence there exists some $y \in \bigcap P_0$. From the definition of P_0 we have, by Axiom **II**, xRy and, as above, $Q(y) = z$. Thus Q is a homomorphism.

$\Box$

In the context of K-frames, this result may be adapted as follows:

Theorem 1.13.6 *Suppose* $\mathcal{F} = \langle W, R \rangle \models \alpha$, *and* V *is a valuation on* $\mathcal{F}$ *for which the frame* $\mathcal{F}_V = \langle W, R, P_V \rangle$ *of 1.3.6 is descriptive. Then* $\mathcal{F}$ *is isomorphic to a subframe of* $\mathcal{F}^K_{K\alpha}$.

Proof. Define $Q : W \to W_{K\alpha}$ by $Q(x) = \{\beta : x \in V(\beta)\}$. Since $\mathcal{F} \models \alpha$, $Q(x)$ is $K\alpha$-maximal as required. The proof of 1.13.5 shows that Q is a homomorphism into $W_{K\alpha}$. Now if $x \neq y \in W$, then by Axiom **I**, since $\mathcal{F}_V$ is descriptive there is some $\beta \in \Phi$ such that $x \in V(\beta), y \notin V(\beta)$. Then $\beta \in Q(x) - Q(y)$. Hence Q is injective, and so maps $\mathcal{F}$ isomorphically into $\mathcal{F}^K_{K\alpha}$.

$\Box$

In view of 1.13.6 we introduce

Definition 1.13.7 *A logic* Λ *is **super-complete** for a class* $\mathfrak{C}$ *of K-frames iff*

(i) $\vdash_\Lambda \alpha$ *only if* $\mathfrak{C} \models \alpha$,
    and

(ii) *every* Λ*-consistent set of wffs is satisfiable on some* $\mathcal{F} \in \mathfrak{C}$ *by a valuation* V *for which* $\mathcal{F}_V$ *is descriptive.*

Theorem 1.13.8 *If* $K\alpha$ *is super-complete, then* $\mathcal{F}^K_{K\alpha} \models \alpha$.

Proof. Suppose $K\alpha$ is super-complete for a class $\mathfrak{C}$ of K-frames. If $\mathcal{F}^K_{K\alpha} \not\models \alpha$ then for some V on $\mathcal{F}^K_{K\alpha}$ and $x \in W_{K\alpha}$, $x \notin V(\alpha)$. But then by 1.4.11, $\mathcal{F}_x \not\models \alpha$, where $\mathcal{F}_x$ is the subframe of $\mathcal{F}^K_{K\alpha}$ generated by x. Now x is $K\alpha$-consistent, so for some $\mathcal{F} \in \mathfrak{C}$ there is a V on $\mathcal{F}$ such that $\mathcal{F}_V$ is descriptive, and for some $t \in \mathcal{F}$, $t \in V(x)$. By 1.13.7(i) $\mathcal{F} \models \alpha$, hence by 1.13.6 $\mathcal{F}$ is isomorphic to $\mathcal{F}_Q \subseteq \mathcal{F}^K_{K\alpha}$ by a map Q. By 1.5.6, $\mathcal{F}_Q \models \alpha$. But by the maximality of x, $x = Q(t) \in \mathcal{F}_Q$, and so $\mathcal{F}_x$ is a subframe of $\mathcal{F}_Q$. Thus by 1.4.10, $\mathcal{F}_x \models \alpha$, a contradiction.

$\Box$

Note that the above proof still works if Q is merely a homomorphism (by 1.5.5). Hence 1.13.8 still holds if super complete is replaced by a weaker notion obtained from 1.13.7 by requiring that $\mathcal{F}_V$ satisfy only Axioms **II** and **III** for descriptive frames.

1.14   d-Persistent Formulae

Much of the early work with possible worlds semantics was devoted to establishing that various modal logics were determined by particular classes of K-frames that were specifiable by some reasonably simple mathematical property, e.g. a first-order condition on binary relations. This program has now been carried out for all the significant logics that had been developed at the time of Kripke's work, and many more besides (cf. [47, 59, 86]). Recently Thomson [97] and Fine [13] demonstrated that there exist normal logics that are not determined by any class of K-frames at all (thereby refuting a conjecture of Lemmon and Scott [59]). Subsequently there has been a shift of emphasis in research into modal logic. One of the major topics of current concern would seem to be the adequacy of the second-order semantics. Which formulae axiomatise logics with characteristic K-frames? In this section we provide a partial answer to that question.

Definition 1.14.1 $\alpha \in \Phi$ is **d-persistent** iff for all frames $\mathcal{F} = \langle W, R, P \rangle$, if $\mathcal{F} \models \alpha$ and $\mathcal{F}$ is descriptive, then $\langle W, R \rangle \models \alpha$. Let

$$D = \{\alpha \in \Phi : \alpha \text{ is d-persistent}\}.$$

For d-persistent formulae we have the following extension of the compactness results of Section 1.8.

Theorem 1.14.2 If $\Gamma \subseteq D$ and $\mathfrak{K}$ is the class of all K-frames, then $\Gamma \models_0 \alpha(\mathfrak{K})$ only if $\Gamma' \models_0 \alpha(\mathfrak{K})$ for some finite $\Gamma' \subseteq \Gamma$.

Proof. If $\Gamma' \not\models_0 \alpha(\mathfrak{K})$, all finite $\Gamma' \subseteq \Gamma$, then using full frames we have for all such Γ', $\Gamma' \not\models_0 \alpha(\mathfrak{C})$, where $\mathfrak{C}$ is the class of all first order frames. Hence, by 1.8.4(2), there is $\mathcal{F} \in \mathfrak{C}$ such that $\mathcal{F} \models \Gamma$ and not $\mathcal{F} \models \alpha$. By 1.10.6 it follows that there is a descriptive $\mathcal{F}' = \langle W', R', P' \rangle$ such that $\mathcal{F}' \models \Gamma$ and $\mathcal{F}' \not\models \alpha$, hence $\langle W', R' \rangle \not\models \alpha$. But $\Gamma \subseteq D$, so, by 1.14.1, $\langle W', R' \rangle \models \Gamma$. Thus $\Gamma \not\models_0 \alpha(\mathfrak{K})$. $\square$

Theorem 1.14.3 If $\Gamma \subseteq D$ then the logic $K\Gamma$ is strongly determined by the K-frame $\mathcal{F}_{K\Gamma}^K$.

Proof. We have seen previously that $\Delta \not\vdash_{K\Gamma} \alpha$ only if $V_{K\Gamma}(\Delta) \not\subseteq V_{K\Gamma}(\alpha)$, whence $\Delta \not\models \alpha(\mathcal{F}_{K\Gamma}^K)$. But $\mathcal{F}_{K\Gamma} \models \Gamma$ (1.13.4), $\mathcal{F}_{K\Gamma}$ is descriptive (1.13.3), and $\Gamma \subseteq D$, so by 1.14.1 $\mathcal{F}_{K\Gamma}^K \models \Gamma$. Thus all the axioms of $K\Gamma$ are valid on $\mathcal{F}_{K\Gamma}^K$. Since the rules of inference of normal logics are validity preserving on K-frames, all the theorems of $K\Gamma$ are valid on $\mathcal{F}_{K\Gamma}^K$. So if $\Delta \vdash_{K\Gamma} \alpha$, then $\vdash_{K\Gamma} \beta \rightarrow \alpha$, where β is a conjunction of members of Δ. Then $\mathcal{F}_{K\Gamma}^K \models \beta \rightarrow \alpha$, from which it follows, as in the proof of 1.13.4, that $\Delta \models \alpha(\mathcal{F}_{K\Gamma}^K)$. $\square$

1.14.3 does not completely encompass our problem, since there are logics with characteristic K-frames and non-persistent axioms (cf. Section 1.18). Nevertheless, the concept of d-persistence is a useful one, for the properties of descriptive frames allow us to find wide-ranging syntactic criteria that are sufficient for a formula to be valid on its associated canonical frame.

Definition 1.14.4

(1) $\alpha \in \Phi$ is **constant** iff $h_\alpha^{\mathfrak{A}}$ is a constant function on any MA $\mathfrak{A}$, i.e. iff $V(\alpha) = V'(\alpha)$ for any valuations on the same frame. α is **atomic** iff it is either a variable or a constant wff.

(2) α is **positive** iff it is formed from atomic wffs using only $\wedge, \vee, \square, \diamond$. Π denotes the set of positive wffs.

(3) α is a $\square$**-string** iff for some $k \in \mathbb{N}$ and some variable p, $\alpha = \square^k p$. Similarly, a $\diamond$**-string** is a wff of the form $\diamond^k p$. (Note that with $k = 0$, each variable is both a $\square$-string and a $\diamond$-string). $\alpha \in \Pi$ is $\diamond$**-positive** iff the only occurrences of $\square$ in α are within $\square$-strings. $\Pi_\diamond$ denotes the class of $\diamond$-positive wffs. Similarly, the class $\Pi_\square$ of $\square$**-positive** wffs consists of those $\alpha \in \Pi$ whose only occurrences of $\diamond$ are within $\diamond$-strings.

The class of constant formulae can be described syntactically as follows:

Theorem 1.14.5 *Let C be the smallest subset of Φ satisfying*

(i) *every instance of a PC-tautology is in C,*

(ii) *$\alpha \in C$ only if $\neg\alpha, \square\alpha \in C$,*

(iii) *$\alpha, \beta \in C$ only if $\alpha \wedge \beta \in C$.*

Then $\beta \in \Phi$ is constant iff $\vdash_K \beta \leftrightarrow \alpha$ for some $\alpha \in C$.

Proof. It is clear that each member of C is constant and therefore anything deductively equivalent to a member of C is constant (because $\vdash_K \beta \leftrightarrow \alpha$ only if $h_\beta^{\mathfrak{A}} = h_\alpha^{\mathfrak{A}}$, any $\mathfrak{A}$).

Conversely, for each variable p, let α_p be a PC tautology. Suppose that β is constant. Let β' be obtained by replacing each p in β by α_p. Then clearly $\beta' \in C$. If not $\vdash_K \beta \leftrightarrow \beta'$, by the completeness theorem for K [59, Section 2] there exists a V on some K-frame such that $V(\beta) \neq V(\beta')$. Choose a valuation V' on this frame such that $V'(p) = W = V(\alpha_p)$. Then a simple induction shows that $V'(\beta) = V(\beta') \neq V(\beta)$, so β is not constant, contrary to hypothesis. Thus $\vdash_K \beta \leftrightarrow \beta'$ as required. $\square$

We note that 1.14.5 would seem to be the "best possible" result, for there are constant wffs not actually in C, e.g. $(\square p \wedge \square q) \to \square(p \wedge q)$.

We saw in Section 1.10 that within isomorphism every descriptive frame is $\mathfrak{A}^+$ for some MA $\mathfrak{A}$. Thus the question of d-persistence is equivalent to that of determining which MA polynomial identities are preserved in passing from an MA to the power-set algebra of its associated frame. This problem was considered for BA's with additive operators by Jónsson and Tarski [48] (cf. also Section 2.8 of [42]). Theorem 2.18 of [48] may be interpreted as proving d-persistence for any wff constructed from atomic ones using only $\wedge$, $\vee$, $\Diamond$. We now propose to develop and expand the Jónsson-Tarski techniques to show that the property is possessed by a much wider class of wffs.

Theorem 1.14.6 *For any* $\alpha \in \Pi$, h_α *is a monotonic function, i.e.* $a_i \leq b_i$ *only if* $h_\alpha(\ldots a_i \ldots) \leq h_\alpha(\ldots b_i \ldots)$.

Proof. By induction on α. From $a \leq b$ we may infer $a \cap c \leq b \cap c$, $a \cup c \leq b \cup c$, $ma \leq mb$, $la \leq lb$.   □

From now on we assume that $\mathcal{F} = \langle W, R, P \rangle$ is a descriptive frame, τ and ζ are as in 1.9.4 and $\alpha \in \Phi$ has a single variable, i.e. h_α is a one-place function on $\mathcal{F}^+$. The reason for the latter restriction is simply expository clarity. All proofs may be adapted with only technical modifications to wffs with any number of variables.

Theorem 1.14.7 *If* $\alpha \in \Pi$,

(1) *For* $a \in \tau$, $h_\alpha(a) = \bigcup_{a \supseteq b \in P} h_\alpha(b)$,

(2) *For* $a \in \zeta$, $h_\alpha(a) = \bigcap_{a \subseteq b \in P} h_\alpha(b)$.

Proof.

(1) By 1.14.6 the result holds from right to left. We prove the converse by induction on α.

  (i) If $x \in h_p(a) = a$, by definition of τ there is some $b \in P$ such that $b \subseteq a$ and $x \in b = h_p(b)$. If α is a constant wff, with $h_\alpha = b$ identically for some $b \subseteq W$, then $t \in h_\alpha(a) = b$ only if $t \in h_\alpha(\emptyset) = b$ and $a \supseteq \emptyset \in P$.

  (ii) If $t \in h_{\alpha \wedge \beta}(a) = h_\alpha(a) \cap h_\beta(a)$ then by IH there exist $b_1, b_2 \in P$ such that $a \supseteq b_1$, $a \supseteq b_2$, $t \in h_\alpha(b_1)$ and $t \in h_\beta(b_2)$. Then by 1.14.6 $h_\alpha(b_1) \subseteq h_\alpha(b_1 \cup b_2)$, $h_\beta(b_2) \subseteq h_\beta(b_1 \cup b_2)$, so $t \in h_{\alpha \wedge \beta}(b_1 \cup b_2)$ and $a \supseteq b_1 \cup b_2 \in P$.

  (iii) If $t \in h_{\alpha \vee \beta}(a)$, then say $t \in h_\alpha(a)$, so by IH, for some $a \supseteq b \in P$, $t \in h_\alpha(b) \subseteq h_{\alpha \vee \beta}(b)$.

  (iv)
$$\begin{aligned}
h_{\Diamond \alpha}(a) &= m_R(h_\alpha(a)) \\
&= m_R(\bigcup_{a \supseteq b \in P} h_\alpha(b)) \qquad (IH) \\
&= \bigcup_{a \supseteq b \in P}(m_R(h_\alpha(b))) \quad (1.9.6(2)) \\
&= \bigcup_{a \supseteq b \in P}(h_{\Diamond \alpha}(b)).
\end{aligned}$$

  (v) If $t \in h_{\Box \alpha}(a) = l_R(h_\alpha(a))$, then by IH and 1.9.9(1), for some $c \in \zeta$, $t \in l_R(c)$ and $c \subseteq h_\alpha(a) = \bigcup_{a \supseteq b \in P} h_\alpha(b)$. But $b \in P$ only if $h_\alpha(b) \in P$, so by 1.9.4(5) there exist $b_1, \ldots, b_n$

such that $a \supseteq b_i \in P$, and $c \subseteq h_\alpha(b_1) \cup \ldots \cup h_\alpha(b_n) \subseteq h_\alpha(b_1 \cup \ldots \cup b_n)$ (1.14.6) so $t \in l_R(c) \subseteq h_{\square\alpha}(b_1 \cup \ldots \cup b_n)$ and $a \supseteq b_1 \cup \ldots \cup b_n \in P$.

(2) This is proven by a similar induction using 1.9.9(2) and 1.9.4(6). □

Theorem 1.14.8

(1) If $a \in \Pi_\Diamond$, then for all $a \subseteq W$, $h_\alpha(a) = \bigcup_{a \supseteq b \in \zeta} h_\alpha(b)$.

(2) If $a \in \Pi_\square$, then for all $a \subseteq W$, $h_\alpha(a) = \bigcap_{a \subseteq b \in \tau} h_\alpha(b)$.

Proof.

(1) Again by 1.14.6, the inclusion from right to left follows. Conversely
   (i) For □-strings:

$$h_{\square^n p}(a) = l_R^n h_p(a) = l_R^n(a) = \bigcup_{a \supseteq b \in \zeta} l_R^n(b) \quad (1.9.9(1))$$
$$= \bigcup_{a \supseteq b \in \zeta} h_{\square^n p}(b).$$

The case for constant wffs is similar to 1.14.7(1)(i), but using $\emptyset \in \zeta$.

   (ii) If $t \in h_{\alpha \wedge \beta}(a) = h_\alpha(a) \cap h_\beta(a)$, by IH there are $b_1, b_2 \in \zeta$ such that $a \supseteq b_1, b_2$; $t \in h_\alpha(b_1)$ and $t \in h_\beta(b_2)$. Then, as in 1.14.7(1)(ii), $t \in h_{\alpha \wedge \beta}(b_1 \cup b_2)$, and $a \supseteq b_1 \cup b_2 \in \zeta$ by 1.9.4(3).

The cases of $\vee$ and $\square$ follow in the manner of 1.14.7(1).

(2) By a similar induction, using 1.9.9(2) for $\Diamond$-strings. □

Theorem 1.14.9 *If $\alpha \in \Pi$, then for all $a \subseteq W$,*

(1) $h_\alpha(a) \supseteq \bigcup_{a \supseteq b \in \zeta} (\bigcap_{b \subseteq c \in P} h_\alpha(c))$,

(2) $h_\alpha(a) \subseteq \bigcap_{a \subseteq b \in \tau} (\bigcup_{b \supseteq c \in P} h_\alpha(c))$,

(3) $h_{\neg\alpha}(a) \supseteq \bigcup_{a \subseteq b \in \tau} (\bigcap_{b \supseteq c \in P} h_{\neg\alpha}(c))$.

Proof.

(1) 1.14.6 and 1.14.7(2).

(2) 1.14.6 and 1.14.7(1).

(3) By (2), antitonicity of $-$, and De Morgan's Laws. □

Theorem 1.14.10 *For all $a \subseteq W$,*

(1) $\alpha \in \Pi_\Diamond$ only if $h_\alpha(a) = \bigcup_{a \supseteq b \in \zeta} (\bigcap_{b \subseteq c \in P} h_\alpha(c))$,

(2) $a \in \Pi_\square$ only if $h_\alpha(a) = \bigcap_{a \subseteq b \in \tau} (\bigcup_{b \supseteq c \in P} h_\alpha(c))$.

Proof.

(1) 1.14.8(1) and 1.14.7(2).
(2) 1.14.8(2) and 1.14.7(1).

□

Theorem 1.14.11 $\beta \in \Phi$ *is d-persistent if it is equivalent in K to an* $\alpha \in \Phi$ *that satisfies any of the following:*

(1) $\alpha \in \Pi$,
(2) $\alpha = \neg\gamma$ *with* $\gamma \in \Pi$,
(3) $\alpha = \gamma \to \delta$ *with* $\gamma \in \Pi_\Diamond$, $\delta \in \Pi$,
(4) $\alpha = \gamma \to \delta$ *with* $\gamma \in \Pi$, $\delta \in \Pi_\Box$,
(5) α *is a conjunction of wffs satisfying* (1)–(4).

Proof. Let $\mathcal{F} = \langle W, R, P \rangle$ be a descriptive frame with $\mathcal{F} \models \alpha$. Then

$$(*) \qquad h_\alpha(a) = W, \quad \text{all } a \in P.$$

(1) Let $\alpha \in \Pi$. Then if $a \subseteq W$,

$$h_\alpha(a) \supseteq \bigcup_{a \supseteq b \in \zeta}(\bigcap_{b \subseteq c \in P} h_\alpha(c)) = W,$$

by 1.14.9(1) and (*). (Note that for any a, $a \supseteq \emptyset \in \zeta$, $a \subseteq W \in P$).
Thus $\langle W, R, 2^W \rangle \models \alpha$ as required.

(2) Similar to (1), using 1.14.9(3).

(3) If $\mathcal{F} \models \gamma \to \delta$, $h_\gamma(a) \subseteq h_\delta(a)$, all $a \in P$, so for any $a \subseteq W$,

$$\bigcup_{a \supseteq b \in \zeta}(\bigcap_{b \subseteq c \in P} h_\gamma(c)) \subseteq \bigcup_{a \supseteq b \in \zeta}(\bigcap_{b \subseteq c \in P} h_\delta(c)),$$

i.e. by 1.14.10(1) and 1.14.9(1), $h_\gamma(a) \subseteq h_\delta(a)$. Thus $\langle W, R \rangle \models \gamma \to \delta$.

(4) Similar to (3), using 1.14.9(2) and 1.4.10(2).

(5) $\mathcal{F} \models \alpha \wedge \beta$ iff $\mathcal{F} \models \alpha$ and $\mathcal{F} \models \beta$ for any frame $\mathcal{F}$.

□

It should be noted that (2) and (4) are obtainable as corollaries of (1) and (3). If $\alpha \in \Pi$, let $\hat{\alpha}$, the dual of α, be obtained by interchanging $\vee$ and $\wedge$, $\Box$ and $\Diamond$. Let α' be the result of negating all the variables in α. Then α is semantically equivalent to α' $(h_{\alpha'}(a) = h_\alpha(a'))$. But $\neg\alpha$ is equivalent to $(\hat{\alpha})'$, which is equivalent to $\hat{\alpha} \in \Pi$, so (1) implies (2). Since $\alpha \in \Pi_\Diamond$ iff $\hat{\alpha} \in \Pi_\Box$, (4) then follows from (3).

1.15   A General Characterization Theorem

Many of the logics with characteristic K-frames are determined by classes of frames that are definable by a first-order condition on binary relations. Not all logics enjoy this property (Section 1.17). While not all d-persistent formulae are first-order definable (and conversely, cf. Section 1.18), many of them do have elementary characterizations, as we shall show by proving the first-order definability of an axiom schema devised by Lemmon and Scott [59] that forms a special case of 1.14.11(3).

Definition 1.15.1 *Let* $\alpha(p_1, \ldots, p_k)$ *be a positive formula and*

$$n = \langle n_1, \ldots, n_k \rangle$$

a k-tuple of natural numbers. If $\langle W, R \rangle$ is a K-frame and

$$t = \langle t_1, \ldots, t_k \rangle \in W^k,$$

define a first-order condition $R_\alpha(x, t, n)$ on $\langle W, R \rangle$ by recursion as follows:

$$
\begin{array}{lll}
R_{p_i}(x, t, n) & \text{iff} & t_i R^{n_i} x \quad (i \leq k) \\
R_{\alpha \wedge \beta}(x, t, n) & \text{iff} & R_\alpha(x, t, n) \text{ and } R_\beta(x, t, n) \\
R_{\alpha \vee \beta}(x, t, n) & \text{iff} & R_\alpha(x, t, n) \text{ or } R_\beta(x, t, n) \\
R_{\Box \alpha}(x, t, n) & \text{iff} & \text{for all } y, \, xRy \text{ only if } R_\alpha(y, t, n) \\
R_{\Diamond \alpha}(x, t, n) & \text{iff} & \text{for some } y, \, xRy \text{ and } R_\alpha(y, t, n).
\end{array}
$$

*Given a pair $n = \langle n_1, \ldots, n_k \rangle$ and $m = \langle m_1, \ldots, m_k \rangle$ of k-tuples of numbers, we define the **Lemmon-Scott axiom** α_n^m to be the wff*

$$\Diamond^{m_1} \Box^{n_1} p_1 \wedge \ldots \wedge \Diamond^{m_k} \Box^{n_k} p_k \to \alpha(p_1, \ldots, p_k).$$

Corresponding to α_n^m is the condition

$$R\alpha_n^m : \forall x \forall t_1 \ldots \forall t_k (x R^{m_1} t_1 \wedge \ldots \wedge x R^{m_k} t_k \Rightarrow R_\alpha(x, t, n)).$$

Theorem 1.15.2 *If V is a valuation on $\langle W, R \rangle$, $R_\alpha(x, t, n)$ and $t_i \in V(\Box^{n_i} p_i)$ $(i \leq k)$, then $x \in V(\alpha)$.*

Proof. By induction. If $\alpha = p_i$ we have $t_i R^{n_i} x$, $t_i \in V(\Box^{n_i} p_i)$ and hence $x \in V(p_i)$. For the case of $\alpha = \Box \beta$, we have xRy only if $R_\beta(y, t, n)$. But $t_i \in V(\Box^{n_i} p_i)$ so by IH, xRy only if $y \in V(\beta)$. Hence $x \in V(\Box \beta)$. The other cases are equally straightforward. $\square$

Corollary 1.15.3 $\mathcal{F} = \langle W, R \rangle \models \alpha_n^m$ *if $\mathcal{F}$ satisfies $R\alpha_n^m$.*

Proof. Suppose $x \in V(\bigwedge_{i \leq k} \Diamond^{m_i} \Box^{n_i} p_i)$ for $x \in W$ and V any valuation on $\mathcal{F}$. Then there exist t_i $(i \leq k)$ such that $x R^{m_i} t_i$ and $t_i \in V(\Box^{n_i} p_i)$. Since $\mathcal{F}$ satisfies $R\alpha_n^m$, we have $R_\alpha(x, t, n)$. Then by 1.15.2, $x \in V(\alpha)$. Thus $x \in V(\alpha_n^m)$ for any x and V, hence $\mathcal{F} \models \alpha_n^m$. $\square$

Theorem 1.15.4 *For each t_i $(i \leq k)$, let $S_i = \{y : t_i R^{n_i} y\}$. Then $x \in h_\alpha(S_1, \ldots, S_k)$ only if $R_\alpha(x, t, n)$.*

Proof. If $\alpha = p_i$, by the definition of h_{p_i}, we have $x \in S_i$, whence $t_i R^{n_i} x$, i.e. $R_{p_i}(x, t, n)$. If $x \in h_{\alpha \wedge \beta}(S)$, then $x \in h_\alpha(S)$ and $x \in h_\beta(S)$, whence by IH, $R_\alpha(x, t, n)$ and $R_\beta(x, t, n)$, so $R_{\alpha \wedge \beta}(x, t, n)$.

If $x \in h_{\Box\alpha}(S)$, then xRy only if $y \in h_\alpha(S)$ only if $R_\alpha(y,t,n)$ (IH). Hence $R_{\Box\alpha}(x,t,n)$.

The cases of $\vee$, $\Diamond$ are similar. $\qquad\square$

Corollary 1.15.5 $\mathcal{F}$ *satisfies* $R\alpha_n^m$ *if* $\mathcal{F} \models \alpha_n^m$.

Proof. Suppose $xR^{m_i}t_i$ $(i \le k)$. Choose a V such that $V(p_i) = S_i$ as in 1.15.4. Then $t_i \in V(\Box^{n_i}p_i)$, so $x \in V(\Diamond^{m_i}\Box^{n_i}p_i)$ for $i \le k$. Since $\mathcal{F} \models \alpha_n^m$, $x \in V(\alpha) = h_\alpha(V(p_1), \ldots, V(p_k))$ (1.3.2). Hence by 1.15.4, $R_\alpha(x,t,n)$. $\qquad\square$

We shall now see that Corollary 1.15.5 holds for descriptive frames as well as K-frames.

Theorem 1.15.6 *If* $\mathcal{F} = \langle W, R, P \rangle$ *is descriptive and* $\alpha(p_1, \ldots p_k) \in \Pi$,

$$R_\alpha(x,t,n) \text{ iff } \{h_\alpha(\ldots a_i \ldots) : a_i \in P \text{ and } t_i \in l_R^{n_i}(a_i) \ (i \le k)\} \subseteq Px.$$

Proof. We give the proof for $k = 1$.

(i) For $\alpha = p$, we require tR^nx iff $\{a \in P : t \in l_R^n(a)\} \subseteq Px$. But this is given by 1.9.6 and 1.9.7.

(ii) If $R_{\alpha\vee\beta}(x,t,n)$, then say $R_\alpha(x,t,n)$ (the case $R_\beta(x,t,n)$ is analogous). Then $t \in l_R^n(a)$ only if $h_\alpha(a) \in Px$ (IH), only if $h_\alpha(a) \cup h_\beta(a) = h_{\alpha\vee\beta}(a) \in Px$.

Conversely, if not $R_{\alpha\vee\beta}(x,t,n)$ then $R_\alpha(x,t,n)$ and $R_\beta(x,t,n)$ both fail. By IH there are $a,b \in P$ such that $l_R^n(a), l_R^n(b) \in P_t$, but $h_\alpha(a), h_\beta(b) \notin Px$. Let $c = a \cap b \in P$. Then $t \in l_R^n(c) = l_R^n(a) \cap l_R^n(b)$. By 1.14.6 $h_\alpha(c) \subseteq h_\alpha(a)$ and $h_\beta(c) \subseteq h_\beta(b)$. So $h_{\alpha\vee\beta}(c) \subseteq h_\alpha(a) \cup h_\beta(b) \notin Px$ (1.7.1(6)).

(iii) Suppose $R_{\Diamond\alpha}(x,t,n)$. Then for some y, xRy and $R_\alpha(y,t,n)$. Then $t \in l_R^n(a)$ only if $y \in h_\alpha(a)$ (IH), only if $x \in m_R(h_\alpha(a)) = h_{\Diamond\alpha}(a)$.

Conversely, suppose

$$(*) \qquad t \in l_R^n(a) \text{ only if } h_{\Diamond\alpha}(a) \in Px, \text{ all } a \in P.$$

Let $P_0 = \{a \in P : x \in l_R(a)\} \cup \{h_\alpha(b) : b \in P \text{ and } t \in l_R^n(b)\}$. Then if P_0 does not have the fip there exist $a, h_\alpha(b_1), \ldots, h_\alpha(b_n)$ such that $x \in l_R(a)$, $t \in l_R^n(b_i)$ $(i \le n)$ and $a \cap (\bigcap_{i \le n} h_\alpha(b_i)) = \emptyset$. Then by 1.14.6, $a \cap h_\alpha(b) = \emptyset$, where $b = \bigcap_{i \le n} b_i \in P$. Then we have $l_R(a) \subseteq l_R(-h_\alpha(b)) = -h_{\Diamond\alpha}(b)$. But $x \in l_R(a)$, so $x \notin h_{\Diamond\alpha}(b)$. Since $t \in l_R^n(b) = \bigcap_{i \le n} l_R^n(b_i)$, this contradicts $(*)$. Thus P_0 has the fip, so by 1.9.3 **IV** there exists $y \in \bigcap P_0$. Then from the definition of P_0, Axiom **II**, and IH, we have xRy and $R_\alpha(y,t,n)$. Thus $R_{\Diamond\alpha}(x,t,n)$ as required.

The cases of $\wedge$ and $\Box$ are straightforward and will be omitted. $\qquad\square$

Corollary 1.15.7 *If $\mathcal{F} \models \alpha_n^m$, and $\mathcal{F}$ is descriptive, then $\langle W, R \rangle$ satisfies $R\alpha_n^m$.*

Proof. Let $xR^{m_i}t_i$ $(i \le k)$. Then if $t_i \in l_R^{n_i}(a_i)$ $(i \le k)$, where $a_i \in P$, we have $x \in m_R^{m_i} l_R^{n_i}(a_i)$. Since $\mathcal{F} \models \alpha_n^m$, we then get $x \in h_\alpha(\dots a_i \dots)$. Thus by 1.15.6, $R_\alpha(x, t, n)$. $\qquad\square$

Corollary 1.15.8 α_n^m *is d-persistent.*

Proof. 1.15.7 and 1.15.3. $\qquad\square$

Corollary 1.15.9 $K\alpha_n^m$ *is strongly determined by the class of all K-frames satisfying $R\alpha_n^m$.*

Proof. 1.15.3, 1.14.3. $\qquad\square$

Corollary 1.15.9 solves a conjecture of Lemmon and Scott. It is indeed a wide-ranging result. The Lemmon-Scott axioms include as special cases the Hintikka schemata of [59], which Segerberg [86] observes "cover most of the 'ordinary' systems in the literature."

1.16   First-Order Definability

Let $\mathfrak{R}$ be (the set of sentences of) the first-order language with equality and a single binary predicate letter. The appropriate structures for this language are precisely the K-frames $\langle W, R \rangle$. In this section we establish conditions under which modally characterised classes of K-frames can be axiomatised by sentences of $\mathfrak{R}$. The converse problem will be taken up in Section 1.20.

Definition 1.16.1 *If α is a modal wff or an $\mathfrak{R}$-sentence, let $\mathfrak{K}(\alpha) = \{\mathcal{F} \in \mathfrak{K} : \mathcal{F} \models \alpha\}$, where $\mathfrak{K}$ is the class of all K-frames.*

(For $\alpha \in \mathfrak{R}$, $\mathcal{F} \models \alpha$ means that α is true in $\mathcal{F}$ in the standard first-order sense (cf., e.g.,[4, p. 56])).

If Γ is a set of wffs ($\mathfrak{R}$-sentences) let $\mathfrak{K}(\Gamma) = \{\mathcal{F} \in \mathfrak{K} : \mathcal{F} \models \Gamma\} = \bigcap\{\mathfrak{K}(\alpha) : \alpha \in \Gamma\}$.

If $X \subseteq \mathfrak{K}$ we write:

$X \in EC$ *(X is **elementary**) iff $X = \mathfrak{K}(\alpha)$ for some $\alpha \in \mathfrak{R}$;*

$X \in EC_\Delta$ *(X is Δ-**elementary**) iff X is the intersection of a set of elementary classes;*

$X \in EC_\Sigma$ *(X is Σ-**elementary**) iff X is the union of elementary classes;*

$X \in EC_{\Sigma\Delta}$ *(X is $\Sigma\Delta$-**elementary**) iff X is the intersection of Σ-elementary classes, or equivalently iff X is closed under first-order semantic equivalence.*

For modal characterisations, we write $X \in MEC$ iff $X = \mathfrak{K}(\alpha)$ for some $\alpha \in \Phi$, and $X \in MAC$ iff $X = \mathfrak{K}(\Gamma)$ for some $\Gamma \subseteq \Phi$.

In what follows we will need the following facts about model classes (here $-X$ denotes $\mathfrak{K} - X$. Proofs may be found in [4, Chapter 7]).

(A) $X \in EC$ iff X and $-X$ are both closed under isomorphism and ultraproducts.

(B) $X \in EC_\Delta$ iff X is closed under isomorphism and ultraproducts and $-X$ is closed under ultrapowers.

(C) $X \in EC_\Sigma$ iff $-X \in EC_\Delta$ iff X is closed under ultrapowers and $-X$ is closed under isomorphism and ultraproducts.

(D) $X \in EC_{\Sigma\Delta}$ iff X and $-X$ are closed under isomorphism and ultrapowers.

Theorem 1.16.2

(i) $X \in MEC$ only if $-X$ is closed under ultraproducts.

(ii) $X \in MAC$ only if $-X$ is closed under ultrapowers.

Proof.

(i) Suppose $X = \mathfrak{K}(\alpha)$ for some $\alpha \in \Phi$, $\{\mathcal{F}_i : i \in I\} \subseteq -X$, and G is an ultrafilter on I. Then $\{i : \mathcal{F}_i \models \alpha\} = \emptyset \notin G$, so by 1.7.14 $\mathcal{F}_G \not\models \alpha$, whence $\mathcal{F}_G \in -X$.

(ii) Suppose $X = \mathfrak{K}(\Gamma)$ for some $\Gamma \subseteq \Phi$, and $\mathcal{F}^I/G$ is an ultrapower of $\mathcal{F} \in -X$ (i.e. $\mathcal{F}^I/G = \prod \mathcal{F}_i/G$ where $\mathcal{F}_i = \mathcal{F}$ all $i \in I$). Since $\mathcal{F} \notin X$, $\mathcal{F} \not\models \alpha$ for some $\alpha \in \Gamma$. Then as in (i), $\mathcal{F}^I/G \not\models \alpha$, whence $\mathcal{F}^I/G \not\models \Gamma$ and so $\mathcal{F}^I/G \in -X$.

□

Corollary 1.16.3

(i) If $X \in MEC$, then $X \in EC$ iff X is closed under ultraproducts.

(ii) If $X \in MAC$, then $X \in EC_\Delta$ iff X is closed under ultraproducts.

Proof.

(i) If $X \in MEC$ (indeed if $X \in MAC$) then X and $-X$ are closed under isomorphism (1.5.6). The result then follows from 1.16.2(i) and (A) above.

(ii) Similar to (i), using 1.16.2(ii) and (B).

□

Since Δ-elementary classes are closed under ultraproducts, it follows from 1.16.3(i) that any $X \in MEC$ is Δ-elementary only if it is elementary. Recently this result has been improved by van Benthem [101] to replace "Δ-elementary" by "$\Sigma\Delta$-elementary". We will present an alternative proof of this result, but first we undertake an analysis of MAC classes, using

Theorem 1.16.4 *Let X be a class of K-frames closed under isomorphism, subframes, disjoint unions and ultrapowers. Then X is closed under ultraproducts.*

Proof. Suppose $\{\mathcal{F}_i : i \in I\} \subseteq X$ and G is an ultrafilter on I. We define first a map $\theta : \prod W_i \to (\sum W_i)^I$, where

$$\Sigma W_i = \bigcup_{i \in I} W_i \times \{i\}$$

is the disjoint union of the W_i's (cf. Section 1.6). If $f \in \prod W_i$, we put $\theta(f)(i) = \langle f(i), i \rangle$. It is easy to see that $f \sim g$ only if $\theta(f) \sim \theta(g)$ (cf. 1.7.3) and so the correspondence $\hat{\theta} : \hat{f} \mapsto \widehat{\theta(f)}$ is a well defined injection of $\prod W_i/G$ into $(\sum W_i)^I/G$. We leave it to the reader to verify that θ satisfies 1.5.1(1) and (2). Hence (1.5.3) θ establishes an isomorphism from the ultraproduct $\prod \mathcal{F}_i/G$ onto a subframe of the ultrapower $(\sum \mathcal{F}_i)^I/G$ of the disjoint union $\sum \mathcal{F}_i$. The closure conditions assumed to hold for X then give $\prod \mathcal{F}_i/G \in X$. $\qquad\square$

Theorem 1.16.5 *If $X \in MAC$ then the following are equivalent:*

(i) $X \in EC$,
(ii) $X \in EC_\Sigma$.

Proof. Obviously (i) implies (ii). To establish the converse it suffices by (A) and (C) to show that $X \in EC_\Sigma$ only if X is closed under ultraproducts. But $X \in MAC$, so if X is also Σ-elementary it satisfies all of the closure conditions in the hypothesis of 1.16.4. The required conclusion is then given by that Theorem. $\qquad\square$

Theorem 1.16.6 *If $X \in MAC$ then the following are equivalent:*

(i) $X \in EC_\Delta$,
(ii) $X \in EC_{\Sigma\Delta}$,
(iii) X *is closed under ultrapowers,*
(iv) X *is closed under ultraproducts.*

Proof. That (ii) follows from (i), and (iii) from (ii) is standard. (iv) follows from (iii) by 1.16.4, since $X \in MAC$. Finally 1.16.3(ii) yields (i) from (iv). $\qquad\square$

Theorem 1.16.7 *There exists an $X \in MAC$ such that $X \in EC_\Delta$ but $X \notin EC$.*

Proof. Let $X = \mathfrak{K}(A)$, where $A = \{\alpha_n : n \geq 1\}$ and for each positive integer n, α_n is the sentence

$$\Diamond p_1 \wedge \ldots \wedge \Diamond p_n \to \Diamond(\Diamond p_1 \wedge \ldots \wedge \Diamond p_n).$$

Each α_n is a Lemmon-Scott axiom as defined in Section 1.15, and so by 1.15.1, 1.15.3 and 1.15.5, $\mathfrak{K}(\alpha_n) = \mathfrak{K}(\beta_n)$ where β_n is the $\mathfrak{R}$-sentence

$$\forall x, y_1, \ldots, y_n (\bigwedge_{i \leq n} xRy_i \Rightarrow \exists z(xRz \,\&\, (\bigwedge_{i \leq n} zRy_i))).$$

Thus we have $X = \mathfrak{K}(B)$ where $B = \{\beta_n : n \geq 1\}$ and so $X \in EC_\Delta$.

To show that $X \notin EC$ we observe first that α_n is derivable from α_m if $m > n$ by simply identifying variables, from which it follows that

(1) $\mathfrak{K}(\beta_m) \subseteq \mathfrak{K}(\beta_n)$ if $m > n$.

To show that this inclusion is proper, consider for each n the frame $\mathcal{F}_n = \langle W_n, R \rangle$, where $W_n = \{0, 1, \ldots, n+1\}$, and R is the codiagonal relation, i.e. xRy iff $x \neq y$. Then 0 has the $n+1$ R-successors $1, \ldots, n+1$. Since these are the only successors of 0, and none of them is related to itself, we see, with $x = 0$, that $\mathcal{F}_n \not\models \beta_{n+1}$. On the other hand if $x \in W_n$ has n successors $y_1, \ldots, y_n$ then by the cardinality of W_n there exists in W_n some $z \neq x, y_1, \ldots, y_n$. Then xRz and zRy_i for $i \leq n$, whence $\mathcal{F}_n \models \beta_n$. It follows from (1) and the above example that

(2) $X \neq \mathfrak{K}(\beta_n)$,   all n.

To complete the proof, suppose that $X = \mathfrak{K}(B) \in EC$. Then by the Compactness Theorem for first-order logic there is some finite $B_0 \subseteq B$ such that $X = \mathfrak{K}(B_0)$. Let n_0 be the largest index of any any $\beta_n \in B_0$. Then by (1)

$$\mathfrak{K}(B_0) = \bigcap\{\mathfrak{K}(\beta_n) : \beta_n \in B_0\} = \mathfrak{K}(\beta_{n_0}),$$

which is a contradiction in view of (2). $\square$

We thus see that the collection of MAC classes can be partitioned into three categories: those satisfying the conditions of 1.16.5, those satisfying 1.16.6 but not 1.16.5, and those having no first order semantic description at all. However for MEC classes there are only two categories:

Theorem 1.16.8 (van Bentham [101]) *If $X \in MEC$ then the following are equivalent:*

(i) $X \in EC$,

(ii) $X \in EC_\Sigma$,

(iii) $X \in EC_\Delta$,

(iv) $X \in EC_{\Sigma\Delta}$,

(v) *X is closed under ultrapowers,*

(vi) *X is closed under ultraproducts.*

Proof. If $X \in MEC$ then $X \in MAC$, so the equivalence of (i) and (ii) is given by 1.16.5, and the equivalence of (iii)–(vi) is 1.16.6. However (i) implies (vi) in general, and the converse for $X \in MEC$ is given by 1.16.3(i). $\square$

1.17   The Logic KM

We come now to an application of the techniques developed so far to a semantic analysis of the wff

$$M : \quad \Box\Diamond p \to \Diamond\Box p,$$

one that has received considerable attention in the literature. In the field of reflexive linear tense logic, M serves as an axiom for "ending" or "beginning" time [85, p. 320]. The wff was used by Thomason [96] to construct a consistent tense logic with no K-frames at all. Lemmon and Scott [59] showed that the logic $K4M$, where 4 denotes the transitivity axiom $\Box p \to \Box\Box p$, was (strongly) determined by the class of transitive K-frames satisfying

$$m^\infty : \quad \forall x \exists y (xRy \text{ and } \forall w, z(yRw \text{ and } yRz \Rightarrow w = z)).$$

They also established that KM itself was not determined by this condition, and left open the problem as to whether KM had a characteristic class of K-frames. An affirmative solution has recently been found by Fine [14]. Of course a K-frame validates M iff it satisfies

$$\forall S \in 2^W \forall x \in W(Rx \subseteq m_R(S) \Rightarrow Rx \cap l_R(S) \neq \emptyset),$$

so Fine's result may be interpreted as showing KM is simply determined by the class of (finite) K-frames satisfying this condition. In this section we shall show that the condition cannot be replaced by a first-order one, or even a set of first-order ones.

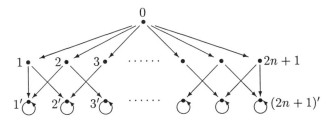

Figure 1.17.1

Consider the class $\{\mathcal{F}_n = \langle W_n, R_n \rangle : n \in \mathbb{N}\}$ of finite K-frames, where $\mathcal{F}_n$ is depicted in Figure 1.17.1. Formally we have $W_n = W_n^1 \cup W_n^2 \cup W_n^3$, where

$$
\begin{aligned}
W_n^1 &= \{0\} \\
W_n^2 &= \{1, \ldots, 2n+1\} \\
W_n^3 &= \{1', \ldots, (2n+1)'\}
\end{aligned}
$$

and R_n holds precisely in the cases

$$0 R_n j \qquad\qquad\qquad\qquad\qquad\qquad\qquad 1 \le j \le 2n+1$$

$$1 R_n 1', \; 1 R_n 2', \; (2n+1) R_n (2n+1)', \; (2n+1) R_n 2'_n$$

$$j R_n (j-1)', j R_n (j+1)' \qquad\qquad\qquad\qquad 1 < j < 2n+1$$

$$j' R_n j' \qquad\qquad\qquad\qquad\qquad\qquad\qquad 1 \le j \le 2n+1$$

Now for any $n \in \mathbb{N}$, $\mathcal{F}_n \models M$. For clearly if $\Diamond p$ is true at $x \in W_n^3$, then so is $\Box p$, hence M cannot be falsified at any point in $W_n^2 \cup W_n^3$. But to have M false at 0 we require p true at one and false at the other of the two R_n-successors of each point in W_n^2. But because of the finiteness and odd order of W_n^3, no such valuation can be constructed.

Now let G be a non-principal ultrafilter on $\mathbb{N}$, and $\mathcal{F}_G$ the resulting ultraproduct[1]. By Łoś's Theorem, any first-order (i.e. $\mathfrak{R}$-expressible) property that is true of almost all of the $\mathcal{F}_n$'s will also be true of $\mathcal{F}_G$. Using this we can show that $\mathcal{F}_G$ is an infinite frame whose base set W_G is partitioned into three disjoint subsets W_G^1, W_G^2, W_G^3, where W_G^1 is a singleton whose member we continue to denote 0, W_G^2 is the set of R_G-successors of 0, and W_G^3 is the set of R_G-successors of members of W_G^2. Each member of W_G^3 has itself as its only R_G-successor, and is an R_G-successor of exactly two points from W_G^2. Each member of W_G^2 has exactly two R_G-successors, and shares each of these with one of two distinct points in W_G^2 that themselves have no R_G-successors in common.

In fact $W_G^2 \cup W_G^3$ consists of an infinite collection of disjoint copies of the graph depicted as follows.

---

[1] The following description of the ultraproduct differs from that given in the original version of this article, since the latter was not strictly correct, as pointed out by Bjarni Jónsson.

It can then be seen that a valuation can be defined on W_G that has p true at one and false at the other of the R_G-successors of each point in W_G^2. In this valuation the wff M is false at 0. Accordingly we have $\{n : \mathcal{F}_n \models M\} = \mathbb{N} \in G$ but $\mathcal{F}_G \not\models M$, and hence:

Theorem 1.17.1 $\mathfrak{K}(M)$ *is not closed under ultraproducts, so is not* EC_Δ, *and* M *is not first-order definable.* □

Note that the above construction gives the promised counter-example to the converse of 1.7.14. Also, if $\mathcal{F}'_n = \langle W_n, R_n, P_n \rangle$ where $P_n = 2^{W_n}$, then $\{n : \mathcal{F}'_n \models M\} = \mathbb{N} \in G$, so by 1.7.13, $\langle W_G, R_G, P_G \rangle \models M$, and so $P_G \neq 2^{W_G}$. Thus an ultraproduct of full frames need not be full. Furthermore, this example is of the kind envisaged in the comments after 1.9.11, and shows that an ultraproduct of descriptive frames can be non-descriptive.

Now any logic has countably many non-theorems (since Φ is denumerable) and so any characteristic class for a logic can be made countable by retaining only one falsifying frame for each non-theorem. Furthermore, by the work on disjoint unions in Section 1.6, we may amalgamate frames and preserve the characterisation. Thus since KM is determined by a class of K-frames, it will certainly be determined by such classes other than $\mathfrak{K}(M)$ itself. We shall now prove that no such class can be axiomatic.

For $n \in \mathbb{N}$, let $\Delta_n \subseteq \Phi$ consist of the following wffs in the variables $p_0, \ldots, p_{2n+1}, q_1, \ldots, q_{2n+1}$.

(1) p_0

(2) $\square^k \neg p_0$ $(1 \leq k \in \mathbb{N})$

(3) $\neg p_1 \wedge \ldots \wedge \neg p_{2n+1}$

(4) $\square^k(\neg p_1 \wedge \ldots \wedge \neg p_{2n+1})$ $(2 \leq k \in \mathbb{N})$

(5) $\square(p_1 \vee \ldots \vee p_{2n+1})$

(6) $\lozenge p_1 \wedge \ldots \wedge \lozenge p_{2n+1}$

(7) $\square \neg (p_i \wedge p_j)$ $(1 \leq i \neq j \leq 2n+1)$

(8) $\square^k(q_1 \vee \ldots \vee q_{2n+1})$ $(2 \leq k \in \mathbb{N})$

(9) $\lozenge^2 q_1 \wedge \ldots \wedge \lozenge^2 q_{2n+1}$

(10) $\square^k \neg (q_i \wedge q_j)$ $(1 \leq i \neq j \leq 2n+1,\ 2 \leq k \in \mathbb{N})$

(11) $\neg q_1 \wedge \ldots \wedge \neg q_{2n+1}$

(12) $\square \neg (q_1 \vee \ldots \vee q_{2n+1})$

(13) $\square(p_1 \rightarrow \lozenge q_1 \wedge \lozenge q_2 \wedge \square(q_1 \vee q_2))$

(14) $\square(p_{2n+1} \rightarrow \lozenge q_{2n} \wedge \lozenge q_{2n+1} \wedge \square(q_{2n} \vee q_{2n+1}))$

(15) $\square(p_j \rightarrow \lozenge q_{j-1} \wedge \lozenge q_{j+1} \wedge \square(q_{j-1} \vee q_{j+1}))$ $(1 < j < 2n+1)$

(16) $\square^k(q_j \rightarrow \lozenge q_j \wedge \square q_j)$ $(1 \leq j \leq 2n+1,\ 2 \leq k \in \mathbb{N})$.

Theorem 1.17.2 *Let $\mathcal{F}_t = \langle W, R \rangle$ be generated by $t \in W$. If there exists a V on $\mathcal{F}_t$ for which $t \in V(\Delta_n)$, then $\mathcal{F}_n \preccurlyeq \mathcal{F}_t$.*

Proof.

Since t generates $\mathcal{F}_t$, by 1.4.4

    (17) for all $x \in W$ there is $k \in \mathbb{N}$ such that $tR^k x$.

By (2), (17), $x \notin V(p_0)$ if $x \neq t$. Hence by (1),

    (18) $V(p_0) = \{t\}$.

Now let $S = \{x : tRx\}$. By (1), (2),

    (19) $t \notin S$.

By (3), (4), (17),

    (20) $V(p_j) \subseteq S$ for $1 \leq j \leq 2n + 1$.

By (5), (20),

    (21) $S = V(p_1) \cup \ldots \cup V(p_{2n+1})$.

By (6),

    (22) $V(p_i) \neq \emptyset$.

By (7), (20),

    (23) $V(p_i) \cap V(p_j) = \emptyset$, for $1 \leq i \neq j \leq 2n + 1$.

Hence by (20)-(23),

    (24) the sets $V(p_1), \ldots, V(p_{2n+1})$ partition S.

By similar reasoning, using (8)-(12) we obtain

    (25) $V(q_1), \ldots, V(q_{2n+1})$ partition $W - (S \cup \{t\})$.

Thus by (19), (24), (25), for each $x \in W$ either

(i) for exactly one $j \leq 2n + 1$, $x \in V(p_j)$,

    or

(ii) for exactly one $j \leq 2n + 1$, $x \in V(q_j)$,

and not both.

    If (i), let $Q(x) = j$. If (ii), let $Q(x) = j'$. Then $Q : W \to W_n$ is a well-defined surjective mapping. Now from (13)-(16) we deduce

    (26) Each $x \in V(p_1)$ has alternatives in and only in (inn) $V(q_1)$ and $V(q_2)$.

    (27) If $1 < j < 2n + 1$, $x \in V(p_j)$ has alternatives inn $V(q_{j-1})$ and $V(q_{j+1})$.

    (28) $x \in V(p_{2n+1})$ has alternatives inn $V(q_{2n})$ and $V(q_{2n+1})$.

    (29) $x \in V(q_j)$ has alternatives inn $V(q_j)$, for $1 \leq j \leq 2n + 1$.

Also by (5), (6),

    (30) t has alternatives inn $V(p_1), \ldots, V(p_{2n+1})$.

A checking of cases, using (26)–(30) shows that Q satisfies 1.5.1(1) and 1.5.1(2), so is a homomorphism from $\mathcal{F}_t$ onto $\mathcal{F}_n$.

□

Theorem 1.17.3 Δ_n *is KM-consistent.*

Proof. Take V on $\mathcal{F}_n$ such that $V(p_j) = \{j\}$ for $0 \leq j \leq 2n + 1$, and $V(q_j) = \{j'\}$ for $1 \leq j \leq 2n + 1$.

Then it is readily shown that $0 \in V(\Delta_n)$. If Δ_n is not KM-consistent, then $\vdash_{KM} \neg\alpha$, for α the conjunction of some finite subset of Δ_n. But $\mathcal{F}_n$ is a KM-frame, so $\mathcal{F}_n \models \neg\alpha$, whence $0 \in V(\neg\alpha)$, which is impossible, as $V(\Delta_n) \subseteq V(\alpha)$, whence $0 \in V(\alpha)$. □

We are now ready for our second main result.

Theorem 1.17.4 *Let $\mathfrak{C}$ be a class of K-frames. If $\mathfrak{C}$ simply determines KM, then $\mathfrak{C}$ is not closed under ultraproducts.*

Proof. Suppose $\mathfrak{C}$ determines KM and is closed under ultraproducts. Then by 1.8.6 $\mathfrak{C}$ strongly determines KM. Hence by 1.8.3(1') and 1.17.3, for each $n \in \mathbb{N}$ there exists $\mathcal{F} \in \mathfrak{C}$, V on $\mathcal{F}$, and some $t \in V(\Delta_n)$. Then $t \in V_t(\Delta_n)$ where V_t is the valuation on $\mathcal{F}_t$ derived from V (1.4.11). Hence by 1.17.2 $\mathcal{F}_n \preccurlyeq \mathcal{F}_t$.

Now let $\mathfrak{C}^*$ be the class of homomorphic images of subframes of frames in $\mathfrak{C}$. Then by the above $\mathcal{F}_n \in \mathfrak{C}^*$, all $n \in \mathbb{N}$. By 1.8.3(2), $\mathfrak{C} \models M$, so by 1.4.10 and 1.5.5, $\mathfrak{C}^* \models M$. Furthermore, since $\mathfrak{C}$ is closed under ultraproducts, it follows from 1.7.9 and 1.7.11 that $\mathfrak{C}^*$ is closed under ultraproducts. Thus all ultraproducts of the $\mathcal{F}_n$'s validate M, which contradicts our earlier work. □

Corollary 1.17.5 *If $\mathfrak{C}$ determines KM then $\mathfrak{C} \notin EC_\Delta$, and hence $\mathfrak{C} \notin EC$.*

Proof. By Łoś's Theorem, $\mathfrak{C} \in EC_\Delta$ only if $\mathfrak{C}$ is closed under ultraproducts. □

Corollary 1.17.6 *If $\mathfrak{C} \subseteq \mathfrak{K}(M)$ and $\mathfrak{C} \in EC_\Delta$, then $\mathcal{F}_{KM}^K \notin \mathfrak{C}$. In particular $\mathcal{F}_{KM}^K$ does not satisfy m^∞.*

Proof. If $\mathfrak{C} \subseteq \mathfrak{K}(M)$ and $\mathcal{F}_{KM}^K \in \mathfrak{C}$, then as each non-theorem of KM is falsifiable on $\mathcal{F}_{KM}^K$ it follows that $\mathfrak{C}$ determines KM. Hence by 1.17.5 $\mathfrak{C} \notin EC_\Delta$. To complete the result we observe that the class of K-frames satisfying m^∞ is an EC_Δ subclass of $\mathfrak{K}(M)$. □

1.18 Some Special Classes of Formulae

In this section we explore the relationships between various classes of wffs that are defined by special model-theoretic properties.

Definition 1.18.1 $\alpha \in \Phi$ *is **r-persistent** iff for every refined frame* $\mathcal{F} = \langle W, R, P \rangle$, $\mathcal{F} \models \alpha$ *only if* $\langle W, R \rangle \models \alpha$.

$E = \{\alpha : \alpha$ *is r-persistent*$\}$

$E_1 = \{\alpha : \alpha \text{ is first-order definable}\}$
$N = \{\alpha : \mathcal{F}^K_{K\alpha} \models \alpha\}$
$SC = \{\alpha : K\alpha \text{ is super-complete for some } \mathfrak{C} \subset \mathfrak{K}\}$
$SF = \{\alpha : K\alpha \text{ is strongly determined by some } \mathfrak{C} \subseteq \mathfrak{K}\}$
$F = \{\alpha : K\alpha \text{ is simply determined by some } \mathfrak{C} \subseteq \mathfrak{K}\}$.

Theorem 1.18.2

(1) $E \subset D \subseteq N = SC \subseteq SF \subset F$.
(2) $E \subset E_1 \cap D$.
(3) $E_1 \cap F \subseteq SF$, $E_1 \cap F \subseteq N$.
(4) E_1 *is not comparable with any of* D, N, SC, SF, F.

Proof.

(1) Recall that D is the class of d-persistent wffs. Since every descriptive frame is refined, it is immediate that $E \subseteq D$ (proper inclusion will be shown in (2)). 1.14.3 gives $D \subseteq N$.

Now let $\mathcal{F}$ be the canonical K-frame for $K\alpha$ and V the canonical valuation. Then $\mathcal{F}_V$ is the canonical first-order frame, which is descriptive (1.13.3) and satisfies every $K\alpha$-consistent set (1.1.3, 1.2.6). So if $\alpha \in N$, $K\alpha$ is supercomplete for $\mathcal{F}^K_{K\alpha}$, so $\alpha \in SC$. The converse is 1.13.8. It follows easily from the definitions that $SC \subseteq SF \subseteq F$. That the latter inclusion is proper is known, but for the sake of completeness we include an example. Let α be the conjunction of the wffs

(i) $\Box p \to \Box\Box p$
(ii) $(\Diamond p \wedge \Diamond q) \to (\Diamond(p \wedge q) \vee \Diamond(p \wedge \Diamond q) \vee \Diamond(q \wedge \Diamond p))$
(iii) $\Box(\Box p \to p) \to \Box p$.

Then $K\alpha$ is the logic $K4.3W$ of [86], which is shown there to be determined by the class of finite strictly linearly ordered K-frames. Thus $\alpha \in F$. Now if $\mathcal{F}$ is a K-frame such that $\mathcal{F} \models \alpha$ and $x \in \mathcal{F}$, then one may show that $\mathcal{F}_x$ is a finite strict linear ordering. Consider $\Delta = \{\Diamond^k p : k \in \mathbb{N}\}$. If $\Delta' \subseteq \Delta$ is finite, and k is the largest number such that $\Diamond^k p \in \Delta'$, then Δ' is satisfiable at 0 on the $K\alpha$-frame $\langle \{0, 1, \ldots, k\}, < \rangle$ when p is true everywhere. Hence every finite subset of Δ is $K\alpha$-consistent, and so Δ itself is $K\alpha$-consistent. But any strict linear ordering that satisfies Δ must be infinite, so no $K\alpha$-frame can satisfy Δ. Hence $\alpha \notin SF$.

(2) Let $\alpha \in E$ and let $\{\mathcal{F}_i : i \in I\} \subseteq \mathfrak{K}(\alpha)$, with $\mathcal{F}_i = \langle W_i, R_i \rangle$. Then if G is an ultrafilter on I, $\{i : \mathcal{F}'_i \models \alpha\} = I \in G$, where $\mathcal{F}'_i = \langle W_i, R_i, P_i \rangle$ and $P_i = 2^{W_i}$. By 1.7.13, $\mathcal{F}_G = \langle W_G, R_G, P_G \rangle \models \alpha$. But each $\mathcal{F}'_i$ is refined (1.9.2), so $\mathcal{F}_G$ is refined (1.9.11). Since

$\alpha \in E$, it follows that $\langle W_G, R_G \rangle \models \alpha$. This shows that $\mathfrak{K}(\alpha)$ is closed under ultraproducts, so by 1.16.3 $\alpha \in E_1$. Together with (1), we thus have $E \subseteq E_1 \cap D$. To prove the inclusion proper, let $\alpha = $ (i) $\wedge$ (ii) $\wedge$ (iii) where (i) and (ii) are as in (1), and (iii) $= M$. Now, as Thomason [96] observes, if $\mathcal{F}$ is refined, $\mathcal{F} \models$ (i) iff R is transitive, and $\mathcal{F} \models$ (ii) iff R is connected, (i.e. (xRy and xRz) only if $y = z$, or yRz, or zRy).

Consider $\langle \mathbb{N}, <, P \rangle$, where P is the class of finite or cofinite subsets of $\mathbb{N}$. This frame is refined and validates α, but $\langle \mathbb{N}, < \rangle \not\models M$, whence $\langle \mathbb{N}, < \rangle \not\models \alpha$ [96, p. 154]. Thus $\alpha \notin E$.

Now the completeness proof of [59, Section 5] for the logic $KM4$ is readily adaptable to show (i) $\wedge$ (iii) $\in D$. But (ii) $\in E \subseteq D$, so $\alpha \in D$. To show finally that $\alpha \in E_1$, we prove that $\mathfrak{K}(\alpha)$ is the class of transitive connected frames satisfying the condition m^∞ mentioned at the beginning of Section 1.17. It is easy to see that any such frame validates α. Conversely if $\mathcal{F} \models \alpha$, then $\mathcal{F} \models$ (i) and $\mathcal{F} \models$ (ii) so R is transitive and connected. Let $x \in \mathcal{F}$. Then $\mathcal{F}_x = \langle W, R \rangle \models \alpha$, where $\mathcal{F}_x \subseteq \mathcal{F}$ is generated by x. Since R is transitive, an equivalence relation is defined on W by

$$y \sim z \text{ iff } (yRz \text{ and } zRy) \text{ or } y = z.$$

Let Cy be the equivalence class of y (called the "cluster" of y in [86]). Let $\mathfrak{C} = \{Cy : y \in W\}$. Defining $Cy < Cz$ iff yRz, $\langle \mathfrak{C}, < \rangle$ is a linear ordering by (i) and (ii). If $\mathfrak{C}$ has no last member we may choose $\mathfrak{B} \subseteq \mathfrak{C}$ such that $\mathfrak{B}$ and its complement are both cofinal in $\mathfrak{C}$. Putting $V(p) = \bigcup \mathfrak{B}$, we then have $x \notin V(M)$ (cf. [96, p. 153]) and so $\mathcal{F}_x \not\models \alpha$, a contradiction. Hence $\mathfrak{C}$ has a last member, Cy say. Since $\mathcal{F}_x$ is transitive and generated by x, xRy. Now suppose yRz and yRw. Put $V(p) = \{z\}$. Then yRt only if tRy (as Cy is last), only if tRz, only if $t \in V(\Diamond p)$. Hence $y \in V(\Box \Diamond p)$. Since $\mathcal{F}_x \models M$ there is some s such that yRs and $s \in V(\Box p)$. But similar reasoning shows sRw, whence $w \in V(p) = \{z\}$, i.e. $w = z$. We have therefore shown that $\mathcal{F}$ satisfies m^∞ as required.

(3) If $\alpha \in E_1 \cap F$, then $\alpha \in F$ so $K\alpha$ is determined by some $\mathfrak{C} \subseteq \mathfrak{K}(\alpha)$. But then clearly $K\alpha$ is determined by $\mathfrak{K}(\alpha)$ which is closed under ultraproducts as $\alpha \in E_1$. By 1.8.6 we then have $K\alpha$ strongly determined by $\mathfrak{K}(\alpha)$, whence $a \in SF$.

That $E_1 \cap F \subseteq N$ has been shown by Fine [15] (cf. 1.20.15).

(4) Thomason [97] exhibits an α such that $\langle W, R \rangle \models \alpha$ iff R is reflexive and transitive, whence $\alpha \in E_1$, but $K\alpha$ is not determined by any class of K-frames. Thus α is not in F, therefore by (1) is not in any of D, N, or SF.

Fine [15] has shown that the wff $\alpha = \Diamond \Box p \rightarrow (\Diamond \Box (p \wedge q) \vee \Diamond \Box (p \wedge \neg q))$ is in N, hence by (1) is in SF and F, but is not in E_1. The proof that $\alpha \in N$ may be carried through on any descriptive frame, so we also have $\alpha \in D$.

□

We note that a proof that $E \subseteq E_1$ was first given by Lachlan [54] using a somewhat different method. The example in that paper which purports to show the inclusion is proper would seem to be in error, since it involves the wff $(\Box p \rightarrow p)$, which is in E. For, by Axiom **II**, any refined frame validating this wff has a reflexive relation, so its full extension validates it as well.

1.19   Replete Frames and Saturated Models

This section introduces a new class of first-order frames that play a special role in the discussion of axiomatic classes of K-frames.

Definition 1.19.1 *If $\mathcal{F} = \langle W, R, P \rangle$ is a frame, then for each $x \in W$ we define $MPx = \{m_R(S) : x \in S\}$. $\mathcal{F}$ is **replete** iff it satisfies the compactness condition **IV** (and hence **III** and **V**) of section 1.9, and also*

VI $MPy \subseteq Px$ *only if for some z, xRz and $Pz = Py$.*

The notion of replete frame adapts that of a "modally saturated model" defined by Fine [15] in terms of valuations on K-frames.

Theorem 1.19.2

(i) *Every finite frame is replete.*

(ii) *Every descriptive frame is replete.*

Proof.

(i) Condition **IV** is trivial if W, hence P, is finite. For **VI**, suppose $MPy \subseteq Px$, and let $S = \bigcap Py$. Since Py is finite, $S \in Py$, whence $m_R(S) \in Px$, i.e. $x \in m_R(S)$. It follows that for some z, xRz and $z \in S$. Then $Py \subseteq Pz$, and so $Py = Pz$.

(ii) If $\mathcal{F}$ is descriptive it satisfies **III**, so to show $\mathcal{F}$ replete we need only consider **VI**. But axiom **II** of descriptive frames may be written as

$$MPy \subseteq Px \text{ only if } xRy,$$

and with $y = z$ this clearly implies **VI**.

□

The converse of 1.19.2(ii) is false (cf. the comments at the end of this section). However a connection between the two kinds of frame is given by

Theorem 1.19.3 *For any replete frame $\mathcal{F}$ there is a descriptive frame $\mathcal{F}'$ such that $\mathcal{F}' \preccurlyeq \mathcal{F}$ and $\mathcal{F}^+ \cong \mathcal{F}'^+$.*

Proof. Let $\mathcal{F}' = (\mathcal{F}^+)_+$. Then $\mathcal{F}'$ is descriptive by 1.10.5, and $\mathcal{F}^+ \cong \mathcal{F}'^+$ by 1.10.3. To establish that $\mathcal{F}' \preccurlyeq \mathcal{F}$ we show that the map $Q : x \mapsto Px$, as used in 1.10.7, is a surjective homomorphism according to 1.5.1. Since $\mathcal{F}$ satisfies **III**, the proof that Q is onto follows exactly as in 1.10.7, as does the proof of 1.5.1(1). 1.5.1(2) may be "read off" from **VI** and the definition of $\mathcal{F}'$ (cf. 1.10.1). Finally to establish 1.5.1(3) we note that every proposition of $\mathcal{F}'$ is of the form $|S|^{\mathcal{F}^+}$ for some $S \in P$ (1.10.1). Then

$$Q^{-1}(|S|^{\mathcal{F}^+}) = \{x : Px \in |S|^{\mathcal{F}^+}\} = \{x : S \in Px\} = \{x : x \in S\} = S,$$

and so $Q^{-1}(|S|^{\mathcal{F}^+}) \in P$.   □

We come now to a method, based on some work of Fine [15], of producing replete frames through a construction from classical model theory. The map $x \mapsto Px$ used in 1.19.3 is not even a modal homomorphism unless its domain has some special properties. Our construction shows in effect that this map may always be factored through an elementary embedding followed by a modal homomorphism. The idea here is to treat a first-order frame as a genuine model for a first-order language.

Definition 1.19.4 *Let $\mathcal{F} = \langle W, R, P \rangle$ be a frame, with $P = \{S_i : i \in I\}$ for some indexing set I. Let $\mathfrak{R}(I)$ be a first-order language obtained from the language $\mathfrak{R}$ of Section 1.16 by the addition of a set $\{\mathbf{S}_i : i \in I\}$ of monadic predicate letters. Then $\mathcal{F}$ is a realisation of $\mathfrak{R}(I)$, and constants may be added to $\mathfrak{R}(I)$ by taking each element of W to be a name for itself.*

*A set Δ of formulae of $\mathfrak{R}(I)$ with at most one free variable is **satisfiable** in $\mathcal{F}$ iff there exists some $a \in W$ such that $\mathcal{F} \models \delta[a]$ for all $\delta \in \Delta$. $\mathcal{F}$ is **2-saturated** iff whenever Δ is a set of formulae with at most one free variable and at most one constant (the same for all members of Δ), and such that every finite subset of Δ is satisfiable in $\mathcal{F}$, then Δ itself is satisfiable in $\mathcal{F}$.*

Theorem 1.19.5 (Fine [15]) *If $\mathcal{F}$ is a 2-saturated realisation of $\mathfrak{R}(I)$, then $\mathcal{F}$ is a replete modal frame.*

Proof. To prove 1.9.3**IV**, let Q be a subset of P and put $\Delta_Q = \{\mathbf{S}_i(v) : S_i \in Q\}$ for v some variable of $\mathfrak{R}$. Then if Q has the fip, each finite subset of Δ_Q is satisfiable in $\mathcal{F}$. Since $\mathcal{F}$ is 2-saturated, it follows that there is some $a \in W$ such that $a \in S_i$ for all $S_i \in Q$ and hence $\bigcap Q \neq \emptyset$

as required. To prove **VI**, suppose $MPb \subseteq Pa$, for $a, b \in W$, and let $\Delta = \{R(a, v)\} \cup \{S_i(v) : S_i \in Pb\}$, where $\boldsymbol{R}$ is the dyadic predicate letter of the language $\mathfrak{R}$. Since Pb is closed under finite intersections, and by hypothesis $b \in S$ only if $a \in m_R(S)$, it follows that each finite subset of Δ is satisfiable in $\mathcal{F}$. But only one constant appears in Δ, so we conclude that Δ is satisfiable by some $c \in W$. Then aRc and $c \in S_i$ for all $S_i \in Pb$, whence $Pc = Pb$. $\qquad\square$

Definition 1.19.6

(i) *For any first-order frame $\mathcal{F} = \langle W, R, P \rangle$, the **reduct** of $\mathcal{F}$ is the K-frame $\mathcal{F}_0 = \langle W, R \rangle$.*

(ii) *Two structures $\mathcal{F}, \mathcal{F}'$ are **elementarily equivalent**, $\mathcal{F} \equiv \mathcal{F}'$, iff they satisfy precisely the same first-order sentences.*

Theorem 1.19.7 $\mathcal{F}_0 \models \alpha$ *only if $\mathcal{F} \models \alpha$.*

Proof. By 1.4.1. $\qquad\square$

Theorem 1.19.8 *For any frame $\mathcal{F}$ there is a replete frame $\mathcal{F}'$ such that $\mathcal{F}^+ \cong \mathcal{F}'^+$ and $\mathcal{F}_0 \equiv \mathcal{F}'_0$.*

Proof. By [4, Chapter 11], $\mathcal{F}$ as a realisation of $\mathfrak{R}(I)$ has a 2-saturated elementary extension $\mathcal{F}' = \langle W', R', P' \rangle$, with $P' = \{S'_i : i \in I\}$. Now for any realisation $\mathfrak{G}$ of $\mathfrak{R}(I)$ we have the following $\mathfrak{R}(I)$-definability of MA operations.

$$
\begin{aligned}
S_i = S_j &\quad \text{iff} \quad \mathfrak{G} \models \forall v(\boldsymbol{S}_i(v) \leftrightarrow \boldsymbol{S}_j(v)) \\
S_i \cap S_j = S_k &\quad \text{iff} \quad \mathfrak{G} \models \forall v(\boldsymbol{S}_i(v) \wedge \boldsymbol{S}_j(v) \leftrightarrow \boldsymbol{S}_k(v)) \\
S_i = -S_j &\quad \text{iff} \quad \mathfrak{G} \models \forall v(\boldsymbol{S}_i(v) \leftrightarrow \neg \boldsymbol{S}_j(v)) \\
S_i = \boldsymbol{m}_R(S_j) &\quad \text{iff} \quad \mathfrak{G} \models \forall v(\boldsymbol{S}_i(v) \leftrightarrow \exists u(\boldsymbol{R}(v, u) \wedge \boldsymbol{S}_j(u))).
\end{aligned}
$$

Now $\mathcal{F}$ and $\mathcal{F}'$ satisfy the same sentences, i.e. $\mathcal{F} \equiv \mathcal{F}'$, and so $\mathcal{F}_0 \equiv \mathcal{F}'_0$. But the above sentences show that P' is closed under $\cap$, $-$, $\boldsymbol{m}_{R'}$ (i.e. $\mathcal{F}'$ is indeed a frame) and that the map $S_i \mapsto S'_i$ is an MA isomorphism of $\mathcal{F}^+$ and $\mathcal{F}'^+$. Finally, 1.19.5 gives $\mathcal{F}'$ replete. $\qquad\square$

Corollary 1.19.9 *For any frame $\mathcal{F}$ there exists a descriptive frame $\mathcal{F}'$ such that $\mathcal{F}^+ \cong \mathcal{F}'^+$ and $\mathcal{F}'_0$ is a homomorphic image of some K-frame that is elementarily equivalent to $\mathcal{F}_0$.*

Proof. By 1.19.8 and 1.19.3. $\qquad\square$

It should be noted that the construction of 1.19.8 will not in general produce a descriptive frame, and so the homomorphism of 1.19.9 cannot be avoided. To see this let $\mathcal{F} = \langle \mathbb{N}, < \rangle$. Then if $\mathcal{G}$ is descriptive and $\mathcal{F}^+ \cong \mathcal{G}^+$ we cannot have $\mathcal{F}_0 \equiv \mathcal{G}_0$. For by 1.10.7, if $\mathcal{F}^+ \cong \mathcal{G}^+$ then the reduct of $(\mathcal{F}^+)_+$ is isomorphic, and hence elementarily equivalent, to $\mathcal{G}_0$.

But Thomason [98] has shown that the former satisfies the $\mathfrak{R}$-sentence $\forall v \exists u (\boldsymbol{R}(v, u) \wedge \boldsymbol{R}(u, u))$, and clearly this is not true of $\mathcal{F}$.

1.20 $\Sigma\Delta$–Elementary Classes of Kripke Frames

Throughout this section X will denote a class of K-frames. We write $X \in EC_{\Sigma\Delta}$ (X is $\Sigma\Delta$-elementary) iff X is closed under the relation $\equiv$ of elementary equivalence (1.19.6(ii)).

Our present aim is to use the results just obtained to establish some significant properties of $EC_{\Sigma\Delta}$ classes. The key fact, immediate from 1.19.9, is

Theorem 1.20.1 *If $X \in EC_{\Sigma\Delta}$ and X is closed under homomorphic images, then $\mathcal{F}_0 \in X$ only if there is a descriptive frame $\mathcal{F}'$ such that $\mathcal{F}^+ \cong \mathcal{F}'^+$ and $\mathcal{F}_0' \in X$.*

Definition 1.20.2

(i) If $\mathfrak{A}$ is an MA, the **completion** of $\mathfrak{A}$ is the K-frame $\mathfrak{A}_\# = (\mathfrak{A}_+)_0$, i.e. the reduct of the descriptive frame $\mathfrak{A}_+$.

(ii) If $\mathcal{F} = \langle W, R \rangle$ is a K-frame, the **completion** of $\mathcal{F}$ is the K-frame $\mathcal{F}^\# = (\mathcal{F}^+)_\#$. Thus $\mathcal{F}^\# = \langle W^\#, R^\# \rangle$ where $W^\#$ is the set of ultrafilters on W, and $x R^\# y$ iff for all $S \subseteq W$, $y \in S$ only if $x \in m_R(S)$.

Theorem 1.20.3

(i) $\mathfrak{A}_\# \models \alpha$ *only if* $\mathfrak{A} \models \alpha$.

(ii) $\mathcal{F}^\# \models \alpha$ *only if* $\mathcal{F} \models \alpha$.

(iii) *If $\mathcal{F}$ is a descriptive frame, then $\mathcal{F}_0 \cong (\mathcal{F}^+)_\#$.*

Proof.

(i) $\mathfrak{A}_\# \models \alpha$ only if $\mathfrak{A}_+ \models \alpha$ (1.19.7), only if $\mathfrak{A} \models \alpha$ (1.10.4).

(ii) $\mathcal{F}^\# \models \alpha$ only if $\mathcal{F}^+ \models \alpha$ (part (i)), only if $\mathcal{F} \models \alpha$ (1.3.3).

(iii) $\mathcal{F} \cong (\mathcal{F}^+)_+$ (1.10.7), whence $\mathcal{F}_0 \cong ((\mathcal{F}^+)_+)_0 = (\mathcal{F}^+)_\#$.

$\square$

If $\mathfrak{N}$ is a class of MA's then $\mathbf{HS}(\mathfrak{N})$ denotes the class of all homomorphic images of subalgebras of members of $\mathfrak{N}$. For X a class of K-frames we put

$$X^+ = \{\mathfrak{A} \in \mathfrak{M} : \mathfrak{A} \cong \mathcal{F}^+ \text{ for some } \mathcal{F} \in X\}.$$

Theorem 1.20.4 *If X is closed under disjoint unions then $\mathbf{HS}(X^+)$ is the smallest equational class of MA's containing X^+.*

Proof. By [39, pp. 152 and 171] the equational class generated by X^+ is $\mathbf{HS}(\mathbf{P}(X^+))$, where $\mathbf{P}(X^+)$ is the closure of X^+ under direct products. But since X is closed under disjoint unions, Theorem 1.6.5 gives $\mathbf{P}(X^+) = X^+$.   □

Theorem 1.20.5

(i) $\mathfrak{A}_\# \in X$ *only if* $\mathfrak{A} \in \mathbf{HS}(X^+)$.

(ii) *If* $X \in EC_{\Sigma\Delta}$ *and* X *is closed under homomorphic images and subframes, then* $\mathfrak{A} \in \mathbf{HS}(X^+)$ *only if* $\mathfrak{A}_\# \in X$.

Proof.

(i) $\mathfrak{A} \cong (\mathfrak{A}_+)^+ \subseteq (\mathfrak{A}_\#)^+$ (1.10.3, 1.20.2, 1.19.6).

(ii) Since $\mathfrak{A} \cong (\mathfrak{A}_+)^+$, if $\mathfrak{A} \in \mathbf{HS}(X^+)$ then $(\mathfrak{A}_+)^+$ is a homomorphic image of a subalgebra of $\langle W, R \rangle^+$, for some $\langle W, R \rangle \in X$. Thus there is some $P \subseteq 2^W$ such that $(\mathfrak{A}_+)^+$ is a homomorphic image of $\langle W, R, P \rangle^+$. By 1.20.1 we may presume that $\langle W, R, P \rangle$ is descriptive. But $\mathfrak{A}_+$ is descriptive, so by 1.10.9 and 1.10.7, $\mathfrak{A}_+$ is embeddable in $\langle W, R, P \rangle$. Then $\mathfrak{A}_\# = (\mathfrak{A}_+)_0$ is isomorphic to a subframe of $\langle W, R \rangle \in X$, whence our hypothesis gives $\mathfrak{A}_\# \in X$.

           □

The above result gives access to the following characterisation of modal axiomatic classes.

Theorem 1.20.6 *If* $X \in EC_{\Sigma\Delta}$, *then* $X \in MAC$ *iff*

(i) X *is closed under disjoint unions, homomorphic images and subframes, and*

(ii) $\mathcal{F}^\# \in X$ *only if* $\mathcal{F} \in X$, *for any* K-*frame* $\mathcal{F}$.

Proof. If $X \in MAC$, (i) holds by 1.6.4, 1.5.5, and 1.4.10, and (ii) by 1.20.3(ii).

Conversely, suppose $X \in EC_{\Sigma\Delta}$ and that (i), (ii) hold. By 1.20.4 $\mathbf{HS}(X^+) = \Delta^*$ for some set Δ of MA polynomial identities. Let Γ_Δ be the set of modal wffs determined by Δ as in the proof 1.12.2. Then if $\mathcal{F} \in X$, $\mathcal{F}^+ \in X^+$, so $\mathcal{F}^+ \models \Delta$ and thus $\mathcal{F} \models \Gamma_\Delta$. On the other hand if $\mathcal{F} \models \Gamma_\Delta$, then $\mathcal{F}^+ \models \Delta$, so $\mathcal{F}^+ \in \mathbf{HS}(X^+)$. By 1.20.5(ii) we then have $(\mathcal{F}^+)_\# = \mathcal{F}^\# \in X$, whence by hypothesis (ii) $\mathcal{F} \in X$. Thus $X = \Gamma_\Delta^*$, and X is modal axiomatic.   □

In order to discuss classes of frames characterised by a single modal wff we need to introduce a new construction.

Definition 1.20.7 *If* $\{\mathcal{F}_i : i \in I\}$ *is a family of* K-*frames, and* G *an ultrafilter on* I, *then the* **completed ultraproduct** *of the* $\mathcal{F}_i$'s *over* G *is the* K-*frame*

$$\mathcal{F}_G^{\#} = (\prod_{i \in I} \mathcal{F}_i^+ / G)_{\#} \,.$$

Theorem 1.20.8

(i) $\mathcal{F}_G^{\#} \models \alpha$ *only if* $\{i : \mathcal{F}_i \models \alpha\} \in G$.

(ii) $\mathcal{F}_G^{\#}$ *contains a substructure (not necessarily a subframe) isomorphic to* $\prod \mathcal{F}_i / G$ *(cf. 1.7.3)*.

Proof.

(i) $\qquad \mathcal{F}_G^{\#} \models \alpha \quad$ only if $\quad \prod \mathcal{F}_i^+ / G \models \alpha \qquad$ (1.20.7, 1.20.3(i))

$\qquad\qquad\qquad\qquad$ only if $\quad \{i : \mathcal{F}_i^+ \models \alpha\} \in G \qquad$ (Łoś's Thm)

$\qquad\qquad\qquad\qquad$ only if $\quad \{i : \mathcal{F}_i \models \alpha\} \in G \qquad$ (1.3.3).

(ii) For each $i \in I$ take $\mathcal{F}_i$ to be $\langle W_i, R_i, P_i \rangle$ where $P_i = 2^{W_i}$, and consider the ultraproduct $\mathcal{F}_G = \langle W_G, R_G, P_G \rangle$ as defined in 1.7.6. Now if $(\prod \mathcal{F}_i^+ / G)_+ = \langle W, R, P \rangle$ then by 1.7.8 $\langle W, R, P \rangle \cong (\mathcal{F}_G^+)_+$. But $\mathcal{F}_G^{\#} = \langle W, R \rangle$, and so it suffices to find an appropriate map from $\mathcal{F}_G$ into $(\mathcal{F}_G^+)_+$. Now by 1.9.2 and 1.9.11, $\mathcal{F}_G$ is a refined frame and so satisfies axioms **I** and **II** of Section 1.9. But then the proof of 1.10.7 shows that the map given there from $\mathcal{F}_G$ to $(\mathcal{F}_G^+)_+$ has the required properties for our theorem. $\qquad\qquad\qquad\qquad\qquad\square$

Corollary 1.20.9 *If* $X \in UC$ *(i.e.* $X = \delta^*$ *for some universal sentence* δ *of* $\mathfrak{R}$*), then* $\mathfrak{K} - X$ *is closed under completed ultraproducts.*

Proof. By 1.20.8(ii), since UC classes are closed under substructures and their complements are closed under ultraproducts. $\qquad\qquad\qquad\square$

Theorem 1.20.10 *If* $X \in EC_{\Sigma\Delta}$*, then* $X \in MEC$ *iff*

(i) $X \in MAC$ *and*

(ii) $\mathfrak{K} - X$ *is closed under completed ultraproducts.*

Proof. If $X \in MEC$ then (i) is immediate and (ii) follows from 1.20.8(i) by a similar argument to that used in 1.16.2. For the converse, if (i) and (ii) hold and $X \in EC_{\Sigma\Delta}$, then we know from the proof of 1.20.6 that $X = \Gamma_\Delta^*$, where $\mathbf{HS}(X^+) = \Delta^*$. Let I be the set of all finite subsets of Γ_Δ, and suppose for all $i \in I$ that $\Gamma_\Delta^* \neq i^*$. Then for each i, since $\Gamma_\Delta^* \subseteq i^*$, there is a K-frame $\mathcal{F}_i$ such that $\mathcal{F}_i \models i$ and $\mathcal{F}_i \nvDash \Gamma_\Delta$, whence $\mathcal{F}_i \notin X$. Then for each $i \in I$ we have $\mathcal{F}_i^+ \models i$, so by the argument of 1.8.4 we may construct an ultrafilter G on I such that $\prod \mathcal{F}_i^+ / G \models \Gamma_\Delta$. But then $\prod \mathcal{F}_i^+ / G \in \mathbf{HS}(X^+)$, so by 1.20.5(ii) $\mathcal{F}_G^{\#} = (\prod \mathcal{F}_i^+ / G)_{\#} \in X$,

contrary to the closure of $\mathfrak{K} - X$ under completed ultraproducts. We
therefore conclude that for some finite $i \subseteq \Gamma_\Delta$, $i^* = \Gamma_\Delta^* = X$, so $X = \alpha^*$
where α is the conjunction of the members of i. Hence $X \in MEC$. □

Since every EC class is $EC_{\Sigma\Delta}$, 1.20.10 characterises those EC classes
that are MEC. Whether EC classes enjoy properties that allow the
conditions of 1.20.10 to be simplified is not known. We can only offer

Theorem 1.20.11 *If $X \in UC$, then $X \in MEC$ iff $X \in MAC$.*

Proof. By 1.20.9 and 1.20.10.   □

As well as leading to a discussion of axiomatic classes, 1.20.5(ii) yields
some interesting consequences concerning logics whose models are $\Sigma\Delta$-
elementary.

Definition 1.20.12 *A normal modal logic Λ is **canonical** iff it is deter-
mined by its canonical Kripke frame $\mathcal{F}_\Lambda^K$ (cf. 1.2.5). Λ is an $EC_{\Sigma\Delta}$ **logic**
iff the class Λ^* of all K-frames on which Λ is valid is $\Sigma\Delta$-elementary.*

We recall from Section 1.2 that any non-theorem of Λ is falsified by
$\mathcal{F}_\Lambda^K$, and so to show that Λ is canonical it suffices to prove $\mathcal{F}_\Lambda^K \models \Lambda$, or
even that $\mathcal{F}_\Lambda^K \models \Gamma$ where Γ is some axiom set that generates Λ.

Theorem 1.20.13 *If X is $\Sigma\Delta$-elementary and closed under disjoint
unions, homomorphic images, and subframes, then X determines a logic
that is canonical.*

Proof. By 1.20.4 $\mathbf{HS}(X^+) = \Delta^*$ for some Δ. Let Λ be the logic $K\Gamma_\Delta$.
Now if $\mathcal{F} \in X$, $\mathcal{F}^+ \models \Delta$ and so $\mathcal{F} \models \Gamma_\Delta$, whence $X \subseteq \Gamma_\Delta^* = \Lambda^*$. If
$\mathcal{F}_\Lambda$ is the canonical Λ-frame of 1.13.1, then by 1.13.4 $\mathcal{F}_\Lambda \models \Gamma_\Delta$ so $\mathcal{F}_\Lambda^+ \in$
$\mathbf{HS}(X^+)$, and thus by 1.20.5(ii), $(\mathcal{F}_\Lambda^+)_\# \in X$. But $\mathcal{F}_\Lambda$ is descriptive
(1.13.3) so by 1.20.3(iii) $(\mathcal{F}_\Lambda)_0 = \mathcal{F}_\Lambda^K \in X$. Since $X \subseteq \Lambda^*$, and $\alpha \notin \Lambda$
only if $\mathcal{F}_\Lambda^K \not\models \alpha$, this implies that $\alpha \in \Lambda$ iff $X \models \alpha$, i.e. X determines Λ.
In particular $\mathcal{F}_\Lambda^K \models \Lambda$, and Λ is canonical.   □

Corollary 1.20.14 *For any $EC_{\Sigma\Delta}$ logic Λ_1 there is a canonical logic
Λ_2 such that $\Lambda_1 \subseteq \Lambda_2$ and $\Lambda_1^* = \Lambda_2^*$.*

Proof. Let $X = \Lambda_1^*$. Then $X \in MAC$ by definition, and $X \in EC_{\Sigma\Delta}$
by hypothesis. Let Λ_2 be the logic $K\Gamma_\Delta$, where $\Delta^* = \mathbf{HS}(X^+)$. Then
by the proof of 1.20.13, Λ_2 is canonical and determined by X. Thus if
$\alpha \in \Lambda_1$, $X \models \alpha$ and so $\alpha \in \Lambda_2$. Hence $\Lambda_1 \subseteq \Lambda_2$. But the proof of 1.20.6
shows in fact that $X = \Gamma_\Delta^* = \Lambda_2^*$, i.e. $\Lambda_1^* = \Lambda_2^*$.   □

We note that if Λ_1 is the incomplete logic of Thomason [97] then Λ_2
above is the logic S4.

Corollary 1.20.15 *Any $EC_{\Sigma\Delta}$ logic that is determined by some class
of Kripke frames is canonical.*

Proof. If Λ_1 is determined at all by a class of K-frames then it is certainly determined by Λ_1^*. Suppose further that Λ_1 is $EC_{\Sigma\Delta}$ and let Λ_2 be the canonical logic then given by 1.20.14. If $\alpha \in \Lambda_2$, $\Lambda_2^* \models \alpha$, so $\Lambda_1^* \models \alpha$ and therefore $\alpha \in \Lambda_1$. Thus $\Lambda_1 = \Lambda_2$, so Λ_1 is canonical.   □

Corollary 1.20.15, which yields the second part of 1.18.2(3), was first proven by Fine [15].

2

Semantic Analysis of Orthologic

Some physicists maintain that from a quantum-theoretic standpoint the propositions pertaining to a physical system exhibit a non-standard logical structure, and indeed that their associated algebra is an orthomodular lattice, rather than a Boolean algebra as in the case of classical systems. Consequently a new area of logical investigation has grown up under the name of "quantum logic", of which one aspect is the study of the propositional logic characterised by the class of orthomodular lattices.

In recent years we have seen the development of a powerful alternative to algebraic semantics for formal systems—namely the "possible worlds" model theory initiated by Saul Kripke. This approach was originally used to analyse modal systems, but it was soon realised that its ramifications were far wider than that. It has subsequently been applied to many other kinds of intensional logic, including tense, deontic, epistemic, intuitionist, and entailment systems, and is currently proving relevant to the study of natural languages.

The purpose of this paper is to lay a foundation for an intensional model theory for quantum logic. To do this we broaden the inquiry to encompass what we shall call Orthologic. The minimal calculus in this area is the system O, characterised by the class of ortholattices. This logic will be shown to have a semantics reminiscent of the relational structures of normal modal logic. These models are then used to establish a connection between O and the Brouwerian modal system that parallels the McKinsey-Tarski translation of intuitionist logic into S4. A discussion of filtration theory shows that O is decidable, and in the final section of the paper we extend the analysis to obtain a characterisation of quantum logic itself.

2.1 Syntax

The primitive symbols of our object language are (i) a denumerable collection $\{p_i : i < \omega\}$ of propositional variables, (ii) the connectives $\sim$ and $\wedge$ of negation and conjunction, (iii) parentheses (and). The set Φ of well-formed formulae (wffs) is constructed from these in the usual way. The letters A, B, C etc. are used as metavariables ranging over Φ. Parentheses may be omitted where convenient, the convention being that $\sim$ binds more strongly than $\wedge$. The disjunction connective $\vee$ is introduced by the definitional abbreviation $A \vee B =_{df} \sim (\sim A \wedge \sim B)$.

Our concern is to explore the relationships between two quite different ways of studying formulae. The *semantical* approach, to be explained in detail in the next section, has as its goal the assignment of meanings or interpretations to wffs, and the setting out of conditions under which a wff is to be true or false. The *syntactical* approach examines formal relationships between wffs, and focuses on the notion of *consequence* or *derivability* of formulae. In this context we can distinguish between *axiomatic systems*, and *logics*. Given a formal language, an axiom system S can be defined as an ordered pair $\langle \mathfrak{A}, \mathfrak{R} \rangle$ where $\mathfrak{A}$ is a set of wffs of the language, called *axioms*, and $\mathfrak{R}$ is a set of *rules of inference* that govern operations allowing certain formulae to be derived from others. A wff A is said to be a *theorem* of S, written $\vdash_S A$, if there exists in S a *proof* of A, i.e. a finite sequence of wffs whose last member is A, and such that each member of the sequence is either an axiom, or derivable from earlier members by one of the rules in $\mathfrak{R}$.

A logic on the other hand can be thought of as a set L of formulae closed under the application of certain inferential rules to its members. The members of L are called L-theorems, and in this case the symbolism $\vdash_L A$ indicates merely that $A \in L$.

For example, if $S = \langle \mathfrak{A}, \mathfrak{R} \rangle$ is an axiom system, then an S-logic can be defined as any set of wffs that includes the axiom set $\mathfrak{A}$ and is closed under the rules of $\mathfrak{R}$. In general the intersection Ls of all S-logics will be an S-logic, whose members are precisely those wffs for which there are proofs in S. This is often described by saying that S is an *axiomatisation* of Ls, or that Ls is *generated* by S.

Thus each axiom system has a corresponding logic (the set of its theorems) and in some formal treatments little or no distinction is made between the two. The converse however is not true. Not every logic is axiomatisable. In any semantical framework the set of wffs true in a particular model will be a logic of some kind, for which, in some cases, there may be no effectively specifiable generating procedure. A classic example is the first-order theory of the standard model of arithmetic.

Now it is sometimes the case that, in studying the concept of de-ducibility, we are concerned not so much with which formulae are the-orems, but rather with which formulae can be derived when others are taken as hypotheses. This, for example, would seem to be part of the motivation behind systems of so-called *natural deduction*. From this point of view one might conceive of a logic, not as a set of wffs, but as a collection L of ordered pairs of wffs that satisfies certain closure condi-tions, the idea being that the presence of the pair ⟨A, B⟩ in L indicates that B can be inferred from A in L. To preserve the distinction we shall call logics of this kind *binary* logics, and refer to logics of the first kind (sets of wffs closed under certain rules) as *unary* logics.

Again by way of example, if S is an axiom system (or even a unary logic) whose language includes an implication operator → for which the Deduction Theorem holds, then we may associate with S the binary logic of all pairs ⟨A, B⟩ such that $\vdash_S$ A → B. For languages of the kind to be considered in this paper, that have no implication connective at all, the notion of a binary logic becomes an extremely useful one for syntactically generating classes of formulae.

In general, if L is a binary logic, we write A $\vdash_L$ B when ⟨A, B⟩ ∈ L.

2.1.1 DEFINITION

An *orthologic* is a binary logic L such that, for all A, B, C ∈ Φ,

#1.	A $\vdash_L$ A
#2.	A∧B $\vdash_L$ A
#3.	A∧B $\vdash_L$ B
#4.	A $\vdash_L$ $\sim\sim$A
#5.	$\sim\sim$ A $\vdash_L$ A
#6.	A ∧ $\sim$A $\vdash_L$ B
#7.	if A $\vdash_L$ B and B $\vdash_L$ C, then A $\vdash_L$ C
#8.	if A $\vdash_L$ B and A $\vdash_L$ C, then A $\vdash_L$ B ∧ C
#9.	if A $\vdash_L$ B, then $\sim$ B $\vdash_L$ $\sim$ A.

It is easy to see that the intersection of any family of orthologics is an orthologic, and hence that there is a smallest logic, which we call O, that satisfies #1, . . . , #9. O is characterised by the class of ortholattices, in the sense that A $\vdash_O$ B iff $v(A) \leq v(B)$ for all valuations v on all or-tholattices (a valuation on an ortholattice is a function from Φ into the lattice under which $\sim$ and ∧ are interpreted as orthocomplement and lattice meet respectively). The necessity part of the above biconditional is proved by showing that it holds for #1, . . . , #6 and is preserved by #7, . . . , #9. Sufficiency may be established by showing that the Linden-baum Algebra for O is an ortholattice (if Φ is thought of as an algebra, then the Lindenbaum Algebra for an orthologic L is the quotient algebra $\Phi/\equiv_L$, where $A \equiv_L B$ iff $A \vdash_L B$ and $B \vdash_L A$).

2.1.2 Definitions

Let L be an orthologic and Γ a non-empty set of wffs. A wff A is said to be L-*derivable from* Γ, $\Gamma \vdash_L A$, if there exist $B_1, \ldots, B_n \in \Gamma$ such that $B_1 \wedge \ldots \wedge B_n \vdash_L A$. If A is L-derivable from $\{A \vee \sim A\}$ then we simply say that A is L-*derivable*, or is an L-*theorem*, and write $\vdash_L A$. Γ is L-*consistent* if there is at least one wff not L-derivable from Γ, and L-*inconsistent* otherwise. (It can be shown that Γ is L-consistent iff for no A do we have both $\Gamma \vdash_L A$ and $\Gamma \vdash_L \sim A$.) Γ is L-*full* iff it is L-consistent and closed under conjunction and L-derivability i.e. iff

- (i)   1or some A, not $\Gamma \vdash_L A$,
- (ii)   if $A \in \Gamma$ and $A \vdash_L B$ then $B \in \Gamma$,
- (iii) $A, B \in \Gamma$ only if $A \wedge B \in \Gamma$.

2.1.3 Lemma

If $x \subseteq \Phi$ is L-full, then
- (i) $A \wedge B \in x$ iff $A \in x$ and $B \in x$,
- (ii) $x \vdash_L A$ iff $A \in x$,
- (iii) $A \vee \sim A \in x$, for all wffs A.

Proof.

(i) The 'if' part is 2.1.2(iii), and the converse follows from #2 and #3 by 2.1.2(ii).

(ii) Since $A \vdash_L A$ (#1), sufficiency follows from the definition of L-derivability. Necessity uses 2.1.2(ii) and (iii).

(iii) By definition x is non-empty, so there exists $B \in x$. But

$$B \vdash_L A \vee \sim A$$

(use #6, #9, #4 and #7), so the result follows by 2.1.2(ii).   □

The basic result linking full sets and derivability is the following version of Lindenbaum's Lemma.

2.1.4 Theorem

$\Gamma \vdash_L A$ iff A belongs to every L-full extension of Γ.

Proof. If $\Gamma \vdash_L A$, then there exist $B_1, \ldots, B_n \in \Gamma$ such that

$$B_1 \wedge \ldots \wedge B_n \vdash_L A.$$

If x is L-full and $\Gamma \subseteq x$, we have $B_1, \ldots, B_n \in x$. Applying 2.1.2(iii) and then 2.1.2(ii) we obtain $A \in x$.

For the converse, suppose A is not L-derivable from Γ. Let $x = \{B : \Gamma \vdash_L B\}$. From #1, $\Gamma \subseteq x$, and by hypothesis $A \notin x$. Our proof will therefore be complete if we can show that x is L-full. Now if $B \in x$ and $B \vdash_L C$, there exist $B_1, \ldots, B_n \in \Gamma$ such that $B_1 \wedge \ldots \wedge B_n \vdash_L B$, hence by #7, $B_1 \wedge \ldots \wedge B_n \vdash_L C$ and so $\Gamma \vdash_L C$, i.e. $C \in x$. If on the other hand $B, C \in x$ then there exist $B_1, \ldots, B_n, C_1, \ldots, C_m \in \Gamma$ such that $B_1 \wedge \ldots \wedge B_n \vdash_L B$ and $C_1 \wedge \ldots \wedge C_m \vdash_L C$. Letting

$$D = B_1 \wedge \ldots \wedge B_n \wedge C_1 \wedge \ldots \wedge C_m$$

we have by #2, #3 and #7 that $D \vdash_L B$, $D \vdash_L C$, and so by #8, $D \vdash_L B \wedge C$. Thus $\Gamma \vdash_L B \wedge C$ and therefore $B \wedge C \in x$. This shows that x is closed under L-derivability and conjunction. Hence, since $A \notin x$, A is not L-derivable from x, and so x is L-consistent, and Theorem 2.1.4 is established. $\qquad\square$

It is interesting to note how direct the proof of 2.1.4 is, in that we are able to define explicitly the required set x. In all other logical systems that the author knows of the proof of Lindenbaum's Lemma involves a complex induction over a fixed enumeration of Φ, or even worse, an application of Zorn's Lemma. On the lattice theoretic level it can be said that our full sets correspond to (proper) filters, whereas for systems that include the distributive law the analogies are with prime or maximal filters, the proof of whose existence requires some variant of the Axiom of Choice.

The last result of this section will be used in our characterisation of negation.

2.1.5 Lemma

If x is L-full and $\sim A \notin x$, then there exists an L-full set y such that $A \in y$, and for all B, either $\sim B \notin x$ or $B \notin y$.

Proof. Let $y = \{B : A \vdash_L B\}$. By #1, $A \in y$. Now let $\sim B \in x$. Then $B \notin y$, or else $A \vdash_L B$, whence $\sim B \vdash_L \sim A$ by #9, and so by 2.1.2(ii), $\sim A \in x$, contrary to hypothesis. By 2.1.3(iii), $A \vee \sim A \in x$, i.e. $\sim(\sim A \wedge \sim\sim A) \in x$. By what we just proved it follows that $\sim A \wedge \sim\sim A \notin y$. Proceeding in a similar manner to 2.1.4 we can show that y is closed under conjunction and L-derivability, and hence that $\sim A \wedge \sim\sim A$ is not L-derivable from y, i.e. y is L-consistent, and therefore L-full as required. $\qquad\square$

2.2   Semantics

2.2.1   Definition

$\mathcal{F}=\langle X, \perp\rangle$ is an *orthoframe* iff X is a non-empty set, the *carrier* of $\mathcal{F}$, and $\perp$ is an *orthogonality relation* on X, i.e. $\perp \subseteq X \times X$ is irreflexive and symmetric.

If $x \perp y$ then we say that x *is orthogonal to* y. If x is orthogonal to every member of a subset Y of X then we say x *is orthogonal to* Y and write $x \perp Y$. $Y \subseteq X$ is said to be $\perp$-*closed* iff for all $x \in X$, $x \notin Y$ only if there exists $y \in X$ such that $y \perp Y$ and not $x \perp y$ (the converse is always true by the symmetry of $\perp$).

2.2.2   Definition

$\mathcal{M}=\langle X, \perp, V\rangle$ is an *orthomodel* on the frame $\langle X, \perp\rangle$ iff V is a function assigning to each propositional variable p_i a $\perp$-closed subset $V(p_i)$ of X. The truth of a wff A at x in $\mathcal{M}$ is defined recursively as follows. (Read "A is true (holds) at x in $\mathcal{M}$" for $\mathcal{M} \models_x A$).

(1) $\mathcal{M} \models_x p_i$ iff $x \in V(p_i)$
(2) $\mathcal{M} \models_x A \wedge B$ iff $\mathcal{M} \models_x A$ and $\mathcal{M} \models_x B$
(3) $\mathcal{M} \models_x \sim A$ iff   for all y, $\mathcal{M} \models_y A$ only if $x \perp y$.

Denoting the set $\{x \in X : \mathcal{M} \models_x A\}$ by $\|A\|^{\mathcal{M}}$, we can rewrite the above as

(1′) $\|p_i\|^{\mathcal{M}}=V(p_i)$
(2′) $\|A \wedge B\|^{\mathcal{M}} = \|A\|^{\mathcal{M}} \cap \|B\|^{\mathcal{M}}$
(3′) $\|\sim A\|^{\mathcal{M}} = \{x : x \perp \|A\|^{\mathcal{M}}\}$.

If Γ is a non-empty set of wffs, then we say Γ *implies* A at x in $\mathcal{M}$, denoted $\mathcal{M}: \Gamma \models_x A$, iff either there exists $B \in \Gamma$ such that not $\mathcal{M} \models_x B$, or else $\mathcal{M} \models_x A$. Γ $\mathcal{M}$-*implies* A, $\mathcal{M}:\Gamma \models A$, iff Γ implies A at all x in $\mathcal{M}$. If $\mathcal{F}$ is a frame, Γ $\mathcal{F}$-*implies* A, $\mathcal{F}:\Gamma \models A$, iff $\mathcal{M}:\Gamma \models A$ for all models $\mathcal{M}$ on $\mathcal{F}$. If $\mathfrak{C}$ is a class of frames, Γ $\mathfrak{C}$-*implies* A, $\mathfrak{C}:\Gamma \models A$, iff $\mathcal{F}:\Gamma \models A$ for all $\mathcal{F} \in \mathfrak{C}$. If $\Gamma=\{A \vee \sim A\}$ then we may simply write $\mathcal{M} \models A$, $\mathcal{F} \models A$ and so on, and speak of *truth* of A in $\mathcal{M}$, $\mathcal{F}$-*validity* of A etc.

Let L be an orthologic. A class $\mathfrak{C}$ of orthoframes is said to *determine* L iff for all A, B $\in \Phi$, A $\vdash_L$ B iff $\mathfrak{C}:A \models B$. $\mathfrak{C}$ *strongly determines* L iff for all Γ and A, $\Gamma \vdash_L A$ iff $\mathfrak{C}:\Gamma \models A$.

The structures that we call orthoframes are not in fact new. They are described as "orthogonality spaces" in Foulis and Randall [18]. That the $\perp$-closed subsets of an orthogonality space form an ortholattice under the partial ordering of set inclusion is a result of long standing (cf. Birkhoff [6, Section V.7]). What appears to be novel is the idea of using such structures to provide models for a propositional language. Furthermore, an algebraic version of the completeness theorem set out below shows that every ortholattice is, within isomorphism, a subortholattice of the lattice of ⊥-closed subsets of some orthogonality space. Previous results in this direction have either been confined to complete ortholattices (cf. [18]) or else have involved a somewhat different notion of orthogonality relation (MacLaren [62]).

Our models can be understood in the following way. X denotes a set of possible outcomes of a number of operations carried out in the performance of some experiment. Elements x and y are orthogonal iff they are *distinct* outcomes of the *same* operation. Thus ⊥ has the character of an "alternativeness" relation as in modal logic. Whereas in the latter a proposition is identified with the set of possible worlds in which it is true, in our present context a proposition A, describing a physical event, is identified with the set $\|A\|^{\mathcal{M}}$ of outcomes that *verify* A. Our truth stipulations then require that an outcome x verifies the conjunction of two propositions iff it verifies each of them separately, and negates a proposition iff only outcomes orthogonal to x verify that proposition. The requirement that $V(p_i)$ be ⊥-closed constitutes a restriction on what sets of outcomes may be identified with propositions or "events".

This interpretation, based largely on ideas expounded in Randall and Foulis [73], is for the present intended merely as a convenient way of thinking about models. Whether or not a description can be found that has significance for quantum theory remains to be seen. Perhaps the structures will contribute to a philosophical clarification of the notion of an event. At any rate these issues are taken to be outside the scope of this paper. Our emphasis throughout is on formal developments. We believe that this attitude is justified, if only by the conviction that the techniques available for handling frames and models are somewhat more elegant, and often more directly applicable to technical problems, than the corresponding algebraic methods. Hopefully the rest of the paper will bear this out.

2.3   The Characterisation of O

In this section we show that the logic O is strongly determined by the class θ of all orthoframes. To this end we need a preliminary lemma.

2.3.1   Lemma

If $\mathcal{M}$ is an orthomodel, then for any A, the set $\|A\|^{\mathcal{M}}$ is ⊥-closed. *Proof.* By induction on the length of A. The result holds for $\|p_i\|^{\mathcal{M}}$ by the definition of orthomodel. Since the intersection of ⊥-closed sets is

$\perp$-closed, it holds for A $\wedge$ B under the hypothesis that it holds for A and B. For the case of negation, suppose $x \notin \|\sim A\|^{\mathcal{M}}$, i.e. not $\mathcal{M} \models_x \sim A$. Then by 2.2.2(3) there exists y such that $\mathcal{M} \models_y$ A and not $x \perp y$. Now if $\mathcal{M} \models_z \sim A$, again by 2.2.2(3), and symmetry of $\perp$, we have $y \perp z$. Thus $y \perp \|\sim A\|^{\mathcal{M}}$, and it follows that $\|\sim A\|^{\mathcal{M}}$ is $\perp$-closed. $\square$

There are a number of results in this paper that are established by induction on the length of formulae. In most instances the case of conjunction is straightforward and so we will present only that part of the proof that involves negation. We also find it convenient to indicate the application of an induction hypothesis by the letters IH.

2.3.2   Soundness Theorem for O

$\Gamma \vdash_O$ A only if $\theta : \Gamma \models$ A.

Proof. The proof, by induction on L-derivability, proceeds by showing that the result holds for #1, ..., #6, and is preserved by applications of #7,..., #9. We consider only the less obvious cases.

#4. Let $\mathcal{M} \models_x$ A. Then if $\mathcal{M} \models_y$ A, by 2.2.2(3) $y \perp x$ and hence (symmetry) $x \perp y$. 2.2.2(3) again gives $\mathcal{M} \models_x \sim \sim A$.

#5. Let $\mathcal{M} \models_x \sim \sim A$. Then $\mathcal{M} \models_y$ A only if $x \perp y$, i.e. $y \perp \|A\|^{\mathcal{M}}$ only if $x \perp y$. But $\|A\|^{\mathcal{M}}$ is $\perp$-closed (2.3.1) and therefore $x \in \|A\|^{\mathcal{M}}$, i.e. $\mathcal{M} \models_x$ A.

#6. If $\mathcal{M} \models_x$ A $\wedge \sim A$ then $\mathcal{M} \models_x \sim A$ and $\mathcal{M} \models_x$ A whence $x \perp x$, contrary to the irreflexivity of $\perp$. Thus not $\mathcal{M} \models_x$ A $\wedge \sim A$ for any x in any $\mathcal{M}$, so $\mathcal{M} : A \wedge \sim A \models$ B for any B.

#9. Suppose $\theta : A \models$ B, and further that $\mathcal{M} \models_x \sim B$. Then $\mathcal{M} \models_y$ A only if $\mathcal{M} \models_y$ B (IH), only if $x \perp y$. This shows that $\mathcal{M} \models_x \sim A$. $\square$

2.3.3   Definition.

If L is an orthologic then the *canonical orthomodel* for L is the structure

$$\mathcal{M}_L = \langle X_L, \perp_L V_L \rangle,$$

where

$X_L = \{ x \subseteq \Phi : x \text{ is L-full} \}$,
$x \perp_L y$ iff there exists A such that $\sim A \in x$, $A \in y$,
$V_L(p_i) = \{ x \in X_L : p_i \in x \}$.

2.3.4   Lemma

$\mathcal{M}_L$ is indeed an orthomodel.

Proof. Let $x \in X_L$. Then for no A do we have $\sim$A, A $\in x$ or else by #6 x would be L-inconsistent. Hence not $x \perp_L x$. If $x \perp_L y$, for some A we have $\sim$A $\in$ x, A $\in y$. Using #4 we conclude $\sim$B $\in y$, B $\in x$, where B=$\sim$A. Thus $y \perp_L x$. Hence $\perp_L$ is an orthogonality relation. To show $V_L(p_i)$ is $\perp_L$-closed, suppose $x \notin V_L(p_i)$, i.e. $p_i \notin x$. By #5, $\sim\sim p_i$ $\notin$ x, whence by 2.1.5 there exists $y \in X_L$ such that not $x \perp_L y$ and $\sim p_i \in y$. Then if $z \in V_L(p_i)$, $p_i \in z$ and so $y \perp_L z$. Thus $y \perp_L V_L(p_i)$ as required.   □

2.3.5 Fundamental Theorem for Orthologics

For all wffs A, and all $x \in X_L$, $\mathcal{M}_L \models_x$ A iff A $\in x$.

Proof. By induction on the length of A. The case A=B∧C is taken care of by 2.1.3(i). Now suppose A=$\sim$B, and the result holds of B. Let $\sim$B $\in x$. Then if $\mathcal{M}_L \models_y$ B, B $\in$ y (IH) and so $x \perp_L y$. This, with 2.2.2(3), yields $\mathcal{M}_L \models_x \sim$B. On the other hand, if $\sim$B $\notin x$, by 2.1.5 there exists $y \in X_L$ such that B $\in y$, whence $\mathcal{M}_L \models_y$ B (IH), but not $x \perp_L y$. From 2.2.2(3) again we conclude that not $\mathcal{M}_L \models_x \sim$B.   □

2.3.6 Corollary

$\Gamma \vdash_L$ A iff $\mathcal{M}_L : \Gamma \models$ A.

Proof. If $\Gamma \vdash_L$ A, $B_1 \wedge \ldots \wedge B_n \vdash_L$ A for some $B_1, \ldots, B_n \in \Gamma$. If $\mathcal{M}_L \models_x$ B for all B $\in \Gamma$ then in particular, by 2.3.5, $B_1, \ldots, B_n \in x$. By 2.1.2(ii) and (iii) it follows that A $\in x$, hence by 2.3.5 $\mathcal{M}_L \models_x$ A.

Conversely, if A is not L-derivable from Γ, by 2.1.4 there exists $x \in X_L$ such that $\Gamma \subseteq x$ and A $\notin x$. By 2.3.5, $\mathcal{M}_L \models_x$ B for all B $\in \Gamma$, but not $\mathcal{M}_L \models_x$ A.   □

2.3.7 Strong Completeness for O

$\theta : \Gamma \models$ A only if $\Gamma \vdash_O$ A.

Proof. Since, by 2.3.4, $\mathcal{M}_O$ is an orthomodel, $\theta : \Gamma \models$ A only if $\mathcal{M}_O : \Gamma \models$ A, and the result follows by 2.3.6.   □

2.4 Translation into $\mathcal{B}$

Let $\{q_i : i < \omega\}$ be a new collection of propositional variables and Φ_M the set of wffs constructible from these by the Boolean connectives ¬ and · (negation and conjunction) and the modal □ (necessity). Material implication → and possibility ◇ are introduced by the usual definitions.

The Brouwerian modal logic $\mathcal{B}$ can be defined as the smallest unary logic, based on Φ_M, that includes all classical PC tautologies, all instances of the schemata

$$\square(A \rightarrow B) \rightarrow (\square A \rightarrow \square B)$$
$$\square A \rightarrow A$$
$$A \rightarrow \square \lozenge A$$

and is closed under Modus Ponens (from A and A → B to infer B) and Necessitation (from A to infer $\square$A).

2.4.1 Definition

$\mathcal{G} = \langle X, R \rangle$ is a *B-frame* if R is a *proximity* relation on X, i.e. $R \subseteq X \times X$ is reflexive and symmetric.

$\mathcal{N} = \langle X, R, V \rangle$ is a *B-model* if V is a function assigning to each q_i a subset $V(p_i)$ of X. The truth stipulations for q_i and conjunction are as in 2.2.2(1), (2). For negation and necessity we have

(i) $\mathcal{N} \models_x \neg A$ iff not $\mathcal{N} \models_x A$

(ii) $\mathcal{N} \models_x \square A$ iff for all y, xRy only if $\mathcal{N} \models_y A$.

A number of basic metalogical facts are known about $\mathcal{B}$, e.g. that it is strongly determined by the class of $\mathcal{B}$-frames, has the finite model property for these structures, has infinitely many non-equivalent modalities, and is not finitely axiomatisable with modus ponens as its sole rule of inference. Apart from that, $\mathcal{B}$ represents an area of modal logic that has attracted little attention. The results of this section may help to alter that situation.

2.4.2 Definition

We recursively define a translation that associates with each $A \in \Phi$ a modal wff $A^* \in \Phi_M$ as follows:

$$p_i^* = \square \lozenge q_i \qquad \text{all } i < \omega$$
$$(A \wedge B)^* = A^* \cdot B^*$$
$$(\sim A)^* = \square \neg (A^*)$$

2.4.3 Lemma

Let $\mathcal{F} = \langle X, \perp \rangle$ and $\mathcal{G} = \langle X, R \rangle$ be an orthoframe and a $\mathcal{B}$-frame respectively, such that $x \perp y$ iff not xRy (clearly $\perp$ is an orthogonality relation iff R is a proximity relation). If $\mathcal{M}$ and $\mathcal{N}$ are models on $\mathcal{F}$ and $\mathcal{G}$ such that for all $i < \omega$, and all $x \in X$, $\mathcal{M} \models_x p_i$ iff $\mathcal{N} \models_x p_i^*$, then for all $A \in \Phi$, and all $x \in X$, $\mathcal{M} \models_x A$ iff $\mathcal{N} \models_x A^*$.

Proof. Let A=∼B and suppose the Lemma holds for B. Then

$$\mathcal{M} \models_x \sim B \quad \text{iff} \quad \text{for all } y, \; \mathcal{M} \models_y B \text{ only if } x \perp y$$

$$\text{iff} \quad \text{for all } y, \; \mathcal{N} \models_y B^* \text{ only if not } xRy \qquad \text{(IH)}$$

$$\text{iff} \quad \text{for all } y, \; xRy \text{ only if not } \mathcal{N} \models_y B^*$$

$$\text{iff} \quad \models_x \Box\neg(B^*) \qquad\qquad (2.4.1(i),(ii))$$

2.4.4 Lemma

$\theta : \Gamma \models A$ only if $\mathfrak{B} : \Gamma^* \models A^*$, where $\Gamma^* = \{B^* : B \in \Gamma\}$ and $\mathfrak{B}$ is the class of B-frames.

Proof. If not $\mathfrak{B}:\Gamma^* \models A^*$, there is a B-model $\mathcal{N} = \langle X, R, V\rangle$ and some $t \in X$ such that $\mathcal{N} \models_t B^*$, all $B \in \Gamma$, but not $\mathcal{N} \models_t A^*$. Let $\mathcal{M} = \langle X, \perp, V'\rangle$, where $V'(p_i) = V(p_i^*)$, and $x \perp y$ iff not xRy. To show that $V'(p_i)$ is $\perp$-closed, suppose $x \notin V'(p_i)$. Then not $\mathcal{N} \models_x \Box\Diamond q_i$ and so there exists y such that xRy and not $\mathcal{N} \models_y \Diamond q_i$. Then if yRw, wRy and so not $\mathcal{N} \models_w \Box\Diamond q_i$, i.e. $w \notin V'(p_i)$. Thus $w \in V'(p_i)$ only if not yRw, i.e. $y \perp w$. Hence $y \perp V'(p_i)$ and, since xRy, not $x \perp y$ as required. Thus $\mathcal{M}$ is an orthomodel satisfying the hypothesis of 2.4.3. We therefore conclude that $\mathcal{M} \models_t B$ for all $B \in \Gamma$, and not $\mathcal{M} \models_t A$, hence not $\theta : \Gamma \models A$. $\qquad\square$

2.4.5 Lemma

$\mathfrak{B} : \Gamma^* \models A^*$ only if $\theta : \Gamma \models A$.

Proof. If not $\theta : \Gamma \models A$, there is an orthomodel $\mathcal{M} = \langle X, \perp, V\rangle$ and some $t \in X$ such that $\mathcal{M} \models_t B$, all $B \in \Gamma$, but not $\mathcal{M} \models_t A$. Let $\mathcal{N} = \langle X, R, V'\rangle$ be a B-model, where xRy iff not $x \perp y$, and $V'(q_i)=V(p_i)$. Then using the fact that $V(p_i)$ is $\perp$-closed we deduce

$$\mathcal{M} \models_x p_i \quad \text{iff} \quad \text{for all } y, \; y \perp \|p_i\|^{\mathcal{M}} \text{ only if } x \perp y$$

$$\text{iff} \quad \text{for all } y, \; xRy \text{ only if for some } z, \text{ not } y \perp z$$
$$\text{and } \mathcal{M} \models_z p_i$$

$$\text{iff} \quad \text{for all } y, \; xRy \text{ only if for some } z, \; yRz$$
$$\text{and } \mathcal{N} \models_z q_i$$

$$\text{iff} \quad \mathcal{N} \models_x \Box\Diamond q_i$$

$$\text{iff} \quad \mathcal{N} \models_x p_i^*.$$

Thus the hypothesis of 2.4.3 is satisfied, and we conclude $\mathcal{N} \models_t B^*$ for all $B \in \Gamma$, but not $\mathcal{N} \models_t A^*$, hence not $\mathfrak{B} : \Gamma^* \models A^*$. $\qquad\square$

Using Lemmata 2.4.4, 2.4.5, our strong determination result for O, and that of B, we arrive at

2.4.6 Theorem

$\Gamma \vdash_O A \quad \text{iff} \quad \Gamma^* \vdash_B A^*.$

2.5   Filtrations and Decidability

In this section we show that O has the *finite model property* (FMP) and, being finitely axiomatisable, is therefore decidable. This result can in fact be obtained indirectly from the results of the previous section, using the fact that $\mathcal{B}$ has the FMP. We wish however to give a direct proof in terms of orthoframes. Part of the reason for this is simply a desire to see how filtration theory works in our present context. More significant however is the recognition that some of the more complex schemata of modal logic have only admitted to successful analysis through filtrations of canonical models, and it seems quite likely that this approach will play an important role in further studies of ortholLogic.

2.5.1   Definition

A set ψ of wffs is *admissible* iff

   (i) if $A \in \psi$ and B is a subformula of A, then $B \in \psi$; and
  (ii) for all $i < \omega$, $p_i \in \psi$ only if $\sim p_i \in \psi$.

2.5.2   Definition

Let $\mathcal{M} = \langle X, \perp, V \rangle$ be an orthomodel and ψ an admissible set of wffs. We define an equivalence relation on X as follows:

$x \approx y$ iff for all $A \in \psi$, $\mathcal{M} \models_x A$ iff $\mathcal{M} \models_y A$.

Putting $[x] = \{y : x \approx y\}$ we define

   $X' = \{[x] : x \in X\}$;

   $[x] \perp' [y]$ iff there exists A such that $\sim A \in \psi$, and either

      (i) $\mathcal{M} \models_x \sim A$ and $\mathcal{M} \models_y A$,   or
     (ii) $\mathcal{M} \models_y \sim A$ and $\mathcal{M} \models_x A$;

   $V'(p_i) = \{[x] : p_i \in \psi$ and $x \in V(p_i)\}$.

The structure $\mathcal{M}' = \langle X', \perp', V' \rangle$ is then called the *filtration of $\mathcal{M}$ through* ψ.

2.5.3   Lemma

   $\mathcal{M}'$ is an orthomodel.

Proof. We note first that the above definitions are correct—independent of the choice of equivalence class representative. $\perp'$ is symmetric by definition. Furthermore by 2.2.2(3) and the symmetry of $\perp$, $[x] \perp' [y]$ only if $x \perp y$. From this, and the irreflexivity of $\perp$, it follows that $\perp'$ is irreflexive, and thus is an orthogonality relation. Now if $p_i \notin \psi$, $V'(p_i) = \emptyset$ and so is $\perp'$-closed. On the other hand if $p_i \in \psi$ and $[x] \notin$

$V'(p_i)$, then $x \notin V(p_i)$, so there exists $y \in X$ such that $y \perp V(p_i)$ and not $x \perp y$. Thus $\mathcal{M} \models_y \sim p_i$ and not $[x] \perp' [y]$. Now if $[z] \in V'(p_i)$, then $\mathcal{M} \models_z p_i$. But $\sim p_i \in \psi$ (2.5.1(ii)) and so by 2.5.2, $[y] \perp' [z]$. Hence $[y] \perp' V'(p_i)$ and $V'(p_i)$ is $\perp'$-closed. □

2.5.4   Filtration Theorem

For all $A \in \psi$ and $x \in X$, $\mathcal{M} \models_x A$ iff $\mathcal{M}' \models_{[x]} A$.

Proof. Let $A = \sim B$ and suppose $\mathcal{M}' \models_{[x]} \sim B$. Then if $\mathcal{M} \models_y B$, $\mathcal{M}' \models_{[y]} B$ (IH) and so $[x] \perp' [y]$, whence $x \perp y$. By 2.2.2(3) this implies $\mathcal{M} \models_x \sim B$. Conversely let $\mathcal{M} \models_x \sim B$. Then if $\mathcal{M}' \models_{[y]} B$, it follows by IH that $\mathcal{M} \models_y B$. But $\sim B \in \psi$ and so by 2.5.2, $[x] \perp' [y]$. Hence $\mathcal{M} \models_{[x]} \sim B$. □

2.5.5   Finite Model Property for O

Let A and B be wffs which together have k subwffs, including l propositional variables. Then $A \vdash_O B$ iff $\mathcal{F}{:}A \models B$ for all orthoframes $\mathcal{F}$ having at most 2^n elements, where $n = k + l$.

Proof. Necessity follows from the Soundness Theorem 2.3.2. Conversely, suppose that not $A \vdash_O B$. By 2.3.6, there exists $x \in X_O$ such that $\mathcal{M}_O \models_x A$ and not $\mathcal{M} \models_x B$. Let ψ be the smallest admissible set including A and B, and $\mathcal{M}'$ the filtration of $\mathcal{M}_O$ through ψ. By 2.5.4, $\mathcal{M}' \models_{[x]} A$ and not $\mathcal{M}' \models_{[x]} B$, hence not $\mathcal{M}' : A \models B$. But clearly ψ consists precisely of the subwffs of A and B, together with the negations of those propositional variables occurring in A or B, i.e. ψ has at most n members. Since each equivalence class x is determined uniquely by the set $\psi_x = \{A \in \psi : \mathcal{M}_O \models_x A\}$, and ψ has at most 2^n subsets, it follows that the frame of $\mathcal{M}'$ has at most this many members, and our proof is complete. □

2.5.6   Corollary

O is decidable. □

2.6   An Approach to Quantum Logic

2.6.1   Definition

A *quantum logic* is an orthologic L such that, for any $A, B \in \Phi$,

#10. $A \wedge (\sim A \vee (A \wedge B)) \vdash_L B$.

We denote by Q the smallest quantum logic i.e. the intersection of all quantum logics. It can be shown that Q bears the same relation to orthomodular lattices that O bears to ortholattices.

In extending our analysis to quantum logics, one desirable outcome would be a first-order condition on orthogonality relations such that Q is strongly determined by the class of orthoframes satisfying that condition[1]. An alternative approach would be to restrict the definition of "model", to impose a constraint on the sets $\|A\|^{\mathcal{M}}$ that characterises the system. Rather than make such a restriction dependent on the object language, we will build it into the frames themselves. Thus a frame now has associated with it a particular collection ξ of $\perp$-closed subsets of its carrier set, and models are obtained by selecting the values of propositions from ξ. Similar structures have been considered by Fine [12] in connection with languages having propositional quantifiers, and there are analogies with the "first-order" tense logic semantics of Thomason [96].

Our next definition provides a natural refinement of the concept of $\perp$-closure.

2.6.2   Definition

If Y and Z are subsets of a frame $\langle X, \perp \rangle$, then Y *is $\perp$-closed in* Z iff for all $x \in Z$, $x \notin Y$ only if there exists $y \in Z$ such that $y \perp Y$ and not $x \perp y$. (Thus $\perp$-closed sets are precisely those that are $\perp$-closed in X.)

2.6.3   Lemma

If $\mathcal{M}$ is any orthomodel, then $\|A\|^{\mathcal{M}}$ is $\perp$-closed in $\|B\|^{\mathcal{M}}$ iff
$\mathcal{M} : B \wedge (\sim B \vee A) \models A$.

Proof. Suppose that $\mathcal{M} : B \wedge (\sim B \vee A) \models A$, $x \in \|B\|^{\mathcal{M}}$, but $x \notin \|A\|^{\mathcal{M}}$. Using 2.2.2(2), we deduce that not $\mathcal{M} \models_x \sim B \vee A$, i.e. not $\mathcal{M} \models_x \sim(B \wedge \sim A)$. By 2.2.2(3) there exists y such that $\mathcal{M} \models_y B \wedge \sim A$ and not $x \perp y$. Then $\mathcal{M} \models_y B$, i.e. $y \in \|B\|^{\mathcal{M}}$, and $\mathcal{M} \models_y \sim A$, hence $y \perp \|A\|^{\mathcal{M}}$ as required.

Conversely, suppose that $\|A\|^{\mathcal{M}}$ is $\perp$-closed in $\|B\|^{\mathcal{M}}$. If we have $\mathcal{M} \models_x B \wedge \sim(B \wedge \sim A)$, then by 2.2.2(2),(3) we conclude $x \in \|B\|^{\mathcal{M}}$ and $(y \in \|B\|^{\mathcal{M}}$ and $y \perp \|A\|^{\mathcal{M}})$ only if $x \perp y$. From 2.6.2 it follows that $x \in \|A\|^{\mathcal{M}}$, i.e. $\mathcal{M} \models_x A$. $\square$

2.6.4   Definition

$\mathcal{F} = \langle X, \perp, \xi \rangle$ is a *quantum frame* if $\langle X, \perp \rangle$ is an orthoframe and ξ is a non-empty collection of $\perp$-closed subsets of X such that

(i) ξ is closed under set intersection, and the operation * defined by
$Y^* = \{x : x \perp Y\}$

---

[1]After this article was first published I discovered that no such condition exists. The proof is given in the next chapter.

(ii) if Y, Z $\in \xi$, then Y $\subseteq$ Z only if Y is $\perp$-closed in Z.

$\mathcal{M} = \langle X, \perp, \xi, V \rangle$ is a *quantum model* if V is a function assigning to each p_i a member of ξ. The truth conditions remain as in 2.2.2.

2.6.5   Lemma

For any quantum model $\mathcal{M}$, and any A$\in \Phi$, $\|A\|^{\mathcal{M}} \in \xi$.

Proof. By induction on the length of A, using 2.6.4(i) and 2.2.2(2′), (3′).
$\square$

2.6.6   Lemma

All quantum models verify #10.

Proof. For any $\mathcal{M}$ we have $\mathcal{M} : A \wedge B \models A$, hence $\|A \wedge B\|^{\mathcal{M}} \subseteq \|A\|^{\mathcal{M}}$. Thus if $\mathcal{M}$ is a quantum model it follows by 2.6.5 and 2.6.4(ii) that $\|A \wedge B\|^{\mathcal{M}}$ is $\perp$-closed in $\|A\|^{\mathcal{M}}$, whence by 2.6.3,

$$\mathcal{M} : A \wedge (\sim A \vee (A \wedge B)) \models A \wedge B.$$

By the verification of #3 and #7, this yields

$$\mathcal{M} : A \wedge (\sim A \vee (A \wedge B)) \models B.$$

$\square$

2.6.7   Corollary (Soundness Theorem for Q)

$\Gamma \vdash_Q A$ only if $\Omega : \Gamma \models A$, where Ω is the class of quantum frames. $\square$

If L is an orthologic and A $\in \Phi$, we denote by $|A|^L$ the set $\{x \in X_L : A \in x\}$. Thus by 2.3.5, $|A|^L = \|A\|^{\mathcal{M}_L}$, where L is the canonical orthomodel for L.

2.6.8   Definition

Let L be a quantum logic. The *canonical quantum frame for* L is the structure $\mathcal{G}_L = \langle X_L, \perp_L, \xi_L \rangle$ where X_L and $\perp_L$ are as in 2.3.3, and $\xi_L = \{|A|^L : A \in \Phi\}$. $\mathcal{N}_L = \langle X_L, \perp_L, \xi_L, V_L \rangle$ is the *canonical quantum model for* L, where, as before, $V_L(p_i) = |p_i|^L$.

2.6.9   Lemma

$\mathcal{G}_L$ is a quantum frame.

Proof. It is plain from our earlier work that $|A|^L \cap |B|^L = |A \wedge B|^L$ and $(|A|^L)^* = |\sim A|^L$, so 2.6.4(i) holds. To establish 2.6.4(ii), suppose that $|A|^L \subseteq |B|^L$, i.e. that $\|A\|^{\mathcal{M}_L} \subseteq \|B\|^{\mathcal{M}_L}$. By #10 and 2.3.6 we have $\mathcal{M}_L : B \wedge (\sim B \vee (B \wedge A)) \models A$. But clearly $\|B \wedge A\|^{\mathcal{M}_L} = \|A\|^{\mathcal{M}_L}$, and so $\mathcal{M}_L : B \wedge (\sim B \vee A) \models A$. From 2.6.3 we conclude that $|A|^L = \|A\|^{\mathcal{M}_L}$ is $\perp_L$-closed in $\|B\|^{\mathcal{M}_L} = |B|^L$.
$\square$

2.6.10 Corollary (Fundamental Theorem for Quantum Logics)

$|A|^L = \|A\|^{\mathcal{M}_L} = \|A\|^{\mathcal{N}_L}$, for all $A \in \Phi$. □

We could at this point produce analogues to 2.3.6 and 2.3.7, and together with 2.6.7 conclude that Q is strongly determined by the class of all quantum frames. However with our new structures, much stronger results are possible. Our earlier methods provided each ortologic with a characteristic model ($\mathcal{M}_L$) but did not produce a characteristic frame, or class of frames, except in the case of O. Whether every ortologic is (strongly) determined by a class of orthoframes is as yet unknown, although recent work by Thomason [97] on incompleteness in modal logic suggests that the answer will probably be negative. For quantum logics and quantum frames however the matter can be settled completely in the affirmative.

2.6.11 Theorem

Let L be a quantum logic. Then $\Gamma \vdash_L A$ iff $\mathcal{G}_L : \Gamma \models A$.

Proof. If $\mathcal{G}_L : \Gamma \models A$, then $\mathcal{N}_L : \Gamma \models A$, so $\mathcal{M}_L : \Gamma \models A$ by 2.6.10. $\Gamma \vdash_L A$ then follows by 2.3.6. Conversely, if $\Gamma \vdash_L A$, there exist $A_1, \ldots, A_n \in \Gamma$ such that $A_1 \wedge \ldots \wedge A_n \vdash_L A$. Now let $\mathcal{M}$ be any model on $\mathcal{G}_L$. For each $i < \omega$, $\|p_i\|^{\mathcal{M}} \in \xi_L$, so there exists B_i such that $\|p_i\|^{\mathcal{M}} = |B_i|^L = \|B_i\|^{\mathcal{N}_L}$. For any wff C, let C' be the result of uniformly replacing each p_i occurring in C by B_i. Clearly we then have $A_1' \wedge \ldots \wedge A_n' \vdash_L A'$ and so by 2.3.6 and 2.6.10, $\mathcal{N}_L : A_1' \wedge \ldots \wedge A_n' \models A'$. But a simple induction shows that $\|C\|^{\mathcal{M}} = \|C'\|^{\mathcal{N}_L}$ and so $\mathcal{M} : A_1 \wedge \ldots \wedge A_n \models A$, whence $\mathcal{M} : \Gamma \models A$. Since this holds for all models $\mathcal{M}$ on $\mathcal{G}_L$, we conclude $\mathcal{G}_L : \Gamma \models A$. □

2.6.12 Corollary (Strong Completeness for Q)

$\Omega : \Gamma \models A$ only if $\Gamma \vdash_Q A$. □

Theorem 2.6.11 shows in fact that every quantum logic L is strongly determined by the class of quantum frames whose only member is $\mathcal{G}_L$.

We could of course, by deleting condition (ii) from 2.6.4, obtain a new concept of frame for which every ortologic is strongly determined. However the most interesting problem of all remains as yet unresolved. Is there a class of orthoframes that determines Q?

Notes

I am indebted to my supervisor, Dr. M. J. Cresswell, for some very helpful discussions and comments on the composition of this paper. I would also like to acknowledge a debt to Mr. K. E. Pledger, through whose involvement with quantum logic I first became interested in the subject. He had earlier established algebraically a connection between the logic of orthomodular lattices and an extension of $\mathcal{B}$. The blame for the techniques and results of this paper however lies solely with its author.

I have a proof that any finite quantum frame is semantically equivalent to one for which ξ is the class of all $\bot$-closed sets. Thus if Q has the FMP for quantum frames it will be determined by a class of finite orthoframes.

3

Orthomodularity is not Elementary

In this article it is shown that the property of *orthomodularity* of the lattice of orthoclosed subspaces of a pre-Hilbert space $\mathcal{P}$ is not determined by any first-order properties of the relation $\perp$ of orthogonality between vectors in $\mathcal{P}$. Implications for the study of quantum logic are discussed at the end of the article.

The key to this result is the following:

(1)   *If $\mathcal{H}$ is a separable Hilbert space, and $\mathcal{P}$ is an infinite-dimensional pre-Hilbert subspace of $\mathcal{H}$, then $(\mathcal{P}, \perp)$ and $(\mathcal{H}, \perp)$ are elementarily equivalent in the first-order language L_2 of a single binary relation.*

Choosing $\mathcal{P}$ to be a pre-Hilbert space whose lattice of orthoclosed subspaces is not orthomodular, we obtain our desired conclusion. In this regard we may note the demonstration by Amemiya and Araki [2] that orthomodularity of the lattice of orthoclosed subspaces is necessary *and* *sufficient* for a pre-Hilbert space to be metrically complete, and hence be a Hilbert space. Metric completeness being a notoriously non-elementary property, our result is only to be expected (note also the parallel with the elementary L_2-equivalence of the natural order $(\mathbb{Q}, <)$ of the rationals and its metric completion to the reals $(\mathbb{R}, <)$).

To derive (1), something stronger is proved, viz. that $(\mathcal{P}, \perp)$ is an elementary substructure of $(\mathcal{H}, \perp)$. This is done by showing that any element of $\mathcal{H}$ can be moved inside $\mathcal{P}$ by an automorphism of $\mathcal{H}$ that leaves fixed a prescribed finite subset of $\mathcal{P}$. Familiarity is assumed with the basic theory of Hilbert spaces, and for this purpose the very accessible exposition of Berberian [5] has been followed.

Theorem 3.1. *Let $\mathcal{H}$ be a separable Hilbert space, and $\mathcal{P}$ an infinite-dimensional linear subspace of $\mathcal{H}$. Then if $a_1, \ldots, a_n \in \mathcal{P}$ and $b \in \mathcal{H}$, there exists an isomorphism $T : \mathcal{H} \to \mathcal{H}$ such that $T(a_i) = a_i$ for $1 \leq i \leq n$, and $T(b) \in \mathcal{P}$.*

Proof. (By an isomorphism is meant a bijective linear transformation that preserves inner products and hence leaves the orthogonality relation invariant.)

Suppose that the a_i's are ordered so that for some $k \leq n$, $a_1, \ldots, a_k$ is a linearly independent set with the same linear span as $a_1, \ldots, a_n$. Then orthonormalise to get an orthonormal set $\{x_1, \ldots, x_k\} \subseteq \mathcal{P}$ with this same linear span [5, p. 47].

As $\mathcal{P}$ is infinite-dimensional, there exists some $x \in \mathcal{P}$ that is linearly independent of $x_1, \ldots, x_k$. Then $y = x - (\sum_1^k (x|x_i)x_i)$ is a nonzero vector in $\mathcal{P}$ that is orthogonal to each of $x_1, \ldots, x_k$. Putting $c = \|y\|^{-1}y$, it follows that $\{x_1, \ldots, x_k, c\}$ is an orthonormal subset of $\mathcal{P}$.

Now if $b \in \mathcal{P}$, the Theorem follows with T as the identity map on $\mathcal{H}$. Hence we may assume $b \notin \mathcal{P}$. But then b is linearly independent of $x_1, \ldots, x_k$ and by the same process that produced c we may orthonormalise to obtain $d \in \mathcal{H}$ with $\{x_1, \ldots, x_k, d\}$ an orthonormal set having the same span as $\{x_1, \ldots, x_k, b\}$.

Now let $\mathcal{M}$ = the linear subspace of $\mathcal{H}$ generated by $\{x_1, \ldots, x_k, d\}$, and $\mathcal{N}$ = the linear subspace of $\mathcal{H}$ generated by $\{x_1, \ldots, x_k, c\}$. Then $\mathcal{N} \subseteq \mathcal{P}$ and $\mathcal{M}$ and $\mathcal{N}$, being of the same finite dimension, are isomorphic by an isomorphism $U : \mathcal{M} \to \mathcal{N}$ that has $U(x_i) = x_i$ for $1 \leq i \leq k$, and $U(d) = c$. But for $1 \leq i \leq n$, a_i is a linear combination of $x_1, \ldots, x_k$, and so $U(a_i) = a_i$. Also, by construction $b \in \mathcal{M}$, so that $U(b) \in \mathcal{P}$.

Now let

$$\mathcal{M}^{\perp} = \{z \in \mathcal{H} : z \perp y \text{ for all } y \in \mathcal{M}\}$$

be the annihilator of $\mathcal{M}$ in $\mathcal{H}$. Then $\mathcal{M}^{\perp}$ is topologically closed [5, p. 59] and hence is a separable Hilbert space. Since $\mathcal{M}$ is finite dimensional, its orthogonal sum with $\mathcal{M}^{\perp}$ is $\mathcal{H}$ [5, p. 66], i.e. $\mathcal{H} = \mathcal{M} \oplus \mathcal{M}^{\perp}$, and so as $\mathcal{H}$ is infinite dimensional, $\mathcal{M}^{\perp}$ must be infinite dimensional too.

Similarly, the annihilator $\mathcal{N}^{\perp}$ of $\mathcal{N}$ in $\mathcal{H}$ is an infinite-dimensional separable Hilbert space. Since any two such spaces are isomorphic [5, p. 55], there exists an isomorphism $V : \mathcal{M}^{\perp} \to \mathcal{N}^{\perp}$. Our desired map T is then realised as the orthogonal sum of U and V. For, each $w \in \mathcal{H}$ has a unique representation $w = y + z$ with $y \in \mathcal{M}$ and $z \in \mathcal{M}^{\perp}$ [5, p. 61]. We put $T(w) = U(y) + V(z)$. Using the fact that, likewise, w has a unique representation $y' + z'$ with $y' \in \mathcal{N}$ and $z' \in \mathcal{N}^{\perp}$, T may be shown to be an isomorphism. Moreover, for $1 \leq i \leq n$, $a_i \in \mathcal{M}$ and so

$$T(a_i) = U(a_i) = a_i,$$

and finally, $T(b) = U(b) \in \mathcal{P}$.   □

Application to Quantum Logic.

A set-theoretic semantics for the propositional logic of ortholattices was developed in [33], using the notion of an *orthoframe* $(X, \perp)$ as a nonempty set X carrying an irreflexive symmetric relation $\perp$. Each sentence is interpreted as a subset Y of X that is $\perp$-*closed*, i.e. satisfies $Y^{\perp\perp} = Y$, where in general $Z^\perp = \{x : x \perp z \text{ all } z \in Z\}$. The set of $\perp$-closed subsets of X is a complete lattice under the partial ordering of set inclusion, with set intersection as lattice meet, and $Z^\perp$ as orthocomplement of Z (Birkhoff [6, Section V.7]). The lattice is *orthomodular* if it satisfies

(2) $Y \subseteq Z$ only if $Z \cap Y^\perp \neq \emptyset$.

If (2) holds, we will say that $(X, \perp)$ is an *orthomodular frame*. A pre-Hilbert space $\mathcal{P}$ gives rise naturally to the orthoframe $(\mathcal{P}^+, \perp)$, where $\mathcal{P}^+$ is the set of nonzero vectors of $\mathcal{P}$, and $x \perp y$ iff $(x|y) = 0$. Now an isomorphism $T : \mathcal{P} \to \mathcal{P}$ preserves the zero vector, and hence acts on $\mathcal{P}^+$ as a bijection that preserves inner products $(x|y)$ and so has

$$x \perp y \quad \text{iff} \quad T(x) \perp T(y).$$

Thus if φ is any formula of the first-order language L_2 of a single binary relation, it follows that for any $x_1, \ldots, x_m$ in $\mathcal{P}$,

$$(\mathcal{P}^+, \perp) \models \varphi[x_1, \ldots, x_m] \quad \text{iff} \quad (\mathcal{P}^+, \perp) \models \varphi[T(x_1), \ldots, T(x_m)].$$

Applying this observation to Theorem 3.1 gives

Theorem 3.2. *Let $\mathcal{P}$ be an infinite-dimensional subspace of a separable Hilbert space $\mathcal{H}$. Then if φ is any L_2 formula, and $a_1, \ldots, a_n$ are elements of $\mathcal{P}$ such that for some $b \in \mathcal{H}$, $(\mathcal{H}^+, \perp) \models \varphi[a_1, \ldots, a_n, b]$, then there is some $a \in \mathcal{P}$ such that $(\mathcal{H}^+, \perp) \models \varphi[a_1, \ldots, a_n, a]$.*   □

Theorem 3.2 gives a well-known criterion [4, p. 76] that ensures that if $\mathcal{P}$ and $\mathcal{H}$ are as stated, then $(\mathcal{P}^+, \perp)$ is an elementary substructure of $(\mathcal{H}^+, \perp)$. Hence the two structures satisfy exactly the same L_2-sentences.

Now let $\mathcal{P}$ be the *incomplete* pre-Hilbert space of finitely nonzero sequences of complex numbers, and $\mathcal{H}$ the separable Hilbert space l^2 of absolutely square-summable sequences [5, Chapter 11]. Then $\mathcal{P}$ is an infinite-dimensional subspace of $\mathcal{H}$, so $(\mathcal{P}^+, \perp)$ is elementarily equivalent to $(\mathcal{H}^+, \perp)$. But the latter is orthomodular, while the former is not. To see this, observe that adjunction of the zero vector to a $\perp$-closed subset of $\mathcal{H}^+$ turns it into a $\perp$-closed subspace of $\mathcal{H}$, and this process gives an isomorphism between the lattices of $\perp$-closed subsets of $\mathcal{H}^+$ and closed subspaces of $\mathcal{H}$ (in $\mathcal{H}$, "$\perp$-closed" and "(topologically) closed" are equivalent). A proof that (2) holds for closed subspaces in $\mathcal{H}$ is given by Halmos [41, p. 23], who describes it as a result which "our geometric intuition makes obvious and desirable." The proof is indeed conceptually natural: it obtains $z \in Z \cap Y^\perp$ as $y - x$, where x is an arbitrary member of $Z - Y$, and y is a member of Y that minimizes the distances from x to vectors in Y. The argument that shows such a y to exist uses the metric completeness of $\mathcal{H}$. Amemiya and Araki [2] showed such a use of completeness to be unavoidable (cf. also Maeda and Maeda [64, Theorem 34.9] or Varadarajan [104, pp. 182–183] for details), by proving that if the lattice of $\perp$-closed subspaces of $\mathcal{P}$ is orthomodular, then $\mathcal{P}$ is complete. Their proof works for any pre-Hilbert space $\mathcal{P}$, but for our present purpose it suffices to apply it to the particular $\mathcal{P}$ cited above to conclude that the class $\{(X, \perp) : (X, \perp)$ is orthomodular$\}$ of L_2-structures is not closed under elementary equivalence, and so is not the class of models of any set of L_2-sentences.

Implications for Quantum Logic.

The semantics based on orthoframes is inspired by the Kripke semantics for propositional modal logics that uses *frames* (X, R), with R a binary relation on X. The development of this type of model has greatly enhanced the study of modal logics, because frames are easier to visualise and manipulate than algebraic models (lattices with operators). One of the more powerful techniques used is the construction for any logic of a *canonical* model falsifying all its nontheorems by taking X as the set of (maximal) theories of the logic.

But the real success of Kripke semantics is perhaps due to the simplicity of its model characterisations. Most of the more important modal logics were shown to have their frames defined by first-order conditions on the relation R (reflexivity, symmetry, transitivity, linearity etc., cf., e.g., [59]). To show that a logic is characterised by a certain class of frames it suffices to show that the class includes the frame of the canonical model. If the class is L_2-definable, then the question of characterisation boils down to showing that the canonical frame satisfies a certain first-order condition, or set of such conditions. This approach is in fact quite general: it was shown by Fine [15] that if the class of all frames of a modal logic is closed under elementary L_2-equivalence, and the logic is characterised by *some* class of frames, then it is characterised by its canonical frame.

Thus the results of this article indicate that the standard approach is unavailable for the logic of orthomodular lattices, as there is no first-order characterisation of orthomodularity for orthoframes. This is further evidence of the intractability of quantum logic. It is perhaps the first example of a natural and significant logic that leaves the usual methods defeated. There are some very basic questions about orthomodular logic which, to my knowledge, remain unanswered:

Is it characterised by the class of orthomodular orthoframes?

Is it characterised by its canonical frame?

Does it have the finite-model property?

Is it decidable?

4

Arithmetical Necessity, Provability and Intuitionistic Logic

4.1 Motivation

The interpretation of the modal operator $\Box$ as "it is provable that" seems to have been first considered by Gödel [20], who observed that there is a theorem-preserving translation of Heyting's intuitionistic logic IL into the modal system S4. He "presumed" further that the translation is deducibility-invariant, i.e., that a sentence is an IL-theorem precisely when its translate is an S4-theorem. This was later verified by McKinsey and Tarski [68].

A recent paper by Solovay [91] considers a number of provability interpretations of modality, the most significant being "it is provable in Peano arithmetic that". The basic idea is that if α is a sentence of Peano Arithmetic (P), then $\Box\alpha$ denotes the sentence

(1) $\text{Bew}(\ulcorner\alpha\urcorner)$,

where $\ulcorner\alpha\urcorner$ denotes the numeral of the Gödel number of α, and $\text{Bew}(x)$ is the formula that expresses "x is the Gödel number of a theorem of P."

Now the most well known reading of $\Box$ is the alethic modality "it is necessarily true that", and the most well known account of necessity is the Leibnizian dictum that a necessary truth is one that is true in all possible worlds. We can relate this to (1) by defining (as seems eminently reasonable) a possible world for arithmetic to be a model of P. A sentence α is then *arithmetically necessary* when it is true in all P-models, which, by the Completeness Theorem, holds precisely when $\text{P} \vdash \alpha$. Since the latter holds just in case $\text{P} \vdash \text{Bew}(\ulcorner\alpha\urcorner)$, we obtain the equivalence of

(2) α is arithmetically necessary

and

(3) $P \vdash \Box \alpha$.

There is however a major inadequacy in this analysis. A necessarily true statement is in particular a true statement, and so for the alethic interpretation we require the validity of the schema

(T) $\Box A \to A$.

But if validity is taken to mean derivability in P, as is done in [91], then not all instances of this schema are valid. Indeed $P \vdash \text{Bew}(\ulcorner \alpha \urcorner) \to \alpha$ only in the event that $P \vdash \alpha$, as was shown by Löb [60]. The purpose of this article is to offer a modified interpretation of $\Box$ that leaves (2) and (3) equivalent, but makes T valid. The modification is simply to take $\Box \alpha$ as

(4) $\alpha \wedge \text{Bew}(\ulcorner \alpha \urcorner)$,

which has the apparent meaning "α is true and provable". Read like that, (4) seems equivalent to "α is provable", i.e., to (1). However the two are not the same, in view of the existence of true but unprovable sentences of arithmetic. The precise situation is that each of (1) and (4) is a P-theorem precisely when the other is (which is why (2) and (3) remain equivalent), the two are materially equivalent in the standard P-model, but this material equivalence is not in general itself provable in P.

An interpretation of the language of modal logic will be developed on the basis of (4), and then by axiomatising the resulting class of valid sentences, and invoking the Gödel-McKinsey-Tarski translations mentioned above, we will obtain a provability interpretation of IL in P, in which an *intuitionistic implication asserts the truth and provability of a material implication*, and an *intuitionistic negation asserts that an arithmetical sentence is false and refutable (inconsistent)*. Subsequently we shall show that the arithmetically necessary non-modal sentences are just the IL-theorems.

These results were obtained while the author held a position as Visiting Scientist at Simon Fraser University. He would like to thank Dr. S. K. Thomason for the hospitality afforded him at that time.

4.2   Method

Let Φ be a modal propositional language based on propositional letters $p_0, p_1, p_2, \ldots$, the connectives $\wedge$, $\vee$, $\sim$, $\to$ (all taken as primitive), and the modal operator $\Box$. Define a translation $A \mapsto A^\circ$ from Φ to Φ by stipulating

(5) $p_i{}^\circ = p_i$,

(6) $(A \wedge B)^\circ = A^\circ \wedge B^\circ,$

(7) $(A \vee B)^\circ = A^\circ \vee B^\circ,$

(8) $(\sim A)^\circ = \sim(A^\circ),$

(9) $(A \to B)^\circ = A^\circ \to B^\circ,$

(10) $(\Box A)^\circ = A^\circ \wedge \Box(A^\circ).$

We presume the reader is familiar with the notions of a frame $\mathcal{F} = \langle U, R \rangle$, a model $\mathcal{M} = \langle U, R, V \rangle$ based on $\mathcal{F}$, the validity of a sentence A on $\mathcal{F}$, and the truth of A at x in $\mathcal{M}$, $\mathcal{M} \models_x A$. We recall only the key clause

(11) $\mathcal{M} \models_x \Box A$ iff for all y, xRy only if $\mathcal{M} \models_y A$.

(The details of these definitions may be found, e.g., in [86]).

Lemma 4.1. *Let $\mathcal{M} = \langle U, R, V \rangle$ be a model with R reflexive. Define $\mathcal{M}' = \langle U, S, V \rangle$, where xSy if and only if xRy and $x \neq y$. Then for any sentence $A \in \Phi$, and any $x \in U$*

$$\mathcal{M} \models_x A \quad \textit{iff} \quad \mathcal{M}' \models_x A^\circ.$$

Proof. By induction on the formation of A, the only non-trivial case being $A = \Box B$, under the inductive hypothesis that the result holds for B. Suppose $\mathcal{M} \models_x \Box B$. Then if xSy, we have xRy, so by (11) $\mathcal{M} \models_y B$, whence by hypothesis $\mathcal{M}' \models_y B^\circ$. The analogue of (11) for $\mathcal{M}'$ then gives $\mathcal{M}' \models_x \Box(B^\circ)$. But also, since R is reflexive we have $\mathcal{M} \models_x B$, hence $\mathcal{M}' \models_x B^\circ$. Altogether then $\mathcal{M}' \models_x B^\circ \wedge \Box(B^\circ)$, i.e., $\mathcal{M}' \models_x A^\circ$.

Conversely let $\mathcal{M}' \models_x A^\circ$ and xRy. If $x = y$, then $\mathcal{M}' \models_x B^\circ$ and so by hypothesis $\mathcal{M} \models_y B$. But if $x \neq y$, xSy, so as $\mathcal{M}' \models_x \Box(B^\circ)$ again we get $\mathcal{M} \models_y B$. Thus, by (11), $\mathcal{M} \models_x \Box B$ as required. $\quad\Box$

Now the modal logic known in the terminology of Segerberg [86] as K4W is axiomatisable by adjoining to a basis for PC the rule

(RN)   *From A to infer $\Box A$,*

and the axiom schemata

   (K) $\Box(A \to B) \to (\Box A \to \Box B),$

   (4) $\Box A \to \Box\Box A,$

   (W) $\Box(\Box A \to A) \to \Box A.$

It is shown in Chapter II §2 of [86] (and it follows also from the results of [91]) that the K4W theorems are precisely those modal sentences valid on all finite strictly ordered frames.

     The system S4Grz (also known as K1.1) has in addition to RN, K, 4, and T (which together define S4), the schema

(Grz) $\Box(\Box(A \to \Box A) \to A) \to A$.

The S4Grz theorems are precisely those sentences valid on all finite partially ordered frames [86, Chapter II, §3].

Theorem 4.1. *For any modal sentence A,*

$$\text{S4Grz} \vdash A \quad \text{iff} \quad \text{K4W} \vdash A°.$$

Proof. If S4Grz $\nvdash A$ then there is some $\mathcal{M} = \langle U, R, V \rangle$ and some $x \in U$, with $\mathcal{M} \nvDash_x A$, U finite, and R a partial ordering. Let $\mathcal{M}'$ be as in Lemma 4.1. Then $\mathcal{M}' \nvDash_x A°$, so $A°$ is not valid on the finite frame $\langle U, S \rangle$, for which S is in fact a strict ordering. Hence K4W $\nvdash A°$.

Conversely if K4W $\nvdash A°$, then $A°$ is false at some point x in a finite strictly ordered model $\mathcal{N} = \langle U, S, V \rangle$. Let $\mathcal{M} = \langle U, R, V \rangle$, where xRy iff xSy or $x = y$. Then clearly $\mathcal{M}' = \mathcal{N}$, so by Lemma 4.1 A is not valid on the frame $\langle U, R \rangle$, which is a finite partial ordering. Hence S4Grz $\nvdash A$.

$\Box$

Now the provability interpretation of Solovay [91] is as follows: a *-*interpretation* of Φ in Peano arithmetic is an assignment to each $A \in \Phi$ of a P-sentence A^* that satisfies

(12) $(A \wedge B)^* = A^* \wedge B^*$,
(13) $(A \vee B)^* = A^* \vee B^*$,
(14) $(\sim A)^* = \sim(A^*)$,
(15) $(A \to B)^* = A^* \to B^*$,
(16) $(\Box A)^* = \text{Bew}(\ulcorner A^* \urcorner)$.

(Actually the connectives are not all treated as primitive in [91] – we have done so here in order to later consider IL-interpretations.) A sentence A is *-*valid*, denoted $\models^* A$, iff under any *-interpretation P $\vdash A^*$. The major result of [91] is that for any $A \in \Phi$,

(17)   K4W $\vdash A$ iff $\models^* A$.

Combining Theorem 4.1 with (17) we then have

Theorem 4.2. *For any $A \in \Phi$,*

$$\text{S4Grz} \vdash A \quad \text{iff} \quad \models^* A°. \qquad \Box$$

Instead of dealing with the translation ° and the above definition of interpretation we could, by combining the two, obtain a characterisation of S4Grz-derivability directly in terms of interpretations in P. The new kind of interpretation would satisfy (12)-(15), and in place of (16) would have

(18) $(\Box A)^* = A^* \wedge \mathrm{Bew}(\ulcorner A^* \urcorner)$.

If $A \mapsto A^*$ satisfies (12)–(16), then the assignment $A \mapsto (A^\circ)^*$ is an interpretation of this new kind. In the other direction, if $A \mapsto A^{*\prime}$ satisfies (12)–(15) and (18), then putting $p_i^* = p_i^{*\prime}$ and extending inductively using (12)–(16) gives a $*$-interpretation $A \mapsto A^*$ that has $A^{*\prime} = (A^\circ)^*$. We leave it to the reader to verify the details of this.

We turn now to the interpretation of IL in P. Let Ψ be the sublanguage of Φ consisting of those sentences containing no occurrence of $\Box$. Define a translation $A \mapsto A^-$ of Ψ into Φ by

(19) $p_i^- = \Box p_i$,

(20) $(A \wedge B)^- = A^- \wedge B^-$,

(21) $(A \vee B)^- = A^- \vee B^-$,

(22) $(\sim A)^- = \Box \sim (A^-)$,

(23) $(A \to B)^- = \Box(A^- \to B^-)$.

Then a result of McKinsey and Tarski [68] is that for any $A \in \Psi$,

(24)   IL $\vdash A$ iff   S4 $\vdash A^-$.

This can be strengthened, as observed by Grzegorczyk [40], to

(25)   IL $\vdash A$ iff   S4Grz $\vdash A^-$.

A straightforward method of proving (24), similar to our derivation of Theorem 4.1 from Lemma 4.1, is given by Fitting [17]. The proof uses the fact that IL and S4 are both determined by the class of frames whose relation R is reflexive and transitive. That this approach yields (25) is immediate from the observation that IL and S4Grz are both determined by the class of finite partially ordered frames (it is the definition of model based on a frame that distinguishes them – cf. Segerberg [84]). Combining Theorem 4.2 with (25) we now have

Theorem 4.3. *For any* $A \in \Psi$,

$$\text{IL} \vdash A \quad \textit{iff} \quad \models^* (A^-)^\circ. \qquad \Box$$

Again the result may be analysed further and expressed in terms of direct interpretations of Ψ in P. This time the interpretations satisfy (12), (13) and

(26) $p_i^* = \alpha_i \wedge \mathrm{Bew}(\ulcorner \alpha_i \urcorner)$,   for some P-sentence α_i,

(27) $(\sim A)^* = \sim (A^*) \wedge \mathrm{Bew}(\ulcorner \sim (A^*) \urcorner)$,

(28) $(A \to B)^* = (A^* \to B^*) \wedge \mathrm{Bew}(\ulcorner A^* \to B^* \urcorner)$.

There are other ways of translating IL into modal logic, and each yields its own mode of interpreting Ψ in P. All of them lead to the same class of valid sentences, viz., the IL-theorems. In the case of the original translation of [20] and the two others in [68], p_i^* can be any P-sentence at all. We leave it to the reader to consult these works to determine how the connectives are treated in these approaches.

4.3   The Case of the Standard Model

A modal sentence $A \in \Phi$ will be called ω^*-*valid*, $\omega \models^* A$, if under all $*$-interpretations A^* is true in the standard P-model $\langle \omega, +, \cdot \rangle$, where ω is the set of non-negative integers. The connection between this notion and $*$-validity is

Lemma 4.2.

(i) $\models^* A$ *only if* $\omega \models^* A$,

(ii) $\models^* A$ *if and only if* $\omega \models^* \Box A$.

Proof.

(i) $P \vdash A^*$ only if A^* is true in ω.

(ii) $P \vdash A^*$ iff $\mathrm{Bew}(\ulcorner A^* \urcorner)$ is true in ω.

$\Box$

Solovay [91] proves that

(29) $\omega \models^* A$ iff $G' \vdash A$,

where G$'$ is the modal system whose theorems form the smallest set of sentences containing all K4W-theorems, and all instances of T, that is closed under Modus Ponens. Combining (29) with Lemma 4.2(ii) and (17) gives the apparently new

Theorem 4.4. K4W $\vdash A$ *iff* $G' \vdash \Box A$. $\Box$

The reader has probably anticipated the next question – which Ψ-sentences are valid in ω when interpreted according to (26)–(28)? He or she may, however, be a little surprised at the answer: precisely the IL-theorems are thus valid. In other words, any Ψ- sentence that is valid in ω is valid in all P-models, and the only Ψ-sentences that are arithmetically necessary in this latter sense are the intuitionistic theorems. To see this we need

Lemma 4.3.

(i) *For any* $A \in \Phi$, $G' \vdash (A \wedge \Box A) \leftrightarrow \Box A$.

(ii) *For any $A \in \Psi$, S4Grz $\vdash A^- \leftrightarrow \Box(A^-)$.*

Proof.

(i) By the schema T, and PC.

(ii) That S4Grz $\vdash \Box(A^-) \to A^-$ is immediate from T. To show that S4Grz $\vdash A^- \to \Box(A^-)$ one can prove by induction that if $\mathcal{M}$ is an S4Grz-model, then if $\mathcal{M} \models_x (A^-)$ and xRy then $\mathcal{M} \models_y (A^-)$. (Alternatively the reader may develop a syntactic proof – the result depends only on S4 principles.)

$\square$

Corollary. *For any $A \in \Psi$,*

$$G' \vdash ((A^-)^\circ) \leftrightarrow \Box((A^-)^\circ).$$

Proof. By part (ii) of the Lemma, Theorem 4.1, and the fact that K4W $\subseteq$ G', we have

$$G' \vdash [A^- \leftrightarrow \Box(A^-)]^\circ,$$

hence

$$G' \vdash (A^-)^\circ \leftrightarrow (\Box(A^-))^\circ,$$

i.e.,

$$G' \vdash (A^-)^\circ \leftrightarrow [((A^-)^\circ) \wedge \Box((A^-)^\circ)].$$

Part (i) of the Lemma then gives the desired result. $\square$

Theorem 4.5. *For any $A \in \Psi$,*

$$\text{IL} \vdash A \quad \textit{iff} \quad \models^* (A^-)^\circ \quad \textit{iff} \quad \omega \models^* (A^-)^\circ.$$

Proof. In view of Theorem 4.3 and Lemma 4.2(i), we need only prove $\omega \models^* (A^-)^\circ$ implies $\models^* (A^-)^\circ$. But if $\omega \models^* (A^-)^\circ$ then $G' \vdash (A^-)^\circ$ (29), whence by the above Corollary and PC we get $G' \vdash \Box((A^-)^\circ)$. Lemma 4.2(ii) and (29) then yield the result as stated. $\square$

In the case of the three other IL-to-S4 translations mentioned earlier, the analogue of Lemma 4.3(ii) does not hold (in particular it fails for propositional letters, which are translated to themselves). However if $A \mapsto A^+$ is one of these translations we do have

$$\text{IL} \vdash A \quad \text{iff} \quad \text{S4Grz} \vdash A^+,$$

and so by Theorem 4.2 and Lemma 4.2(ii), we have

(30) $\text{IL} \vdash A$ iff $\models^* (A^+)^\circ$ iff $\omega \models^* \Box((A^+)^\circ)$.

By Theorem 4.5 and (30) we then have

(31) $\omega \models^* (A^-)^\circ$ iff $\omega \models^* \square((A^+)^\circ)$.

This last result can also be obtained from the fact, noted in [68], that

$$S4 \vdash (A^-) \leftrightarrow \square(A^+),$$

which, reasoning as in the Corollary to Lemma 4.3, yields

$$G' \vdash (A^-)^\circ \leftrightarrow \square((A^+)^\circ),$$

and hence we get (31) by (29) and PC.

Problem

Let $\Delta = \{A \in \Phi : \omega \models^* A^\circ\} = \{A : G' \vdash A^\circ\}$ be the set of modal sentences valid in the standard model under interpretations satisfying (18) in place of (16). Δ is recursive, since G' is, [91, §5], and by Theorem 4.1, S4Grz $\subseteq \Delta$. The problem is to axiomatise Δ by adjoining a finite number of schemata to S4Grz.

Postscript

A number of people have worked on the problems considered here. George Boolos independently proved Theorem 4.1, while a proof that IL $\vdash A$ iff K4W $\vdash (A^-)^\circ$ was previously given by A. Kuznetsov and A. Muzavitski (cf. *Proceedings of the IVth Soviet Union Conference on Mathematical Logic*, Kishiniev, 1976, p. 73, (in Russian)).

The above Problem has also been solved by Boolos (personal communication, April 1979), by showing that Δ is just S4Grz itself! Thus

$$S4Grz \vdash A \quad \text{iff} \quad \omega \models^* A^\circ,$$

or equivalently

$$K4W \vdash A^\circ \quad \text{iff} \quad G' \vdash A^\circ.$$

5

Diodorean Modality in Minkowski Spacetime

ABSTRACT. The Diodorean interpretation of modality reads the operator $\Box$ as "it is now and always will be the case that". In this article time is modelled by the four-dimensional Minkowskian geometry that forms the basis of Einstein's special theory of relativity, with "event" y coming *after* event x just in case a signal can be sent from x to y at a speed *at most* that of the speed of light (so that y is in the causal future of x).

It is shown that the modal sentences valid in this structure are precisely the theorems of the well-known logic S4.2, and that this system axiomatises the logics of two and three dimensional spacetimes as well. Requiring signals to travel slower than light makes no difference to what is valid under the Diodorean interpretation. However if the "is now" part is deleted, so that the temporal ordering becomes irreflexive, then there are sentences that distinguish two and three dimensions, and sentences that can be falsified by approaching the future at the speed of light, but not otherwise.

The Stoic logician Diodorus Chronus described the *necessary* as being that which both *is* and *will always be* the case. This temporal interpretation of modality has been exhaustively investigated by the methods of contemporary formal logic within the context of *linear* temporal orderings (cf. Chapter II of [72] for a survey of this work). The present paper is a contribution to the study of modalities in branching time, and is concerned with the most significant of all non-linear time structures, viz. the four-dimensional Minkowskian spacetime that forms the basis of Einstein's theory of special relativity. Since the temporal ordering of spacetime points is directed (indeed any two have a *least* upper bound) it follows, as observed by Arthur Prior in [72, p. 203], that the associated Diodorean modal logic contains the system S4.2. We shall prove that it is in fact precisely S4.2, and that this holds also for two and three dimensional spacetime.

The language of propositional modal logic comprises sentences constructed from sentence letters $p, q, r, \ldots$ by Boolean connectives and the modal $\square$ ("it will always be"). The connective $\Diamond$ ("it will (at some time) be") is defined as $\sim\square\sim$.

A *time-frame* is a structure $\mathcal{T} = (T, \leq)$ comprising a non-empty set T of times (moments, instants, events) on which $\leq$ is a *reflexive* and *transitive* ordering. A frame is *directed* if any two elements of it have an upper bound, i.e.

for all $t, s \in T$ there exists a $v \in T$ with $t \leq v$ and $s \leq v$.

A $\mathcal{T}$-valuation is a function V assigning to each sentence letter p a set $V(p) \subseteq T$ (the set of times at which p is true). The valuation is then extended to all sentences via the obvious definitions for the Boolean connectives, together with

$$t \in V(\square A) \quad \text{iff} \quad t \leq s \text{ implies } s \in V(A).$$

Hence $t \in V(\Diamond A)$ iff for some $s \in V(A)$, $t \leq s$.

The reflexivity of $\leq$ gives $\square$ the Diodorean "*is* and always will be" interpretation. A sentence A is *valid* in $\mathcal{T}$, $\mathcal{T} \models A$, iff $V(A) = T$ holds for every $\mathcal{T}$-valuation V.

A function $f : T \to T'$ is a *p-morphism* from a frame $\mathcal{T} = (T, \leq)$ to a frame $\mathcal{T}' = (T', \leq')$ if it satisfies

P1: $t \leq s$ implies $f(t) \leq' f(s)$,
P2: $f(t) \leq' v$ implies that there exists some $s \in T$ with $t \leq s$ and $f(s) = v$.

We write $\mathcal{T} \twoheadrightarrow \mathcal{T}'$ to mean that there is a p-morphism from $\mathcal{T}$ to $\mathcal{T}'$ that is *surjective* (onto).

p-Morphism Lemma. *If $\mathcal{T} \twoheadrightarrow \mathcal{T}'$, then for any sentence A, $\mathcal{T} \models A$ only if $\mathcal{T}' \models A$.*

If $T' \subseteq T$ is *future-closed* under $\leq$, i.e. whenever $t \in T'$ and $t \leq s$ we have $s \in T'$, then $\mathcal{T}' = (T', \leq)$ is called a *subframe* of $\mathcal{T}$. By the transitivity of $\leq$, for each t the set $\{s : t \leq s\}$ is the base of a subframe, called the subframe of $\mathcal{T}$ *generated by* t. In general an element 0 of T is called an *initial point* of $\mathcal{T}$ if $0 \leq s$ holds for all $s \in T$. Thus t is an initial point of the subframe generated by t. A frame with an initial point will be simply called a *generated frame*.

Subframe Lemma. *If $\mathcal{T}'$ is a subframe of $\mathcal{T}$, then for any sentence A, $\mathcal{T} \models A$ only if $\mathcal{T}' \models A$.*

The logic S4.2 may be axiomatised as follows;

Axioms: All instances of tautologies, and the schemata

    I $\Box(A \to B) \to (\Box A \to \Box B)$

    II $\Box A \to A$

    III $\Box A \to \Box \Box A$

    IV $\Diamond \Box A \to \Box \Diamond A$.

Rules: Modus Ponens, and Necessitation: *From A derive* $\Box A$.

Axiom I is valid on all frames, as is the rule of Necessitation, re-gardless of the properties of $\leq$. The validity of II depends on reflexivity of $\leq$, III requires transitivity, while IV is valid if $\leq$ is directed. Thus $\vdash_{S4.2} A$ implies that A is valid on all directed frames. The following strong version of the converse to this statement may be found in [86].

Completeness Theorem. *If* $\nvdash_{S4.2} A$, *then there is a finite generated and directed frame* $\mathcal{T}$ *with* $\mathcal{T} \nvDash A$.

We have not required that a frame be *partially ordered*, i.e. that $\leq$ be antisymmetric (indeed there is no sentence whose validity requires it). Thus the equivalence relation defined on T by

$$t \approx s \quad \text{iff} \quad t \leq s \text{ and } s \leq t$$

will in general be non-trivial. The $\approx$-equivalence classes are called the *clusters* of $\mathcal{T}$. They are ordered by putting

$$\hat{t} \leq \hat{s} \quad \text{iff} \quad t \leq s,$$

(where $\hat{t}$ is the cluster containing t etc.), and *this* is an antisymmetric ordering. Thus we may conveniently visualise a frame as a partially-ordered collection of clusters, with the relation $\leq$ being universal within each cluster.

An element ∞ of T is called *final* in $\mathcal{T}$ if $t \leq \infty$ holds for all t in T. All such final points are $\approx$-equivalent and so they form a single cluster. Notice that if $\mathcal{T}$ is directed and *finite* then it must have at least one final point. A unique final point can be adjoined to any frame $\mathcal{T}$ by forming the frame $\mathcal{T}^{\infty} = (T \cup \{\infty\}, \leq)$ where ∞ is some object not a member of T, and the ordering is that of $\mathcal{T}$ extended by

$$\{\langle s, \infty \rangle : s \in T \cup \{\infty\}\}.$$

Notice that $\mathcal{T}^{\infty}$ is always directed, as the final point serves as upper bound for any two elements.

The key to our characterisation of the logic of spacetime is the struc-ture of the infinite binary-branching frame $\mathcal{B} = (B, \leq)$. The members of $\mathcal{B}$ are the finite sequences of the form $x = x_1 x_2 \ldots x_n$, where each

$x_i \in \{0,1\}$. Such a sequence is of *length* n, denoted $l(x) = n$. We include the case $n = 0$, so that B contains the empty sequence $x = \emptyset$. The ordering is defined by specifying that for sequences $x = x_1 x_2 \ldots x_n$ and $y = y_1 y_2 \ldots y_m$ we have

$$x \leq y \quad \text{iff} \quad x \text{ is an initial segment of } y$$
$$\text{iff} \quad n \leq m \text{ and } y = x_1 x_2 \ldots x_n y_{n+1} \ldots y_m.$$

Thus B is partially-ordered, with $\emptyset$ as initial point. The successors $\{y : x \leq y\}$ of x in B are just the sequences that extend x, and so x has exactly two *immediate* successors, viz. $x0$ and $x1$ (cf. Figure 5.1). We shall also refer to $l(x)$ as the *level* of x in B.

In what follows we shall use the abbreviations

$$1^r \quad \text{for} \quad \underbrace{11 \ldots 1}_{r \text{ times}}, \qquad \text{and}$$

$$0^r \quad \text{for} \quad \underbrace{00 \ldots 0}_{r \text{ times}}, \qquad \text{where } r \geq 0.$$

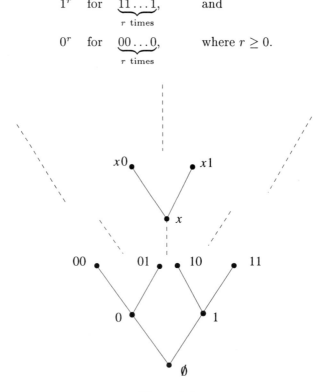

Figure 5.1

The following result is due originally to Dov Gabbay, and was independently discovered by Johan van Benthem. The construction we use in the proof is that devised by the latter.

Theorem 5.1. *If T is any finite generated frame, then $B \twoheadrightarrow T$.*

Proof. We develop inductively an assignment of members of T to the "nodes" of the binary tree B to obtain the desired p-morphism.

Step One: Let 0 be an initial point of the generated frame T. Assign 0 to the initial point $\emptyset$ of B.

Inductive Step: Suppose that $x \in B$ has been assigned an element t of T, but that no B-successor of x has received an assignment. Such a point x that is used to initiate an inductive step will be called a *primary node* of B for the p-morphism being defined.

Now let $t_1, \ldots, t_k$ be all of the $\leq$-future points (i.e. $t \leq t_i$) of t in T. Take the least j such that $k + 1 \leq 2^j$. This j is the *bound* of x: $\beta(x) = j$. Notice that $k \geq 1$, since at least $t \leq t$, and so $j \geq 1$.

Suppose $l(x) = n$. Then we assign t to all B-successors of x up to and including level $n + j$ (cf. Figure 5.2).

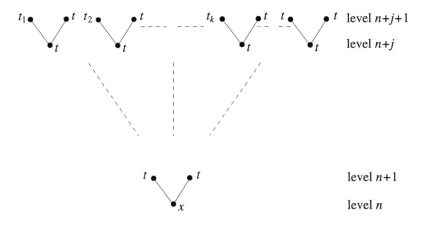

Figure 5.2

Now let $y_1, \ldots, y_k$ be any k of the 2^j successors of x at level $n + j$. Assign t_1 to one of the immediate successors of y_1, and t to the other. Assign t_2 to one of the immediate successors of y_2, and t to the other. Continue this process up to y_k, thereby giving assignments to $2k$ of the 2^{j+1} successors of x at level $n + j + 1$. Let all the other nodes at this level be assigned t (there are such nodes, as $2k < 2^{j+1}$).

The nodes at levels $n+1$ through $n+j$ are designated as *intermediate* nodes for the construction, while the nodes at level $n + j + 1$ are new *primary* nodes. The inductive step is then repeated for each of the latter, and so on. Since $j \geq 1$, the immediate successors of x at level $n + 1$ must receive an assignment (in fact the same one as x). Hence by induction, every member of B gets an assignment, and a function $f : B \to T$ may be defined by letting $f(x)$ be the member of T assigned to x. Since $\{t : 0 \leq t\} = T$, every member of T will be assigned at least once already after the first inductive step, and so f is onto. To prove clause P1 of the p-morphism definition, observe that if $x \leq y$, and $f(x) = t$ say, then y will be assigned a future point of t in $\mathcal{T}$, hence $f(x) \leq f(y)$ (a rigorous argument would proceed by induction on the level of y above x and use the transitivity of $\leq$).

For P2, suppose that $f(x') \leq s$, where $f(x') = t$. If x' is primary at level n, such as the x in Figure 5.2, then there is a point y at level $n + j + 1$ that is assigned s, hence $x' \leq y$ and $f(y) = s$. If however x' is intermediate, then since all points at level $n + j$ have at least one successor at $n + j + 1$ that is assigned t, there will be some such primary node z with $x' \leq z$ and $f(z) = t$. Then by the argument of the previous sentence, there will be a y with $z \leq y$, and hence $x' \leq y$, such that $f(y) = s$. This completes the proof. $\square$

We note in passing that the modal logic S4 has as basis the axioms for S4.2 without the schema IV. It is well known that any non-theorem for S4 is falsifiable on a finite generated (reflexive and transitive) frame, and hence by Theorem 5.1 and the p-Morphism Lemma is falsifiable on $\mathcal{B}$. Thus for any sentence A we have

$$\vdash_{S4} A \quad \text{iff} \quad \mathcal{B} \models A,$$

so that $\mathcal{B}$ is a characteristic frame for S4.

It is apparent that the proof of Theorem 5.1 as given requires only that $k \leq 2^j$. The reason for the stronger constraint is that we have to refine the construction to ensure that f satisfies some combinatorial conditions that will allow us to define a p-morphism on spacetime. In the proof of Theorem 5.1 the chosen nodes $y_1, \ldots, y_k$ at level $n + j$ will be called *special* intermediate points. The other intermediate points are *ordinary*. Then since there are $2^j \geq k + 1$ points above x at level $n + j$;

(a) f can be defined so that for primary x the intermediate node $x0^j$ is ordinary (where $j = \beta(x)$).

We then give the definition of f in the inductive step related to Figure 5.2 quite explicitly as follows:

if z is an intermediate point,

(b) let $f(z) = t = f(x)$;

and if z is at level $n + j$, then

(c) if z is ordinary, let $f(z0) = f(z1) = t = f(x)$, while

(d) if $z = y_i$ is special, let $f(z0) = t_i$ and $f(z1) = t$.

Thus the only case in which an intermediate node has a different assignment to one of its immediate sucessors is when the node is a special point z, and the successor is the primary point $z0$. Moreover in the case of a primary point x the successors $x0$ and $x1$ at level $n + 1$ are intermediate, as $\beta(x) \geq 1$, and so (Figure 5.2) have the same f-value as x. Altogether then we have that for *any* point z in B,

(e) $f(z) = f(z1)$,

and

(f) if z is not a special intermediate point, then $f(z) = f(z0)$.

From (e) we deduce that

(g) for all $z \in B$ and all r, $f(z) = f(z1^r)$.

Next we consider nodes of the form $x0^r$, for primary x. If $r \leq j = \beta(x)$, then $x0^r$ is intermediate and so has the same f-value as x by (b). But by (a), $x0^r$ is not special, so by (f), $f(x0^{j+1}) = f(x0^j) = f(x)$. Since $x0^{j+1}$ is primary, the argument may be repeated up to the next level of primary points, and so by induction,

(h) if x is primary, then $f(x) = f(x0^r)$, for all r.

Lemma 5.2. *For any $x \in B$,*

(i) *if x is special, then $f(x) = f(x10^r)$, all r;   and*

(ii) *otherwise $f(x) = f(x01^r)$, all r.*

Proof. For (i), if x is special then $f(x) = f(x1)$ by (d), and since $x1$ is primary, $f(x1) = f(x10^r)$ by (h).

If however x is not special, then $f(x) = f(x0)$ by (f), and then $f(x0) = f(x01^r)$ by (g).   □

Our next step is to produce a characteristic frame for S4.2 by placing an infinite final cluster at the top of B. Let

$$\Omega = \{\infty_0, \infty_1, \ldots, \infty_n, \ldots\}$$

be an infinite set of objects disjoint from B. Define a frame

$$B^\Omega = (B \cup \Omega, \leq)$$

by taking the ordering $\leq$ to be that of B extended by

$$\{\langle s, \infty_n \rangle : s \in B \cup \Omega \text{ and } n \in \mathbb{N}\}$$

where $\mathbb{N} = \{0, 1, 2, \ldots\}$. Then B^Ω has Ω as its set of final points, with $\infty_n \approx \infty_m$ for all n and m.

Theorem 5.3. *If T is finite directed and generated, then $\mathcal{B}^{\Omega} \twoheadrightarrow T$.*

Proof. By Theorem 5.1 there is a p-morphism $f : \mathcal{B} \twoheadrightarrow T$. We lift this map to $\mathcal{B} \cup \Omega$. Since T is directed it has final points, and these form a (non-empty) cluster, C say. We extend f by mapping Ω *onto* C in any surjective manner. Since the relevant frame orderings are universal within C and Ω, and each of these clusters consists of final points, it is readily seen that such an extension of f yields the desired surjective p-morphism.   □

Applying the Completeness Theorem given above for S4.2 to Theorem 5.3, we deduce

Corollary 5.4. *For any sentence A,*

$$\vdash_{S4.2} A \quad \textit{iff} \quad \mathcal{B}^{\Omega} \models A. \qquad\qquad\qquad □$$

We turn now to the structure of spacetime itself. If $x = \langle x_1, \ldots, x_n \rangle$ is an n-tuple of real numbers, let

$$\mu(x) = x_1^2 + x_2^2 + \ldots + x_{n-1}^2 - x_n^2.$$

Then by *n-dimensional spacetime*, for $n \geq 2$, we mean the frame

$$\mathbb{T}^n = (\mathbb{R}^n, \leq),$$

where $\mathbb{R}^n$ is the set of all real n-tuples, and for x and y in $\mathbb{R}^n$ we have

$$\begin{aligned} x \leq y \quad &\text{iff} \quad \mu(y-x) \leq 0 \text{ and } x_n \leq y_n \\ &\text{iff} \quad \sum_{i=1}^{n-1}(y_i - x_i)^2 \leq (y_n - x_n)^2 \text{ and } x_n \leq y_n. \end{aligned}$$

Then $\mathbb{T}^n$ is a partially-ordered frame, which is directed. As an upper bound of x and y we have, for example, $z = \langle x_1, \ldots, x_{n-1}, z_n \rangle$, where

$$z_n = \sum_{i=1}^{n-1}(x_i - y_i)^2 + |x_n| + |y_n|.$$

Theorem 5.5. $\mathbb{T}^{n+1} \twoheadrightarrow \mathbb{T}^n$.

Proof. Let $f : \langle x_1, \ldots, x_{n+1} \rangle \mapsto \langle x_2, \ldots, x_{n+1} \rangle$ be the (surjective) projection map. Then if $x \leq y \in \mathbb{T}^{n+1}$, we have

$$\sum_{i=1}^{n}(y_i - x_i)^2 \leq (y_{n+1} - x_{n+1})^2 \text{ and } x_{n+1} \leq y_{n+1}.$$

But then as $(y_1 - x_1)^2 \geq 0$,

$$\sum_{i=2}^{n}(y_i - x_i)^2 \leq \sum_{i=1}^{n}(y_i - x_i)^2 \leq (y_{n+1} - x_{n+1})^2$$

and so $f(x) \leq f(y)$ in $\mathbb{T}^n$, establishing P1 for f.

For P2, if $f(x) \leq y$ in $\mathbb{T}^n$, where $x = \langle x_1, \ldots, x_{n+1} \rangle$ and $y = \langle y_2, \ldots, y_{n+1} \rangle$, let $z = \langle x_1, y_2, \ldots, y_{n+1} \rangle \in \mathbb{R}^{n+1}$. Then $z_1 - x_1 = 0$, so

$$\sum_{i=1}^{n}(z_i - x_i)^2 = \sum_{i=2}^{n}(z_i - x_i)^2$$
$$= \sum_{i=2}^{n}(y_i - x_i)^2$$
$$\leq (y_{n+1} - x_{n+1})$$
$$= (z_{n+1} - x_{n+1}),$$

and $x_{n+1} \leq y_{n+1} = z_{n+1}$. Thus $x \leq z$, and by definition $f(z) = y$. Therefore f is the desired p-morphism.   □

Minkowski spacetime is $\mathbb{T}^4$. The intended interpretation of $x \leq y$ is that a signal can be sent from "event" x to "event" y at a speed at most that of the speed of light, and so y is in the "causal" future of x (assuming a choice of coordinates that gives the speed of light as one unit of distance per unit of time).

The frame $\mathbb{T}^2$ is depicted in Figure 5.3. For each point $t = \langle x, y \rangle$ in the plane, the future consists of all points on or above the upwardly directed rays of slopes $+1$ and -1 emanating from t.

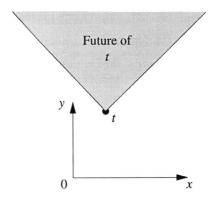

Figure 5.3

By performing the isometry of rotating the plane clockwise through $45°$ about the origin 0, the picture becomes that of Figure 5.4, in which the future points of t are precisely those above and to the right of t. The rotation is a bijective p-morphism (isomorphism of frames) and so from now on we will identify $\mathbb{T}^2$ with the structure of Figure 5.4. This is done largely to make the constructions to follow more tractable, but notice that it reveals $\mathbb{T}^2$ as the direct product of the real linear frame $(\mathbb{R}, \leq)$ with itself, as we now have

(∗) $\langle x_1, y_1 \rangle \leq \langle x_2, y_2 \rangle$ iff $x_1 \leq x_2$ and $y_1 \leq y_2$.

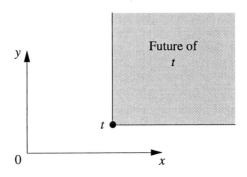

Figure 5.4

Now let $\mathbb{T}_0^2 = \{t : 0 \leq t\}$ be the "first quadrant" of the plane, consisting of all points with non-negative coordinates. A *future-open box* in $\mathbb{T}_0^2$ is a subset of the form $[a, b) \times [c, d)$ (cf. Figure 5.5).

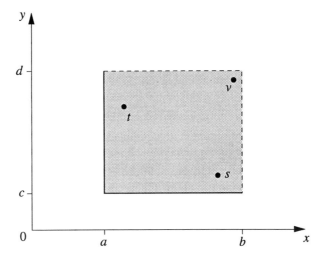

Figure 5.5

Notice that any two members t, s of a future-open box have an upper bound v within the box, and that v may be chosen to lie on the diagonal line joining $\langle a, c \rangle$ to $\langle b, d \rangle$.

Theorem 5.6. *Any future-open box is temporally isomorphic to* $\mathbb{T}_0^2$.

Proof. It is a fact of classical analysis that there is a bijection $f : [a, b) \mapsto [0, \infty) = \{e : 0 \leq e\}$ that preserves order, i.e. has $x \leq y$ iff $f(x) \leq f(y)$. Figure 5.6 displays one method of geometrically constructing f.

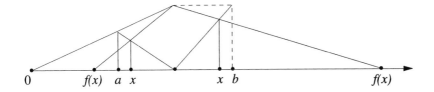

Figure 5.6

Likewise, there is an order-isomorphism $g : [c, d) \to [0, \infty)$. Then the map $\langle x, y \rangle \mapsto \langle f(x), g(y) \rangle$ gives a bijection between $[a, b) \times [c, d)$ and $\mathbb{T}_0^2$ that preserves the temporal ordering defined on each by $(*)$. □

Corollary 5.7. *Any two future-open boxes are temporally isomorphic.*

□

From now on we focus on the structure of the *unit box*

$$I = [0, 1) \times [0, 1).$$

Theorem 5.8. $I \twoheadrightarrow \Omega$.

Proof. Here Ω is considered as a frame in its own right, consisting of an infinite set of points all related to each other by $\leq$. The idea of the proof is that each ∞_n is made to correspond to a subset A_n of I that is *cofinal* with I, i.e.

for each $t \in I$ there is some $s \in A_n$ with $t \leq s$.

We can do this by making rational cofinal assignments up the diagonal of I to $\infty_1, \infty_2, \ldots$, and mapping everything else to ∞_0 (Figure 5.7). Thus we map $\langle \frac{1}{2}, \frac{1}{2} \rangle, \langle \frac{3}{4}, \frac{3}{4} \rangle, \ldots$ to ∞_1; $\langle \frac{2}{3}, \frac{2}{3} \rangle, \langle \frac{8}{9}, \frac{8}{9} \rangle, \ldots$ to ∞_2; $\langle \frac{4}{5}, \frac{4}{5} \rangle, \langle \frac{24}{25}, \frac{24}{25} \rangle, \ldots$ to ∞_3; and so on. Formally, let $\pi_1, \pi_2, \ldots, \pi_n, \ldots$ be a listing without repetition of the prime numbers in order of increasing magnitude, starting with $\pi_1 = 2$. Then if $\langle x, y \rangle \in I$,

(i) if $x = y = 1 - \frac{1}{(\pi_n)^k}$ for some $k \geq 1$, put $f(\langle x, y \rangle) = \infty_n$,

and

(ii) otherwise put $f(\langle x, y \rangle) = \infty_0$.

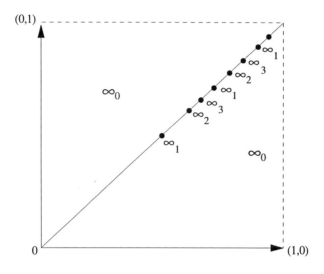

Figure 5.7

That P1 holds for f is immediate, as Ω is a cluster. But the cofinality of the ∞_n-assignments along the diagonal, together with the fact that each point t in I has $\leq$-successors on the diagonal, guarantees that $A_n = f^{-1}(\infty_n)$ is cofinal with I. This cofinality ensures that f satisfies P2. $\qquad \square$

Theorem 5.9. *If $\mathcal{T}$ is finite generated and directed, then $I \twoheadrightarrow \mathcal{T}$.*

Proof. By Theorem 5.3 there exists a p-morphism $f : \mathcal{B}^\Omega \twoheadrightarrow \mathcal{T}$. We define a map $g : I \to B \cup \Omega$ which will compose with f to give the desired result. This is done by assigning each point in $B \cup \{\infty\}$ a future-open box contained in I, through a series of *temporary* and then *permanent* labellings.

Step One: Temporarily assign the initial point $\emptyset$ of $\mathcal{B}$ to I.

Inductive Step: Suppose $x \in B$ has been temporarily assigned a box within I. Divide this box into four equal future-open boxes (Figure 5.8). *Permanently* assign the lower left-hand box to x and the upper right-hand one to ∞. *Temporarily* assign the upper left-hand box to $x0$ and the lower right-hand one to $x1$.

When all members of B have inductively received *permanent* assignments, the picture is as in Figure 5.9.

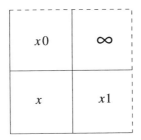

Figure 5.8

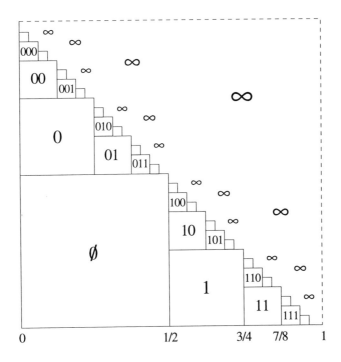

Figure 5.9

It is apparent that

(∗∗) if $z \le y$ in $\mathcal{B}$, then the box permanently assigned y lies inside the one temporarily assigned z.

Lemma 5.10. *If $t \in I$ belongs to the box assigned $x \in B$, then there is some $z \in B$ with $f(x) = f(z)$, and such that the box assigned z lies entirely inside the I-future of t.*

Proof. As indicated in Figure 5.10, by taking r large enough we can ensure that the boxes assigned $z_1 = x01^r$ and $z_2 = x10^r$ both lie inside the future of t.

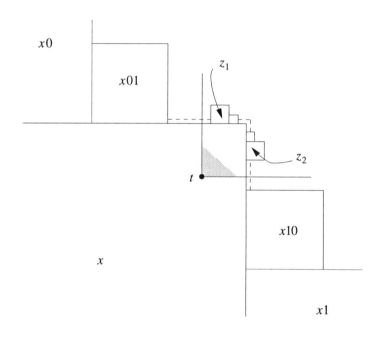

Figure 5.10

Then by Lemma 5.2, if x is a special point for the construction of f as in Theorem 5.1, we may take $z = z_2$ to fulfill Lemma 5.10, while if x is not special, $z = z_1$ meets our requirements.   □

To continue with the proof of Theorem 5.9, we define a map $g : I \to B \cup \Omega$ as follows:

(i) the members of the future-open box permanently assigned $x \in B$ in Figure 5.9 are all mapped to x by g.

(ii) each box assigned ∞ in Figure 5.9 is mapped p-morphically onto Ω by g. This is done by the method of Theorem 5.8, noting Corollary 5.7.

Next a surjective map $h : I \rightarrow \mathcal{T}$ is defined by putting $h(t) = f(g(t))$, for all $t \in I$. To show that h satisfies P1, suppose $t \leq s$ in I. Then if $h(s)$ is final in $\mathcal{T}$, immediately $h(t) \leq h(s)$. If $h(s) = f(g(s))$ is not final, then (cf. proof of Theorem 5.3) $g(s) \notin \Omega$ and so $g(s) \in B$. But since $t \leq s$, the permanent B-assignment to s will be a sequence extending the one assigned to t (Figure 5.8), i.e. $g(t) \leq g(s)$. But then $f(g(t)) \leq f(g(s))$, as f satisfies P1.

For P2, suppose $h(t) = f(g(t)) \leq v$ in $\mathcal{T}$. If $g(t)$ is a member of B, then by Lemma 5.10 there exists some $z \in B$ that is assigned a box entirely inside the future of t and that has $f(z) = f(g(t)) \leq v$. Since f satisfies P2, there is some $y \in B$ with $z \leq y$ and $f(y) = v$. But then (cf. (**)) the box assigned y also lies inside the future of t, and so if we choose an element s from this box, so that $g(s) = y$, we have $t \leq s$ and $h(s) = f(y) = v$. On the other hand, if $g(t) \in \Omega$, then $h(t)$ is final in $\mathcal{T}$, and therefore so is v, hence $v = f(\infty_n)$ for some n. But by the definition of g, t belongs to a box that is assigned ∞ in Figure 5.9, and this box is mapped p-morphically onto Ω by g. Hence there is some s in this box with $t \leq s$ and $g(s) = \infty_n$, so $h(s) = f(\infty_n) = v$ as required.

This completes the proof of Theorem 5.9. $\qquad \Box$

Theorem 5.11. *For any sentence A,*

$$\vdash_{\text{S4.2}} A \quad \textit{iff} \quad \mathbb{T}^n \models A \quad \textit{iff} \quad I \models A.$$

Proof. If $\vdash_{\text{S4.2}} A$, then A is valid on all directed frames and thus in particular on $\mathbb{T}^n$. But if $\mathbb{T}^n \models A$, application $(n - 2$ times$)$ of the p-Morphism Lemma to Theorem 5.5 gives $\mathbb{T}^2 \models A$. The Subframe Lemma then gives $\mathbb{T}^2_0 \models A$, which in turn by Theorem 5.6 yields $I \models A$. To complete the cycle of implications, observe by Theorem 5.9 that if $I \models A$ then A is valid on all finite generated and directed frames, and so by the Completeness Theorem given earlier, A is a theorem of S4.2. $\qquad \Box$

Slower-Than-Light Signals

In $\mathbb{T}^n$, define

$$x \prec y \quad \text{iff} \quad \mu(y - x) < 0 \text{ and } x_n < y_n.$$

Then $x \prec y$ holds just in case a signal can be sent from x to y at less than the speed of light. The reflexive relation

$$x \preccurlyeq y \quad \text{iff} \quad x = y \text{ or } x \prec y$$

yields the same logic as before – we leave it to the reader to analyse the above proof to verify that the valid sentences on $(\mathbb{T}^n, \preccurlyeq)$ are precisely the S4.2 theorems.

The End of Time

Amongst the possible future fates of our universe is that expansion will eventually give way to contraction and collapse to a singularity. In this event, any future-oriented path in spacetime will come to an end (the singularity). Formally, this corresponds to the frame condition

(†) $\forall x \exists y (x \leq y \ \& \ \forall w(y \leq w \rightarrow y = w))$.

In a directed partially-ordered frame there can be only one y as in (†), namely a unique final point, for if y has no successors then an upper bound for y and any other point can only be y itself.

The logic K2 extends the system S4.2 by the additional axiom schema

$$\square \lozenge A \rightarrow \lozenge \square A,$$

which is valid on frames satisfying (†). Conversely, the work of Segerberg [83] may be used to show:

> If A is not a K2-theorem, then A can be falsified on a finite generated directed frame whose final cluster has only one member.

Thus K2 is characterised by the finite generated directed frames with a *unique* final point. Any such frame $\mathcal{T}$ is a p-morphic image of I^∞, as may be deduced from $I \twoheadrightarrow \mathcal{T}$. Indeed any p-morphism $\mathcal{T}_1 \twoheadrightarrow \mathcal{T}$ can be lifted to $\mathcal{T}_1^\infty \twoheadrightarrow \mathcal{T}$ by mapping ∞ to the unique final point of $\mathcal{T}$. We leave it to the reader to use that observation to verify, for any sentence A, that

$$\vdash_{\text{K2}} A \quad \text{iff} \quad (\mathbb{T}^n)^\infty \models A \quad \text{iff} \quad I^\infty \models A \quad \text{iff} \quad \mathcal{B}^\infty \models A.$$

Irreflexive Time

Tense logic, as a branch of modal logic, is generally taken to be concerned with *irreflexive* orderings, so that a point is not considered to be in its own future. In spacetime there are two natural *strict* orderings, viz. the relation

$$x \prec y \quad \text{iff} \quad \mu(y - x) < 0 \text{ and } x_n < y_n$$

defined earlier, and

$$x \alpha y \quad \text{iff} \quad x \neq y \text{ and } x \leq y.$$

(α is the relation "after" axiomatised by Robb in [75]).

The logic of these two orderings can be distinguished in terms of the validity of modal sentences. There may be two propositions A and B that are true in the future at two points that can only be reached by travelling (in opposite directions) at the speed of light (cf. Figure 5.11).

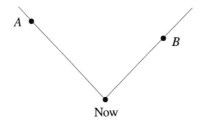

Figure 5.11

In this situation, $\Diamond A \wedge \Diamond B$ will be true now, but never again, and hence the sentence

$$\Diamond A \wedge \Diamond B \rightarrow \Diamond(\Diamond A \wedge \Diamond B)$$

is not valid when α is the temporal ordering. It is however valid under $\prec$, since a slower-than-light journey can always be made to go faster, so we could wait some time and then travel at a greater speed to A and B (Figure 5.12).

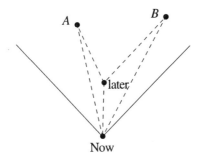

Figure 5.12

These observations apply to $\mathbb{T}^n$ for all $n \geq 2$. However by pushing the idea a little further we can produce a sentence whose truth is dimension-dependent. For, in *three*-dimensional spacetime we can find at least three points that can only be reached by travelling in different directions at the speed of light. In $\mathbb{T}^3$, the future of t is represented by the upper half of a right circular cone centered on t (Figure 5.13).

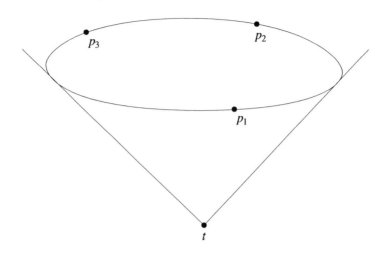

Figure 5.13

Thus in $(\mathbb{T}^3, \alpha)$, and indeed in $(\mathbb{T}^n, \alpha)$ for $n \geq 3$, we can falsify the following sentence (here i and j range over $\{1, 2, 3\}$):

$$(\bigwedge_i \Diamond p_i) \wedge (\bigwedge_{i \neq j} \Box(p_i \rightarrow \sim p_j \wedge \sim \Diamond p_j)) \rightarrow \bigvee_{i \neq j} (\Diamond(\Diamond p_i \wedge \Diamond p_j)).$$

However this sentence is valid under $\prec$ for all $n \geq 2$, and is valid under α in $\mathbb{T}^2$.

Problems

1. Axiomatise the logics corresponding to α and to $\prec$ in the various dimensions.

2. Analyse the logic of *discrete* spacetime, i.e. when $\mathbb{R}$ is replaced by the set $\mathbb{Z}$ of integers .

Notes

I am very much indebted to Johan van Benthem for a stimulating and fruitful dialogue, without which I doubt that I would ever have completed this jigsaw puzzle.

The fact that S4.2 is the logic of the direct product of the real linear frame $(\mathbb{R}, \leq)$ with itself was discovered independently by Valentin Shehtman (cf. [89]). Theorem 5.1 was also proved by A. G. Dragalin. More details of these other works are given in the Editor's footnote to [22].

6

Grothendieck Topology as
Geometric Modality

> *A Grothendieck "topology" appears most naturally as a modal operator of the nature "it is locally the case that".*
>
> F. W. LAWVERE

6.1 Introduction

The language of propositional modal logic extends that of ordinary logic by the additional of a single unary connective. This connective has a long history of investigation in terms of modal interpretations of philosophical interest, such as "it is necessarily the case that" (alethic mode), "it ought to be the case that" (deontic mode), "it is known that" (epistemic), "it will always be that" (temporal), and so on. Recently a number of interpretations have been studied that are of more mathematical concern. Thus we have "it is provable (or true and provable) in Peano arithmetic that" (cf. [91] or [21]), and "whenever a certain program terminates, it is the case that" (cf. [16]).

The present article is a contribution to the study of mathematical modalities, and is concerned with what might be called the *geometric* mode. The above quotation comes from the address [55] at which LAWVERE first announced the results of his work, with M. TIERNEY, on axiomatic sheaf theory—an elementary (first order) treatment of the notion of a Grothendieck topology on a category and its attendant category of sheaves. In order to elucidate this claim we shall interpret the formal modal language within *elementary sites*. An elementary site comprises a topos $\mathcal{E}$ with a topology $\Omega \xrightarrow{j} \Omega$. $\mathcal{E}$ may be thought of as a generalised universe of (perhaps non-extensional) sets, with Ω its "set" of truth-values, and j a unary operator on truth-values—hence a suitable entity for interpreting a modal connective. Using the "logic" of $\mathcal{E}$, a notion of validity is defined, and by adapting the set-theoretic techniques now generally employed in the study of intensional logics we are able to axiomatise the class of sentences valid on all sites and establish that it is recursive.

In order to carry out the program just sketched a Kripke-style semantics will be developed on the basis of a wide-ranging discussion of concepts of "local truth". Since the logic of topoi is in general intuitionistic, we find ourselves working in non-classical modal logic. This area has been explored to some extent before. For instance, BULL [7] provides a philosophically motivated account of the alethic mode in relation to intuitionism. Our approach to intuitionistic logic will however be to see it, not so much as an exegesis of a constructivist theory of the meaning of mathematical statements, but rather as the group of "laws" that arise when one makes the natural generalisation of the "algebra of sets" to the context of the open-set lattices associated with topological spaces. Whether or not the models discussed below have any relevance to intuitionism is another matter. For the present we regard this as an exercise in classical mathematics, dealing with abstract analogues of topological ideas. The main lesson we take from what follows is that whereas GROTHENDIECK's original definition of a "topology" on a category arose by an abstraction of the category-theoretic properties of the category of open subsets of a space, the concept may also be seen to stem from a generalisation of the concepts of "nearness" and "neighbourhood" from classical topology.

The author is indebted to Professors S. K. THOMASON and DANA SCOTT for the hospitality he received during the preparation and writing of this article, which took place initially at the Mathematics Department of Simon Fraser University, and later at the Mathematical Institute, Oxford.

Note on Numbering: As with other chapters, sections are numbered globally, so that the third section of this sixth chapter is numbered 6.3. But we drop the chapter number from items within sections, so that the fourth item of the third section is numbered 3.4, rather than 6.3.4.

6.2   Resume of Propositional Models

We shall be dealing throughout with an object language in which sentences are constructed from a denumerable set Ψ_0 of sentence letters, together with the constant $\perp$ (False), by means of the connectives $\wedge$

(and), $\vee$ (or), $\rightarrow$ (implies) and ∇ (it is locally the case that). Negation, the biconditional, and the constant $\top$ are defined by

$$\sim A = A \rightarrow \bot, \qquad A \equiv B = (A \rightarrow B) \wedge (B \rightarrow A), \qquad \top = \bot \rightarrow \bot.$$

The set of all sentences will be denoted Ψ. Φ denotes the subset of all non-modal sentences, i.e. those with no occurrence of ∇.

The Kripke semantics for Φ-sentences employs as its basic structure a partially-ordered set (poset) $\mathcal{P} = (P, \sqsubseteq)$. For each $p \in P$ we put $[p) = \{q \in P : p \sqsubseteq q\}$. A subset $S \subseteq P$ is $\mathcal{P}$-*hereditary* if it is closed upwards under $\sqsubseteq$, i.e. if $p \in S$ implies $[p) \subseteq S$. We put

$$\mathcal{P}^+ = \{S \subseteq P : S \text{ is } \mathcal{P}\text{-hereditary}\}.$$

A *model based on* $\mathcal{P}$ is a pair $\mathcal{M} = (\mathcal{P}, V)$, where $V : \Psi_0 \rightarrow \mathcal{P}^+$ is a $\mathcal{P}$-*valuation* assigning to each sentence letter $\pi \in \Psi_0$ a $\mathcal{P}$-hereditary set $V(\pi) \subseteq P$. The notion of a non-modal sentence $A \in \Phi$ being *true at a point p in* $\mathcal{M}$, written $\mathcal{M} \models_p A$, is defined inductively by:

(2.1) $\mathcal{M} \models_p \pi$ iff $p \in V(\pi)$,

(2.2) not $\mathcal{M} \models_p \bot$,

(2.3) $\mathcal{M} \models_p A \wedge B$ iff $\mathcal{M} \models_p A$ and $\mathcal{M} \models_p B$,

(2.4) $\mathcal{M} \models_p A \vee B$ iff $\mathcal{M} \models_p A$ or $\mathcal{M} \models_p B$,

(2.5) $\mathcal{M} \models_p A \rightarrow B$ iff $p \sqsubseteq q$ implies that $\mathcal{M} \models_q A$ only if $\mathcal{M} \models_q B$.

Hence we also have

(2.6) $\mathcal{M} \models_p \sim A$ iff $p \sqsubseteq q$ implies that not $\mathcal{M} \models_q A$.

Putting $\mathcal{M}(A) = \{p : \mathcal{M} \models A\}$, these clauses become

(2.7) $\mathcal{M}(\pi) = V(\pi)$,

(2.8) $\mathcal{M}(\bot) = \emptyset$,

(2.9) $\mathcal{M}(A \wedge B) = \mathcal{M}(A) \cap \mathcal{M}(B)$,

(2.10) $\mathcal{M}(A \vee B) = \mathcal{M}(A) \cup \mathcal{M}(B)$,

(2.11) $\mathcal{M}(A \rightarrow B) = \mathcal{M}(A) \Mapsto \mathcal{M}(B)$,

(2.12) $\mathcal{M}(\sim A) = \neg \mathcal{M}(A)$,

where for $S, T \in \mathcal{P}^+$

(2.13) $S \Mapsto T = \{p : [p) \cap S \subseteq T\}$,

and

(2.14) $\neg S = S \Mapsto \emptyset = \{p : [p) \cap S = \emptyset\}$.

The set $\mathcal{P}^+$ contains $\emptyset$ and is closed under the operations $\cap, \cup, \Mapsto$. From this a straightforward induction shows that $\mathcal{M}(A) \in \mathcal{P}^+$, for all $A \in \Phi$ [84, Lemma 2.1]. Sentence A is *true in* $\mathcal{M}$, $\mathcal{M} \models A$, if $\mathcal{M} \models_p A$ for all

p (i.e. $\mathcal{M} \models A$ iff $\mathcal{M}(A) = P$). A is *valid on the poset* $\mathcal{P}$, $\mathcal{P} \models A$, if $\mathcal{M} \models A$ holds for all models $\mathcal{M}$ based on $\mathcal{P}$.

The sentences valid on all posets are precisely the theorems of intuitionistic logic. The requirement that $\mathcal{M}(A)$ be $\mathcal{P}$-hereditary is needed to ensure that all such theorems are $\mathcal{P}$-valid.

Algebraic semantics for Φ uses the notion of a *Heyting algebra* (HA). This is a structure $\mathcal{H} = (H, \sqcap, \sqcup, \Rightarrow, 0)$, where $(H, \sqcap, \sqcup)$ is a lattice with least element 0, and $\Rightarrow$ is an operation of relative pseudo-complementation, satisfying

$$x \sqcap y \sqsubseteq z \quad \text{iff} \quad x \sqsubseteq y \Rightarrow z,$$

where $\sqsubseteq$ is the usual lattice ordering given by

$$x \sqsubseteq y \quad \text{iff} \quad x \sqcap y = x \quad \text{iff} \quad x \sqcup y = y.$$

The element $x \Rightarrow y$ is called the *relative pseudo-complement*(r.p.c.) *of* x *in* y. It is the greatest member of the set $\{z : x \sqcap z \sqsubseteq y\}$.

Any HA has a greatest element 1, with $1 = (x \Rightarrow x)$ for any $x \in H$. An $\mathcal{H}$-*valuation* is a function $V : \Psi_0 \to H$. Such a function is lifted canonically to $V : \Phi \to H$ by stipulating that

(2.15) $V(\bot) = 0$,
(2.16) $V(A \wedge B) = V(A) \sqcap V(B)$,
(2.17) $V(A \vee B) = V(A) \sqcup V(B)$,
(2.18) $V(A \to B) = V(A) \Rightarrow V(B)$,

and hence

$$V(\sim A) = \neg V(A),$$

where $\neg : H \to H$ is defined by $\neg x = x \Rightarrow 0$ (x is called the *pseudo-complement* of x).

Sentence A is $\mathcal{H}$-*valid*, $\mathcal{H} \models A$, if $V(A) = 1$ for every $\mathcal{H}$-valuation V.

Now given a poset $\mathcal{P}$, we obtain the corresponding algebra

$$(\mathcal{P}^+, \cap, \cup, \Rrightarrow, \emptyset)$$

of hereditary sets as above. $\mathcal{P}^+$ is in fact an HA with greatest element P. The $\mathcal{P}^+$-valuations $V : \Psi_0 \to \mathcal{P}^+$ correspond exactly to the models $\mathcal{M} = (\mathcal{P}, V)$ based on $\mathcal{P}$. When such a valuation is extended to $V : \Phi \to \mathcal{P}^+$ by clauses (2.15)–(2.18) we find that for all $A \in \Phi, \mathcal{M}(A) = V(A)$ (using induction on (2.7)–(2.11)). Hence $\mathcal{M} \models A$ iff $V(A) = P = 1$. Since this is true for all V, we have

(2.19) $\mathcal{P} \models A$ iff $\mathcal{P}^+ \models A$, for all $A \in \Phi$.

From this it is immediate that a sentence valid on all HAs will be valid on all posets and hence be an intuitionistic theorem. That all such theorems are HA-valid is established by a routine analysis of the properties of HAs (cf. [74]).

The construction of $\mathcal{P}^+$ assigns to each poset a semantically equivalent HA. In general there is only a partial converse to this. From an HA $\mathcal{H}$ we obtain via the representation theory of Stone [92] the poset $\mathcal{H}_+ = (\mathcal{P}_\mathcal{H}, \subseteq)$, where $\mathcal{P}_\mathcal{H}$ is the set of prime filters in $\mathcal{H}$, and $\subseteq$ is set inclusion. The function $x \mapsto \{p \in \mathcal{P}_\mathcal{H} : x \in p\}$ is an isomorphism between $\mathcal{H}$ and a subalgebra of the HA $(\mathcal{H}_+)^+$ of $\mathcal{H}_+$-hereditary sets. From this it can be shown that $\mathcal{H}_+ \models A$ implies $\mathcal{H} \models A$, but the converse need not obtain. The construction does however give a proof that a sentence valid on all posets will be valid on all HAs.

The class of Heyting algebras includes all Boolean algebras (BAs) amongst its members. In the BA $(2^X, \cap, \cup)$ of all subsets of a set X, the r.p.c. operation is given by $-S \cup T$, where $-S = -S \cup \emptyset = \neg S$ is the usual set complement $\{x \in X : x \notin S\}$. Now if $\mathcal{D}$ is the class of open sets of a topology on X, the lattice $(\mathcal{D}, \cap, \cup)$ will not in general be closed under the operations $-S \cup T$ or $-S$. If however we consider their nearest approximations in $\mathcal{D}$ by defining $S \Rightarrow T = (-S \cup T)^\circ$, and hence $\neg S = (-S)^\circ$, where $(\ )^\circ$ denotes the topological *interior* operation associated with $\mathcal{D}$, then $(\mathcal{D}, \cap, \cup, \Rightarrow, \emptyset)$ proves to be an HA. In this way HA's are seen to be the natural generalisation to the topological context of the Boolean algebra of subsets of a given set.

The collection $\mathcal{P}^+$ is in fact a topology on P, in terms of which we find that $S \mapsto T = (-S \cup T)^\circ$, and hence $\neg S = (-S)^\circ$, where $\mapsto$ and $\neg$ are as defined by (2.13) and (2.14). Thus $S \mapsto T$ is the largest $\mathcal{P}$-hereditary subset of $-S \cup T$ and $\neg S$ is the largest hereditary set disjoint from S.

$\mathcal{P}^+$ is a rather special topology in that it is $\bigcap$-closed, i.e. closed under arbitrary intersections. Also since $p \in [p) \in \mathcal{P}^+$, and $p \sqsubseteq q$ iff $q \in [p)$, we find via the antisymmetry of $\sqsubseteq$ that $\mathcal{P}$ is a T_0 topology (distinct points do not have identical neighbourhoods). Moreover these facts lead us to observe that

$$p \sqsubseteq q \quad \text{iff} \quad p \in S \in \mathcal{P}^+ \text{ implies } q \in S,$$

and so the collection $\mathcal{P}^+$ determines the partial-ordering $\sqsubseteq$.

Conversely, if $\mathcal{D}$ is any $\bigcap$-closed topology on P, by putting $\mathcal{D}_p = \{S \in \mathcal{D} : p \in S\}$ we may *define* a relation $\sqsubseteq$ by

(2.20) $\quad p \sqsubseteq q \quad \text{iff} \quad q \in \bigcap \mathcal{D}_p \quad \text{iff} \quad \mathcal{D}_p \subseteq \mathcal{D}_q,$

so that $[p) = \bigcap \mathcal{D}_p \in \mathcal{D}$. This relation is reflexive and transitive, and will be antisymmetric iff $\mathcal{D}_p = \mathcal{D}_q$ implies $p = q$, which is precisely the T_0 condition. It is clear from (2.20) that the members of $\mathcal{D}$ are

$\sqsubseteq$-hereditary, so that $\mathcal{D} \subseteq \mathcal{P}^+$. But if S is $\mathcal{P}^+$-hereditary, we have $S = \bigcup\{[p) : p \in S\} \in \mathcal{D}$ (since $[p) \in \mathcal{D}$ and $\mathcal{D}$ is closed under arbitrary unions, being a topology), yielding $\mathcal{D} = \mathcal{P}^+$.

Altogether then there is a bijective correspondence between partial-orderings and $\bigcap$-closed T_0 topologies on any set. The T_0 condition (anti-symmetry) is not in fact essential to the model theory of Φ. It is however possessed by most important models, and may always be imposed, by passing to a quotient, without affecting the validity of Φ-sentences. Since it leads to a cleaner theory we shall assume it throughout.

6.3 Frames and Monads

There are two (at least) senses in which the word "local" is employed in topology, and an analysis of these will lead us to extend the modelling of Section 6.2 to provide an interpretation of the connective ∇.

6.3.1 Germs and Monads

Two functions f and g are said to be equivalent, or to have the same germ, at a point p in the intersection of their domains if there is a neighbourhood of p on which f and g assign the same values. Thus f and g have the same germ at p when the statement "$f = g$" is *locally true* at p, i.e. true throughout some neighbourhood of p. Intuitively this conveys the idea that f and g assign the same values to point "close" to p.

Similarly, two sets S and T are equivalent at p if there is some neighbourhood U of p with $S \cap U = T \cap U$. Again this means that S and T are locally equal at p, i.e. that "$S = T$" is true when relativised to points close to p.

From these examples we extract the principles

(3.1) A is locally true at p iff A is true at all points close to p, and

(3.2) A is locally true at p iff A is true throughout some neighbourhood of p.

Informally, (3.1) and (3.2) are equivalent: a neighbourhood of p is any set containing all points close to p, while the points close to p are just those belonging to all neighbourhoods of p. Formally however there are, in any topological space that is at least T_1, *no* points that are close to p in this sense, and so we have to resort to formulation (3.2) (T_1 means that any point distinct from p lies outside at least one p-neighbourhood). Thus whereas the germ of S at p is intuitively a subset of S, namely the collection of points in S that are close to p, formally this germ is defined as the set of all sets that are equivalent to S at p. This situation can be remedied in topological theories that admit infinitesimals. In the work of Robinson [77] the *germ of S at p* is the intersection of S with the *monad* of p, the latter being the set of points infinitely close to p.

6.3.2 Frames

We now introduce the concept of a *frame* as a structure $\mathcal{P} = (P, \sqsubseteq, \mu)$ comprising a poset on which there is a function $\mu : P \to 2^P$ assigning to each $p \in P$ a subset $\mu(p) \subseteq P$ (the *monad* of p) such that

(3.3) $p \sqsubseteq q$ implies $\mu(q) \subseteq \mu(p)$.

Writing $p \prec q$ for $q \in \mu(p)$, (3.3) becomes

(3.4) $p \sqsubseteq q \prec r$ implies $p \prec r$.

The notion of a model $\mathcal{M} = (\mathcal{P}, V)$ based on $\mathcal{P}$ is defined as in Section 6.2, and truth in $\mathcal{M}$ at a point is defined for all Ψ-sentences using (2.1)–(2.5) and the new clause

(3.5) $\mathcal{M} \models_p \nabla A$ iff $\mu(p) \subseteq \mathcal{M}(A)$,

or equivalently

(3.6) $\mathcal{M} \models_p \nabla A$ iff $p \prec q$ implies $\mathcal{M} \models_q A$,

which formalises principle (3.1).

Defining $j_\mu : \mathcal{P}^+ \to \mathcal{P}^+$ by $j_\mu(S) = \{p : \mu(p) \subseteq S\}$—which does indeed make $j_\mu(S)$ $\mathcal{P}$-hereditary by (3.3)—we find that (cf. (3.5))

(3.7) $\mathcal{M}(\nabla A) = j_\mu(\mathcal{M}(A))$,

and from this it can be shown that $\mathcal{M}(A) \in \mathcal{P}^+$, for all $A \in \Psi$.

The notions $\mathcal{M} \models A$ and $\mathcal{P} \models A$ of truth in a model and validity on a frame are again as for posets. Amongst the sentences that are valid on all frames we cite

(3.8) $\nabla(A \to B) \to (\nabla A \to \nabla B)$,
(3.9) $\nabla A \wedge \nabla B \to \nabla(A \wedge B)$,
(3.10) $\nabla A \vee \nabla B \to \nabla(A \vee B)$,

while amongst the validity-preserving rules there are

(3.11) if $\mathcal{P} \models A$ then $\mathcal{P} \models \nabla A$,

and

(3.12) if $\mathcal{P} \models A \to B$ then $\mathcal{P} \models \nabla A \to \nabla B$.

The verification of these facts is left to the reader.

6.3.3 Increasing Frames

We say that μ, or more loosely $\mathcal{P}$, is *increasing* if it satisfies

(3.13) $\mu(p) \subseteq [p)$, for all $p \in P$,

i.e.

(3.14) $p \prec q$ implies $p \sqsubseteq q$.

Then we have

(3.15) $\mathcal{P}$ *is increasing iff* $\mathcal{P} \models A \rightarrow \nabla A$.

Proof. Suppose that $\mathcal{P}$ is increasing and that $\mathcal{M} \models_p A$, i.e. $p \in \mathcal{M}(A)$, for some model $\mathcal{M}$ on $\mathcal{P}$. But $\mathcal{M}(A) \in \mathcal{P}^+$, and so $[p) \subseteq \mathcal{M}(A)$, whence by (3.13) we get $\mu(p) \subseteq \mathcal{M}(A)$ and so (3.5) gives $\mathcal{M} \models_p \nabla A$. Thus $A \rightarrow \nabla A$ cannot be falsified on $\mathcal{P}$.

Conversely suppose $\mathcal{P} \models A \rightarrow \nabla A$ and take a model with $V(\pi) = [p) \in \mathcal{P}^+$. Then $\mathcal{M} \models_p \pi$ and hence $\mathcal{M} \models_p \nabla \pi$, giving $\mu(p) \subseteq \mathcal{M}(\pi) = [p)$.   □

6.3.4   Hereditary Monads

As noted in Section 6.2, the topology $\mathcal{P}^+$ (the *upper order* topology) is $\bigcap$-closed, and so for each set $U \subseteq P$ there is a smallest open (hereditary) set $U^\#$ containing U, given by

(3.16) $U^\# = \bigcap\{S \in \mathcal{P}^+ : U \subseteq S\}$.

A more useful formula however is

(3.17) $U^\# = \bigcup\{[p) : p \in U\}$,

so

(3.18) $q \in U^\#$ iff $\exists p \in U(p \sqsubseteq q)$.

We can use $\#$ to show that there is no sentence whose $\mathcal{P}$-validity is equivalent to the requirement that $\mathcal{P}$ have hereditary monads, i.e. that $\mu(p) \in \mathcal{P}^+$ for all $p \in P$. Given any model $\mathcal{M} = (P, \sqsubseteq, \mu, V)$, put $\mathcal{M}^\# = (P, \sqsubseteq, \mu^\#, V)$, where $\mu^\#(p) = (\mu(p))^\#$, for all p, so that $\mu^\#(p) \in \mathcal{P}^+$. Then the reader may check that $\mathcal{M}^\#$ is indeed a model, i.e. (3.3) holds for $\mu^\#$. But for any $S \in \mathcal{P}^+$ we have

$$\mu^\#(p) \subseteq S \quad \text{iff} \quad \mu(p) \subseteq S$$

in view of (3.16), from which we can show that $\mathcal{M}^\#(A) = \mathcal{M}(A)$ for all $A \in \Psi$.

Thus every frame is semantically equivalent to one with hereditary monads.

6.3.5   Density

Defining $p \prec^2 q$ iff $\exists r(p \prec r \prec q)$, it is evident that

(3.19) $\mathcal{M} \models_p \nabla\nabla A$ iff $p \prec^2 q$ implies $\mathcal{M} \models_q A$.

We shall say that frame $\mathcal{P}$ is *dense* if it satisfies the condition

(3.20) $p \prec q$ implies $p \prec^2 q$.

$\mathcal{P}$ will be *pseudo-dense* if

(3.21) $p \prec q$ implies $\exists r(p \prec^2 r \sqsubseteq q)$.

(3.22) $\mathcal{P}$ *is pseudo-dense iff* $\mathcal{P} \models \nabla\nabla A \rightarrow \nabla A$.

Proof.. The "only if" part is left to the reader. For its converse, let $S = \{r : p \prec^2 r\}$ and take a model on $\mathcal{P}$ that has $\mathcal{M}(\pi) = S^{\#} \in \mathcal{P}^{+}$. By (3.19) it follows that $\mathcal{M} \models_p \nabla\nabla\pi$ so that the $\mathcal{P}$-validity of the schema $\nabla\nabla A \rightarrow \nabla A$ yields $\mathcal{M} \models_p \nabla\pi$. Now if $p \prec q$, it must follow that $q \in \mathcal{M}(\pi)$, so (3.18) gives an $r \in S$ with $r \sqsubseteq q$ as required for pseudo-density.   □

Every dense frame is obviously peudo-dense, and the two notions coincide when all monads are hereditary. A frame on which $A \rightarrow \nabla A$ and $\nabla\nabla A \rightarrow \nabla A$ are valid will be called a *J-frame*. These increasing and (pseudo-)dense structures are our basic models for the logic of Grothendieck topologies.

6.3.6   Limit Points

While "$p \prec q$" is to mean "q is close to p", we will for the most part be interested in cases where we do *not* have $p \in \mu(p)$, i.e. $p \prec p$. Thus whereas the informal classical notion alluded to in relation to (3.1) and (3.2) amounts to

(3.23) $p \prec q$ iff p is a closure point of $\{q\}$ (i.e. iff all neighbourhoods of p intersect $\{q\}$),

we have in mind something more like

(3.24) $p \prec q$ iff p is a *limit point* of $\{q\}$ (i.e. every neighbourhood of p intersects $\{q\}$ at a point *other than* p).

If we use (3.24) as a definition of $\prec$ on a poset by taking neighbourhoods in the sense of the order topology $\mathcal{P}^{+}$, then, as $[p)$ is the smallest of all p-neighbourhoods, we find that

(3.25) $\mu(p) = [p) - \{p\}$,

equivalently

(3.26) $p \prec q$ iff $p \sqsubset q$ (i.e. $p \sqsubseteq q$ and $p \neq q$).

The resulting frame is increasing and has hereditary monads ($\mu(p)^{\#} = \mu(p)$). It will not always be dense however; for instance density of $\sqsubset$ fails when P is finite.

It is noteworthy that on any finite poset with the definition (3.26) the sentence

(3.27) $(\nabla A \rightarrow A) \rightarrow \nabla A$

is valid, since to falsify it at p requires an infinite ascending chain in $[p)$. (3.27) is not however valid on $(\omega, \leq)$, where $\leq$ is the usual numerical ordering of the natural numbers.

Whenever the condition

(3.28) $p \sqsubset q$ implies $p \prec q$

obtains, the sentence

(3.29) $\nabla A \rightarrow (B \vee (B \rightarrow A))$

is valid, and conversely in the presence of hereditary monads. In general, validity of (3.29) is equivalent to

(3.30) $p \sqsubset q$ implies $\exists r (p \prec r \sqsubseteq q)$,

i.e. $[p) - \{p\} \subseteq \mu(p)^{\#}$. To indicate the proof, suppose $p \sqsubset q$, so that $p \notin [q)$, and take a model with A and B being distinct letters having $V(A) = \mu(p)^{\#}$ and $V(B) = [q)$. Then $\mathcal{M} \models_p \nabla A$ but not $\mathcal{M} \models_p B$, so by the validity of (3.29) we obtain $\mathcal{M} \models_p B \rightarrow A$. Since $p \sqsubseteq q$ and $\mathcal{M} \models_q B$, it follows by (2.5) that $\mathcal{M} \models_q A$, hence $q \in \mu(p)^{\#}$ as required. $\square$

Returning now to the question of reflexivity of $\prec$, we observe first that validity of the schema $\nabla A \rightarrow A$ is equivalent to the frame condition

(3.31) $\exists q (p \prec q \sqsubseteq p)$,

which when $\mu(p)^{\#} = \mu(p)$ is itself equivalent to

(3.32) $p \prec p$.

But when $\mu(p)$ is hereditary, $p \prec p$ iff $[p) \subseteq \mu(p)$. Noting that on an increasing frame (3.31) implies (3.32), we conclude that $\nabla A \equiv A$ ("local truth=truth") is valid just in case $\mu(p) = [p)$ for all p, i.e. just in case

(3.33) $p \prec q$ iff $p \sqsubseteq q$.

6.3.7   Continuous Lattices

Let $\mathcal{D}$ be a subtopology of the order topology on a poset $(P, \sqsubseteq)$. The following definition is essentially that made by Scott [81] for a particular $\mathcal{D}$ in his work on continuous lattices:

(3.34) $\mu(p) = [p)^{\circ}$,

where the interior is taken with respect to $\mathcal{D}$, so that

(3.35) $p \prec q$ iff $\exists S \in \mathcal{D}(q \in S \subseteq [p))$.

This yields an increasing frame with hereditary monads. A sufficient (though not necessary) condition for it to be dense, and hence a $\mathcal{J}$-frame, is that $\mathcal{D}$ have a base of sets of the form $[r)$. For if $q \in [r) \subseteq [p)$

with $[r] \in \mathcal{D}$, then $p \prec r \prec q$. Such an example is provided by the model of [82] for Lambda-calculus, where $P = 2^\omega$, $\sqsubseteq$ is set inclusion, and $\mathcal{D}$ is generated by all sets of the form $[r]$ for finite $r \subseteq \omega$. In this case we have

(3.36) $p \prec q$ iff p is a finite subset of q.

Of course if $[r] \in \mathcal{D}$ for *every* r, then $\mathcal{D} = \mathcal{P}^+$ and the construction collapses to (3.33).

6.3.8 Cofinality

There is a standard Grothendieck topology on any topos, namely double negation, which is more appropriately put into words as "it is cofinally the case that".

LAWVERE [55]

In general a subset R is said to be *cofinal* with a subset S of a poset when each element of S has an element of R "above" it, i.e. when

(3.37) $q \in S$ implies $\exists r \in R(q \sqsubseteq r)$.

The treatment of double negation as a modal operator is best considered in two parts. The link with cofinality comes from

(3.38) $\mathcal{M} \models_p \sim\sim A$ *iff* $\mathcal{M}(A)$ *is cofinal with* $[p)$.

Proof.

$$\mathcal{M} \models_p \sim\sim A \quad \begin{aligned} &\text{iff} && p \sqsubseteq q \text{ implies not } \mathcal{M} \models_q \sim A \\ &\text{iff} && p \sqsubseteq q \text{ implies } \exists r(q \sqsubseteq r \text{ and } \mathcal{M} \models_r A) \\ &\text{iff} && q \in [p) \text{ implies } \exists r \in \mathcal{M}(A)(q \sqsubseteq r). \end{aligned}$$

□

Now we have that *the sentence*

(3.39) $\nabla A \rightarrow \sim\sim A$

is valid precisely on those frames satisfying

(3.40) $\mu(p)^\#$ *is cofinal with* $[p)$*, for all* $p \in P$.

Proof. To establish sufficiency of (3.40) suppose that $\mathcal{M} \models_p \nabla A$, and so $\mu(p) \subseteq \mathcal{M}(A)$. As $\mu(p)^\#$ is the smallest hereditary superset of $\mu(p)$, this implies that $\mu(p)^\# \subseteq \mathcal{M}(A)$. But then if $\mu(p)^\#$ is cofinal with $[p)$, so too will be $\mathcal{M}(A)$, so by (3.38) we get $\mathcal{M} \models_p \sim\sim A$. Conversely, putting $V(A) = \mu(p)^\#$, so that $\mathcal{M} \models_p \nabla A$, the validity of (3.39) gives $\mathcal{M} \models_p \sim\sim A$, and from this follows (3.40) by (3.38). □

The converse sentence

(3.41) $\sim\sim A \to \nabla A$

is valid precisely on those frames that are both increasing and satisfy

(3.42) $p \prec q \sqsubseteq r$ implies $q = r$,

i.e. the members of $\mu(p)$ are $\sqsubseteq$-*maximal*. That these conditions imply validity of (3.41) is left to the reader to verify. For the other direction, noting that the schema $A \to \sim\sim A$ is valid on all posets, we have that validity of (3.41) implies that of $A \to \nabla A$, so the frame must be increasing by (3.15). Also the schema $\sim\sim(A \vee \sim A)$ is poset-valid, and so with (3.41) yields validity of

(3.43) $\nabla(A \vee \sim A)$.

But this last sentence is precisely equivalent to (3.42). Once more we shall only prove one direction: Take $p \prec q$ and let $\mathcal{M}(A) = [q) - \{q\} \in \mathcal{P}^+$. If (3.43) is $\mathcal{P}$-valid then $\mathcal{M} \models_p \nabla(A \vee \sim A)$, hence $\mathcal{M} \models_q A \vee \sim A$. But not $\mathcal{M} \models_q A$, so we must have $\mathcal{M} \models_q \sim A$ and thus ((2.12) and (2.14)) $[q) \cap \mathcal{M}(A) = \emptyset$. But $\mathcal{M}(A) = [q) - \{q\} \subseteq [q)$, so this is only possible if $[q) - \{q\} = \emptyset$, i.e. $[q) = \{q\}$ as required.

Confining our attention now to frames with $\mu(p)^{\#} = \mu(p)$, we see that if $\mu(p)$ is cofinal with $[p)$, all $\sqsubseteq$-maximal members of $[p)$ must be in $\mu(p)$. Then any such frame that validates

(3.44) $\nabla A \equiv \sim\sim A$

must by (3.42) satisfy

(3.45) $p \prec q$ iff q is a $\sqsubseteq$-maximal member of $[p)$.

However this condition is not always sufficient for (3.44). It is satisfied vacuously by the frame $(\omega, \leq, \mu)$ where $\mu(p) = \emptyset$ for all p, since there are no $\leq$-maximal elements. But in any model on this frame $\nabla\bot$ is true and $\sim\sim\bot$ is false at all points (alternatively observe that (3.40) fails). On the other hand, on a finite poset every element has a $\sqsubseteq$-maximal successor, and indeed the finite frames on which (3.44) is valid are precisely those satisfying (3.45).

   In Sections 6.6 and 6.8 we shall see how the various schemata we have been discussing can be used to axiomatically generate the logics determined by the structures that satisfy the frame conditions that correspond to those schemata.

6.4   Neighbourhood Spaces and Congruences

In addition to the notion of a sentence or property obtaining locally *at a point,* there is the idea of something holding locally *of an object*— typically a set or a space or a function. In general this refers to the existence of an open cover with the property in question holding of each member of the cover. Thus a topological space is said to be *locally connected* if each of its open sets has an open cover of connected sets, while a function is *locally constant* on its domain if that domain is covered by open sets on each of which the (restricted) function is actually constant.

Suppose then that our poset $(P, \sqsubseteq)$ is the collection of open sets of some topology, with $p \sqsubseteq q$ iff $q \subseteq p$. Let γ be the function assigning to each $p \in P$ the collection $\gamma_p \subseteq 2^P$ of open covers of p, where $S \subseteq P$ is a cover of p iff $p \subseteq \bigcup S$ (i.e. $\bigcup S \sqsubseteq p$). We use γ to interpret the connective ∇ in the light of the above examples by putting

(4.1) $\mathcal{M} \models_p \nabla A$ iff $\exists S \in \gamma_p (S \subseteq \mathcal{M}(A))$.

Since γ_p will be closed under supersets, (4.1) is equivalent to

(4.2) $\mathcal{M} \models_p \nabla A$ iff $\mathcal{M}(A) \in \gamma_p$.

Alternatively, thinking of P simply as a point-set, with γ_p a system of "neighbourhoods" of p, then (4.1) and/or (4.2) formalise the "true throughout some neighbourhood" approach to local truth embodied in (3.2). Clause (4.2) is the basis of the *neighbourhood semantics* used in classical modal logic. It is more general than the *relational semantics* of (3.6). Since all sentences are being interpreted as hereditary sets, we shall confine our attention to neighbourhoods in $\mathcal{P}^+$ (although the theory can be carried through without this requirement).

6.4.1 Spaces

A *neighbourhood space* is a structure $\mathcal{S} = (P, \sqsubseteq, \gamma)$, where γ assigns to each $p \in P$ a collection $\gamma_p \subseteq \mathcal{S}^+$ ($\mathcal{S}^+$ being the $\sqsubseteq$-hereditary sets) such that

(4.3) $p \sqsubseteq q$ implies $\gamma_p \subseteq \gamma_q$.

The notion of a model $\mathcal{M} = (\mathcal{S}, V)$ is evident, and truth in $\mathcal{M}$ is defined as before for the non-modal connectives, with (4.2) being used to interpret ∇. Putting

(4.4) $j_\gamma(S) = \{p : S \in \gamma_p\}$

we have from (4.3) that $j_\gamma(S)$ is hereditary. Clause (4.2) can be re-expressed as

(4.5) $\mathcal{M}(\nabla A) = j_\gamma(\mathcal{M}(A))$

and by induction we establish that, once again, $\mathcal{M}(A) \in \mathcal{S}^+$ for all $A \in \Psi$.

Some of the sentences and rules of Section 6.3 that are frame-valid can be falsified on certain spaces and their validity requires certain constraints on γ_p. The sentence (3.9), i.e. $\nabla A \wedge \nabla B \rightarrow \nabla(A \wedge B)$, corresponds to the requirement that γ_p be closed under finite intersections, i.e.

(4.6) $(R \in \gamma_p$ and $S \in \gamma_p)$ implies $R \cap S \in \gamma_p$,

while the validity of the rule

(3.12) if $A \to B$ is valid then so is $\nabla A \to \nabla B$

on S requires γ_p to be closed under supersets in S^+, i.e.

(4.7) $(R \in \gamma_p$ and $R \subseteq S \in S^+)$ implies $S \in \gamma_p$.

If γ_p is non-empty and satisfies (4.6) and (4.7) then it is a *filter*. The conjunction of (4.6) and (4.7) is equivalent to

(4.8) $(R \in \gamma_p$ and $S \in \gamma_p)$ iff $R \cap S \in \gamma_p$.

S will be called a *filter-space* if each γ_p is a filter.

The S-validity of $A \to \nabla A$ corresponds to the neighborhood condition

(4.9) $[p] \subseteq S$ *implies* $S \in \gamma_p$, *for all* $S \in S^+$.

Proof. Assume (4.9) and suppose $\mathcal{M} \models_p A$. Then $p \in \mathcal{M}(A) \in S^+$, so $[p] \subseteq \mathcal{M}(A)$. (4.9) then gives $\mathcal{M}(A) \in \gamma_p$, whence $\mathcal{M} \models_p \nabla A$. Conversely, choose an $\mathcal{M}$ with $\mathcal{M}(A) = S$. If $[p] \subseteq S$, then $\mathcal{M} \models_p A$, so $S \models A \to \nabla A$ yields $\mathcal{M} \models_p \nabla A$ and then $\mathcal{M}(A) = S \in \gamma_p$ as required. $\square$

Note that (4.9) implies

(4.10) $[p] \in \gamma_p$,

and that the converse is true when γ_p is closed under S^+-supersets (4.7).

The space-condition for the density sentence $\nabla\nabla A \to \nabla A$ is

(4.11) $j_\gamma(S) \in \gamma_p$ implies $S \in \gamma_p$, for all $S \in S^+$,

as the reader may verify using (4.5).

We shall say that γ is a *$\mathcal{J}$-system*, and S is a *$\mathcal{J}$-space*, if each γ_p is a filter and $A \to \nabla A$ and $\nabla\nabla A \to \nabla A$ are S-valid (i.e. (4.10) and (4.11) hold).

6.4.2   Spaces From Frames

Let $\mathcal{P} = (P, \sqsubseteq, \mu)$ be a frame. Define $S^\mu = (P, \sqsubseteq, \gamma^\mu)$, where

(4.12) $\gamma_p^\mu = \{S \in \mathcal{P}^+ : \mu(p) \subseteq S\}$.

Then the frame condition (3.3) implies that γ^μ satisfies (4.3), so S^μ is indeed a space. If $\mathcal{M} = (\mathcal{P}, V)$ is a model, put $\mathcal{M}^\mu = (S^\mu, V)$ to obtain a bijection between $\mathcal{P}$-based and S^μ-based models. Since $\mu(p) \subseteq S$ iff $S \in \gamma_p^\mu$, a straightforward induction shows for all $A \in \Psi$ that

(4.13) $\mathcal{M} \models_p A$ iff $\mathcal{M}^\mu \models_p A$.

Hence it follows that

(4.14) $\mathcal{P} \models A$ iff $\mathcal{S}^\mu \models A$.

It is clear from (4.12) that $\mathcal{S}^\mu$ is always a filter-space.

(4.15) If $\mathcal{P}$ is a $\mathcal{J}$-frame, then $\mathcal{S}^\mu$ is a $\mathcal{J}$-space.

Proof. Let $\mathcal{P}$ be increasing and pseudo-dense. Then $\mu(p) \subseteq [p)$, so by (4.12), $[p) \in \gamma_p^\mu$, giving (4.10). To derive (4.11), suppose $j_{\gamma^\mu}(S) \in \gamma_p^\mu$, i.e. $\mu(p) \subseteq j_{\gamma^\mu}(S)$. We have to show that $S \in \gamma_p^\mu$. But if $q \in \mu(p)$, by pseudo-density there exist r, t with $p \prec r \prec t \sqsubseteq q$. Then $r \in \mu(p)$ and so $r \in j_{\gamma^\mu}(S)$, i.e. $\mu(r) \subseteq S$. Then as $t \in \mu(r)$ we get $t \in S$, and finally $q \in S$ since $S \in \mathcal{P}^+$. Thus we have $\mu(p) \subseteq S$ as required. $\square$

(4.14) and (4.15) together imply

(4.16) a sentence valid on all $\mathcal{J}$-spaces is valid on all $\mathcal{J}$-frames.

In the converse direction, given $\mathcal{S} = (P, \sqsubseteq, \gamma)$, put $\mathcal{P}^\gamma = (P, \sqsubseteq, \mu^\gamma)$, where

(4.17) $\mu^\gamma(p) = \bigcap \gamma_p$.

This time (4.3) implies (3.3), making $\mathcal{P}^\gamma$ a frame. Applying this construction to γ_p^μ (4.12) just gives $\mu(p)$ back again. However applying the construction (4.12) to μ^γ may not recover γ—the new neighbourhood system will always be made up of filters, even if $\mathcal{S}$ is not. But taking $\mathcal{S}$ as a filter-space still does not make it equivalent to $\mathcal{P}^\gamma$. For example, let $\mathcal{S} = (\omega, \leq, \gamma)$, where for all $p \in \omega$,

$$\gamma_p = \{[n) : n \in \omega\}.$$

Since every non-empty $\mathcal{S}$-hereditary set is of the form $[n)$ (i.e. $\mathcal{S}^+ = \gamma_p \cup \{\emptyset\}$), γ_p is a filter in $\mathcal{S}^+$. But $\mu^\gamma(p) = \emptyset$ for all p, so $\mathcal{P}^\gamma \models \nabla\bot$. However since $\mathcal{M}(\bot) = \emptyset \notin \gamma_p$, $\nabla\bot$ is not true at *any* point of any model on $\mathcal{S}$.

The converse to (4.16) will follow from the completeness theorems of Section 6.6.

6.4.3 Grothendieck Topology

Let $\mathcal{D}$ be the collection of open sets of a topological space X, and $\mathcal{D}(X) = (\mathcal{D}, \supseteq)$ the partial-ordering inverse to the set inclusion relation. The *Grothendieck topology* associated with the category of sheaves over X is given by the function assigning to each $\mathcal{D}(X)$-hereditary set S the set

(4.18) $j_\mathcal{D}(S) = \{p \in \mathcal{D} : p \subseteq \bigcup S\}$

of opens that S covers. This notion can be lifted to any *complete Heyting algebra* (CHA). A CHA is by definition an HA $\mathcal{H}$ in which every subset S has a least upper-bound (join) $\bigsqcup S$. CHAs satisfy the infinite distributive law

$$(4.19) \quad (\textstyle\bigsqcup S) \sqcap (\bigsqcup R) = \bigsqcup_{s \in S}(\bigsqcup_{r \in R}(r \sqcap s)).$$

Now if $\mathcal{H}$ is a CHA with lattice ordering $\sqsubseteq$, let $\mathcal{H}_0 = (H, \leq)$ be the inverse ordering to $\sqsubseteq$, i.e. $p \leq q$ iff $q \sqsubseteq p$. Then for $S \in \mathcal{H}_0^+$, put

$$(4.20) \quad j_H(S) = \{p \in H : p \sqsubseteq \textstyle\bigsqcup S\} = [\bigsqcup S) \text{ in } \mathcal{H}_0^+.$$

(4.21) *If* $S, R \in \mathcal{H}_0^+$ *have* $p \sqsubseteq \bigsqcup S$ *and* $p \sqsubseteq \bigsqcup R$, *then* $p \sqsubseteq \bigsqcup(R \cap S)$.

Proof. Let q denote the term on the right hand side of equation (4.19). For each $s \in S$ and $r \in R$ we have $s \leq s \sqcap r$ and $r \leq s \sqcap r$. But S and R are $\leq$-hereditary, and so $s \sqcap r \in R \cap S$, from which it follows that $s \sqcap r \sqsubseteq \bigsqcup(R \cap S)$. Hence we may show that $q \sqsubseteq \bigsqcup(R \cap S)$. But the hypothesis of (4.21) yields $p \sqsubseteq (\bigsqcup S) \sqcap (\bigsqcup R) = q$. (The proof of (4.21) for $\mathcal{D}(X)$ can be done by elementary set theory, without explicit recourse to (4.19).) $\qquad\square$

Now let us define a neighbourhood structure on H by putting

$$(4.22) \quad \gamma_p^H = \{S \in \mathcal{H}_0^+ : p \sqsubseteq \textstyle\bigsqcup S\}.$$

That $p \leq q$ implies $\gamma_p^H \subseteq \gamma_q^H$ is obvious, and so $\mathcal{S}^H = (H, \leq, \gamma^H)$ is a space, in which

$$(4.23) \quad j_{\gamma^H}(S) = j_H(S), \text{ for all } S \in \mathcal{H}_0^+.$$

It is readily seen that γ_p^H is closed under supersets, and the import of (4.21) is that it is closed under finite intersections. Thus $\mathcal{S}^H$ is a filter-space. Moreover, since $[p) = \{q : q \sqsubseteq p\}$ we have

$$(4.24) \quad \textstyle\bigsqcup [p) = p,$$

which ensures that $[p) \in \gamma_p^H$. Finally, if $j_{\gamma^H}(S) \in \gamma_p^H$ then

$$p \sqsubseteq \textstyle\bigsqcup\{q : q \sqsubseteq \bigsqcup S\} = \bigsqcup[\bigsqcup S) = \bigsqcup S \quad (4.24),$$

and so $S \in \gamma_p^H$.

Altogether then, $\mathcal{S}^H$ satisfies (4.10) and (4.11) and so is a $\mathcal{J}$-space. If we construct its associated frame (4.17) by putting $\mu(p) = \bigcap \gamma_p^H$, the definition of $\prec$ becomes

$$(4.25) \quad p \prec q \quad \text{iff} \quad \forall S \in \mathcal{H}_0^+ (p \sqsubseteq \textstyle\bigsqcup S \text{ implies } q \in S).$$

Associated with μ is the operation $j_\mu(S) = \{p : \mu(p) \subseteq S\}$. It can be shown that j_μ is identical with the function j_H, i.e.

$$(4.26) \quad \mu(p) \subseteq S \text{ iff } p \sqsubseteq \textstyle\bigsqcup S \text{ for all } p \in P \text{ and } S \in \mathcal{H}_0^+,$$

precisely when $\mathcal{H}$ satisfies

(4.27) $p = \bigsqcup \{q : p \prec q\}$.

In the case of $\mathcal{D}(X)$ we have $p \prec q$ as in (4.25) when q belongs to every hereditary open cover (decomposition) of p. Intuitively this makes q an irreducible (indecomposable) open subset of p. In the discrete topology $\mathcal{D}(X) = 2^X$ it means that q is a singleton subset of p. In the event that $\mathcal{D}$ is an order topology, it means that q is a set of the form $[x)$ for some $x \in p$.

The condition (4.27) means that p is covered by its irreducible open subsets. It is satisfied by all $\bigcap$-closed topologies: given $x \in p$, choose from each hereditary cover of p a neighbourhood q_x of x. Then the intersection of all these q_x's is both an open set around x and an irreducible subset of p. (4.27) also holds of the topology on 2^ω described in Subsection 6.3.7, where each open set is a union of sets of the form $\{q : r \subseteq q\}$ for finite $r \subseteq \omega$. The latter have the irreducibility property we have been describing.

6.4.4   Cofinality

Let $\mathcal{P} = (P, \sqsubseteq)$ be any poset. Then,

(4.28) *the set* $\gamma_p^c = \{S \in \mathcal{P}^+ : S \text{ is cofinal with } [p)\}$ *is a filter on* $\mathcal{P}^+$.

Proof. $\gamma_p^c \neq \emptyset$, since it contains P, and it is easy to see that cofinality is preserved under supersets. Next, take S and R in γ_p^c. Then if $q \in [p)$ there is some $r \in S$ with $q \sqsubseteq r$. But then $r \in [p)$, so there is some $t \in R$ with $r \sqsubseteq t$. Then $q \sqsubseteq r \sqsubseteq t \in S \cap R$, and so we have shown $S \cap R$ to be cofinal with $[p)$, giving the closure of γ_p^c under finite intersections. □

By (4.28) it follows that $\mathcal{S}^c = (P, \sqsubseteq, \gamma^c)$ is a filter-space (why is (4.3) satisfied?). Since $[p)$ is cofinal with itself, it belongs to γ_p^c. Moreover if $j_{\gamma^c}(S) \in \gamma_p^c$ and $p \sqsubseteq q$ there is some r in $j_{\gamma^c}(S)$ with $q \sqsubseteq r$. But then S is cofinal with $[r)$, so for some $t \in S$ we have $q \sqsubseteq r \sqsubseteq t$. This establishes cofinality of S with $[p)$, i.e. $S \in \gamma_p^c$.

Altogether then we find that $\mathcal{S}^c$ is a $\mathcal{J}$-space. It is left to the reader to verify that

(4.29) $\mathcal{S}^c \models \nabla A \equiv \sim\sim A$.

6.4.5   Topological Neighbourhood Systems

In a topological space $(X, \mathcal{D})$ a set Y is a *neighbourhood* of p when there is an open $S \in \mathcal{D}$ with $p \in S \subseteq Y$. Open sets are characterised as those that are neighbourhoods of all of their points. The axioms for the system γ of topological neighbourhoods are

(4.30) γ_p is a filter,

(4.31) $\forall Y \in \gamma_p \exists S \in \gamma_p [(q \in S \text{ implies } S \in \gamma_q) \text{ and } S \subseteq Y]$,

(4.32) $p \in \bigcap \gamma_p$.

(The set $j_\gamma(Y) = \{p : Y \in \gamma_p\}$ is then the interior Y° of Y with respect to $\mathcal{D}$.)

Now the system γ^μ on a $\mathcal{J}$-frame satisfies (4.30) and (4.31)—for (4.31) take $S = \mu(p)^\#$. However while it is true that $p \in S \in S^+$ implies $S \in \gamma_p$, we do not require that $\mathcal{J}$-spaces satisfy (4.32) (which is equivalent to the validity of $\nabla A \to A$). Thus it may be that $p \notin S \in \gamma_p$ for some S. In this case we may think of S as a *punctured* neighbourhood of p, i.e. a set of the form $R - \{p\}$, where R is a neighbourhood of p. These sets are sometimes used to define limit points: p is a limit point of Y iff every punctured neighbourhood of p meets Y (cf. also Section 6.5).

Notice that in a $\mathcal{J}$-space satisfying (4.32) we have $j_\gamma(S) = S$ for all $S \in \mathcal{P}^+$. Now classical logic is modelled by posets with the discrete ordering (i.e. $p \sqsubseteq q$ iff $p = q$), in which $\mathcal{P}^+ = 2^P$. Thus on a discrete poset (set) the only classically topological neighbourhood system that is also a $\mathcal{J}$-system is the one inducing the discrete topology.

6.4.6   Topological Congruences

On a space $\mathcal{S}$ we can define a weak equality relation between hereditary sets: two sets are *locally equal* if they are neighbourhoods of the same points. Formally we define

(4.33) $S \simeq_\gamma R$ iff $\forall p \in P(S \in \gamma_p$ iff $R \in \gamma_p)$.

Then $\simeq_\gamma$ is an equivalence relation on S^+. If S is a $\mathcal{J}$-space, then $S \subseteq j_\gamma(S)$ (by (4.9), i.e. validity of $A \to \nabla A$). Together with (4.11) and the closure of γ_p under supersets this gives $S \in \gamma_p$ iff $j_\gamma(S) \in \gamma_p$, i.e.

(4.34) $S \simeq_\gamma j_\gamma(S)$.

The relation $\simeq_\gamma$ completely determines the neighbourhood system γ. To see this, we first prove

(4.35) $S \simeq_\gamma R$ *implies* $R \subseteq j_\gamma(S)$.

Proof. If $p \in R$ then $R \in \gamma_p$. But then if $S \simeq_\gamma R$, $S \in \gamma_p$ and so $p \in j_\gamma(S)$.   □

(4.34) and (4.35) together imply that $j_\gamma(S)$ is the largest (union) of all sets that are $\simeq_\gamma$-equivalent to S, i.e.

(4.36) $j_\gamma(S) = \bigcup\{R : S \simeq_\gamma R\}$.

Recalling the definition of $j_\gamma(S)$, this can be re-expressed as

(4.37) $S \in \gamma_p$ iff $\exists R(p \in R \simeq_\gamma S)$, which characterises the p-neighbourhoods in terms of $\simeq_\gamma$. It also shows how to define neighbourhoods in terms of a relation $\simeq$ between hereditary sets. It transpires that the properties required of such a relation in order to produce a $\mathcal{J}$-system are that it be an equivalence relation on $\mathcal{S}^+$ that is stable under the *characteristically topological* operations of forming finite intersections and arbitrary unions, i.e.

(4.38) $(S \simeq R$ and $S' \simeq R')$ implies $S \cap S' \simeq R \cap R'$,

and

(4.39) $(S_i \simeq R_i$ for all $i \in I)$ implies $\bigcup_{i \in I} S_i \simeq \bigcup_{i \in I} R_i$.

We shall call such an equivalence relation a *topological congruence*.

(4.40) $\simeq_\gamma$ *is a topological congruence.*

Proof. The derivation of the property (4.38) for $\simeq_\gamma$ is straightforward, and follows from the filter characterisation (4.8). For the property (4.39), suppose $S_i \simeq_\gamma R_i$ for all $i \in I$. First we observe that $\bigcup S_i \subseteq j_\gamma(\bigcup R_i)$. For, if $p \in \bigcup S_i$ then $p \in S_{i_0}$ for some $i_0 \in I$. But then $S_{i_0} \in \gamma_p$, and since $S_{i_0} \simeq_\gamma R_{i_0}$ this gives $R_{i_0} \in \gamma_p$. But γ_p is closed under $\mathcal{S}^+$-supersets (4.7), so it follows that $\bigcup R_i \in \gamma_p$, i.e. $p \in j_\gamma(\bigcup R_i)$ as required. Now from $\bigcup S_i \subseteq j_\gamma(\bigcup R_i)$ and (4.7) we have that $\bigcup S_i \in \gamma_p$ implies $j_\gamma(\bigcup R_i) \in \gamma_p$ and so (4.11) $\bigcup R_i \in \gamma_p$. Interchanging the R_i and S_i in this whole argument proves conversely that $\bigcup R_i \in \gamma_p$ only if $\bigcup S_i \in \gamma_p$. Thus $\bigcup S_i \simeq_\gamma \bigcup R_i$. $\qquad\square$

Now let $(P, \sqsubseteq)$ be any poset and $\simeq$ be a topological congruence on its class of hereditary sets. Define

(4.41) $S \in \gamma_p^\simeq$ iff $\exists R(p \in R \simeq S)$.

If $p \in R \simeq S$ and $p \sqsubseteq q$, then $q \in R \simeq S$, as R is $\sqsubseteq$-hereditary, making $S \in \gamma_q^\simeq$. Thus (4.3) holds, and $\mathcal{S}^\simeq = (P, \sqsubseteq, \gamma^\simeq)$ is a neighbourhood space in which, by (4.41) and the definition (4.4) of j_γ we see that

(4.42) $j_{\gamma^\simeq}(S) = \bigcup\{R : R \simeq S\}$.

But by property (4.39) of topological congruences (with $S_i = S$ for all i) this implies that

(4.43) $S \simeq j_{\gamma^\simeq}(S)$.

Now if $p \in R \simeq S$ and $p \in R' \simeq S'$, then $p \in R \cap R' \simeq S \cap S'$. Also if $p \in R \simeq S$ and $S \subseteq S' \in \mathcal{S}^+$, then $p \in R \cup S' \simeq S \cup S' = S'$. Thus $\gamma^\simeq$ is closed under finite intersections and supersets, i.e. $\mathcal{S}^\simeq$ is a filter-space. Also $p \in [p) \simeq [p)$, putting $[p)$ in $\gamma_p^\simeq$. Moreover, if $j_{\gamma^\simeq}(S) \in \gamma_p^\simeq$ then for some $R \in \mathcal{S}^+$, $p \in R \simeq j_{\gamma^\simeq}(S)$. Then (4.43) gives $p \in R \simeq S$, i.e. $S \in \gamma_p^\simeq$. Altogether we have proved

(4.44) $S^{\simeq}$ is a $\mathcal{J}$-space.   □

(4.45) $R \simeq S$ iff $j_{\gamma^{\simeq}}(R) = j_{\gamma^{\simeq}}(S)$.

Proof. Suppose $R \simeq S$. Then for any $R' \in S$ we have by symmetry and transitivity of $\simeq$ that $R' \simeq R$ iff $R' \simeq S$. Hence $p \in R' \simeq R$ iff $p \in R' \simeq S$. This yields (4.42) $p \in j_{\gamma^{\simeq}}(R)$ iff $p \in j_{\gamma^{\simeq}}(S)$. Conversely, if $j_{\gamma^{\simeq}}(R) = j_{\gamma^{\simeq}}(S)$ then by (4.43), $R \simeq j_{\gamma^{\simeq}}(R) = j_{\gamma^{\simeq}}(S) \simeq S$.   □

The upshot of (4.45) is that applying definition (4.33) to the system $\gamma^{\simeq}$ just gives back the relation $\simeq$. But (4.37) guarantees that the neighbourhood system constructed from $\simeq_\gamma$ by (4.41) is just γ again. Hence

(4.46) *The constructions $\gamma \mapsto \simeq_\gamma$ and $\simeq \mapsto \gamma^{\simeq}$ give a bijection between $\mathcal{J}$-neighbourhood systems and topological congruences on any poset.*

Examples of topological congruences abound. When $\simeq$ is the diagonal congruence, with $R \simeq S$ iff $R = S$, then γ_p is simply the class of hereditary sets that contain p, i.e. the neighbourhood system for the order topology $\mathcal{P}^+$. In the Grothendieck topology construction of Subsection 6.4.3, $S \simeq R$ means that S and R cover the same open sets. In the double-negation context (6.4.4) it means that S and R are mutually cofinal, i.e. each is cofinal in the other. In the space S^μ constructed from a frame $\mathcal{P} = (P, \sqsubseteq, \mu)$ it means that S and R contain the same monads. In particular, on the frame based on 2^ω described in 6.3.7 it means that S and R have the same finite members.

On any poset $(P, \sqsubseteq)$, if Y is any subset of P the conditions

(4.47) $Y \cup R = Y \cup S$,
(4.48) $Y \cap R = Y \cap S$,
(4.49) $R \mapsto Y^\circ = S \mapsto Y^\circ$

(Y° taken in the sense of the order topology) each define a topological congruence on $\mathcal{P}^+$. We shall return to these examples in the next section.

6.5   Local Algebras

The appropriate algebraic modelling for the language Ψ in our present context is provided by structures $\mathfrak{L} = (\mathcal{H}, j)$, where $\mathcal{H}$ is a Heyting algebra with operator $j : H \to H$, allowing interpretation of Ψ-sentences by clauses (2.15)–(2.18) plus

(5.1) $V(\nabla A) = j(V(A))$.

Then the notion of $\mathfrak{L}$-*validity*, $\mathfrak{L} \models A$, is defined for all modal sentences as for $\mathcal{H}$-validity.

Any frame $\mathcal{P} = (P, \sqsubseteq, \mu)$ gives rise to the structure $\mathfrak{L}^\mu = (\mathcal{P}^+, j_\mu)$. In view of (3.7) we find that

(5.2) $\mathcal{P} \models A$ iff $\mathfrak{L}^\mu \models A$, for all $A \in \Psi$.

Similarly from a space $\mathcal{S} = (P, \sqsubseteq, \gamma)$ we obtain the algebra $\mathfrak{L}^\gamma = (\mathcal{S}^+, j_\gamma)$ (cf. (4.4)) and find from (4.5) that

(5.3) $\mathcal{S} \models A$ iff $\mathfrak{L}^\gamma \models A$.

$\mathfrak{L} = (\mathcal{H}, j)$ will be called a *local algebra* when j is a *local operator* on $\mathcal{H}$. This means that j is

(5.4) *inflationary*: $x \sqsubseteq j(x)$, for all $x \in H$;
(5.5) *idempotent*: $j(j(x)) = j(x)$, for all $x \in H$;
(5.6) *multiplicative*: $j(x \sqcap y) = j(x) \sqcap j(y)$, for all $x, y \in H$.

Clearly for inflationary j to be idempotent it suffices that it satisfy

(5.7) $j(j(x)) \sqsubseteq j(x)$.

The operator j_μ is always multiplicative, since $\mu(p) \subseteq S \cap R$ iff both $\mu(p) \subseteq S$ and $\mu(p) \subseteq R$. If $A \to \nabla A$ is $\mathcal{P}$-valid then j_μ is inflationary, and if $\mathcal{P} \models \nabla\nabla A \to \nabla A$ then j_μ satisfies (5.7). Indeed we have

(5.8) $\mathfrak{L}^\mu$ is a local algebra iff $\mathcal{P}$ is a $\mathcal{J}$-frame,

from which by (5.2) it follows that

(5.9) if A is valid on all local algebras, then A is valid on all $\mathcal{J}$-frames.

On a neighbourhood space $\mathcal{S}$, multiplicativity of j_γ corresponds to the filter property (4.8). In the presence of (4.8), the γ_p's are all non-empty (and hence are filters) just in case

(5.10) $P \in \gamma_p$, for all p,

which is equivalent to

(5.11) $\mathcal{S} \models \nabla\top$,

i.e.

(5.12) $j_\gamma(P) = P$.

(5.12) is implied by the inflationary condition $(\mathcal{S} \models A \to \nabla A)$. We have

(5.13) $\mathfrak{L}^\gamma$ is a local algebra iff $\mathcal{S}$ is a $\mathcal{J}$-space,

and hence

(5.14) a sentence valid on all local algebras is valid on all $\mathcal{J}$-spaces.

Using the fact that $x \sqsubseteq y$ iff $x \sqcap y = x$, it is not hard to see that any multiplicative operator is monotone, i.e. satisfies

(5.15) $x \sqsubseteq y$ only if $j(x) \sqsubseteq j(y)$.

This property is useful in developing an even finer analysis of the relationships between the various structures we have been considering.

(5.16) *There is a bijective correspondence between local operators and topological congruences on $\mathcal{P}^+$, for any poset $\mathcal{P}$.*

Proof. Given a congruence $\simeq$, the neighbourhood system $\gamma^{\simeq}$ of (4.41) yields a $\mathcal{J}$-space $\mathcal{S}^{\simeq}$ (4.44) whose associated operator

(4.42) $j_{\gamma^{\simeq}}(S) = \bigcup\{R : R \simeq S\}$

is local by (5.13). Conversely, given a local operator $j : \mathcal{P}^+ \to \mathcal{P}^+$, then motivated by (4.45) we define

(5.17) $R \simeq_j S$ iff $j(R) = j(S)$.

$\simeq_j$ is obviously an equivalence relation. Multiplicativity of j makes it stable under finite intersections. For unions, suppose $R_i \simeq_j S_i$, all $i \in I$. Then using the inflationary property and (5.15) we have that for each $i_0 \in I$, $R_{i_0} \subseteq j(R_{i_0}) = j(S_{i_0}) \subseteq j(\bigcup S_i)$. Hence we have $\bigcup R_i \subseteq j(\bigcup S_i)$. But then by (5.15) and (5.5) we get $j(\bigcup R_i) \subseteq j(j(\bigcup S_i)) = j(\bigcup S_i)$. Interchanging the R_i and the S_i yields $j(\bigcup S_i) \subseteq j(\bigcup R_i)$, and so $\bigcup R_i \simeq_j \bigcup S_i$ as required. Thus $\simeq_j$ is a topological congruence.

(5.18) $S \simeq_j j(S)$.

Proof. j is idempotent.   □

(5.19) $j(S) = \bigcup\{R : R \simeq_j S\}$.

Proof. By (5.18), $j(S) \in \{R : R \simeq_j S\}$, so $j(S) \subseteq \bigcup\{R : R \simeq_j S\}$. Conversely, if $R \simeq_j S$ then $R \subseteq j(R) = j(S)$.   □

In view of (4.45), (4.42), (5.19), the constructions $\simeq \mapsto j_{\gamma^{\simeq}}$ and $j \mapsto \simeq_j$ are mutually inverse, and this establishes (5.16).   □

The laws (5.4)–(5.6) defining local operators were isolated by LAW-VERE and TIERNEY in developing an abstract characterisation of the notion of a "sheaf over a site" (cf. [55], [56], or [26, Chapter 14]). These laws combine some of the properties of interior and closure operators on standard topological spaces, and lead to a modal logic that differs considerably from those that have been held significant in the context of classical logic. In particular the schema $A \to \nabla A$ is certainly not valid when ∇ is given any of the philosophical interpretations listed in the first paragraph of this paper (cf. the next section for more information about the behaviour of this schema over classical logic). But the naturalness of the combination "inflationary, idempotent, and multiplicative" in the present situation is evinced by the notion of a topological congruence. A local operator is to be found precisely in the presence of such a relation of equivalence (indistinguishability modulo certain properties), with the role of that operator being (5.19) to assign to each open set the largest open set that is equivalent to it.

The notion of topological congruence can be lifted to that of an equivalence relation on a complete lattice that is stable under finite meets $\sqcap$ and arbitrary joins $\sqcup$. We shall call such relations $\sqcup$-*congruences*. On a CHA $\mathcal{H}$ they are in bijective correspondence with the local operators via the definitions

(5.20) $x \simeq_j y$ iff $j(x) = j(y)$, $\qquad j_{\simeq}(x) = \sqcup\{y : x \simeq y\}$.

Given a particular $p \in \mathcal{H}$, then the conditions $p \sqcup x = p \sqcup y$, $p \sqcap x = p \sqcap y$, and $x \Rightarrow p = y \Rightarrow p$ each define a $\sqcup$-congruence on $\mathcal{H}$. The corresponding local operators are, respectively,

(5.21) $j(x) = p \sqcup x \ (= j(0) \sqcup x)$,
(5.22) $j(x) = p \Rightarrow x$,
(5.23) $j(x) = (x \Rightarrow p) \Rightarrow p$.

These three equations define local operators on any HA. In the case of Boolean algebras it can be shown that the equation

(5.24) $j(x) = j(0) \sqcup x$

holds of any local operator [63, Chapter 2], so j must be of the type (5.21).

The theory of local operators is set out in detail by MACNAB [63]. $\sqcup$-congruences on CHAs were studied by DOWKER and PAPERT [10] and shown to correspond to local operators in [63].

Punctured Neighbourhoods

We end this section with a re-examination of the idea of *punctured neighbourhood* mentioned in Subsection 6.4.5.

Let $(X, \mathcal{D})$ be a topological space and

(5.25) $\gamma_p = \{Y \subseteq X : \exists S \in \mathcal{D}(p \in S \subseteq Y)\}$

its associated (topological) neighbourhood system. We form a new system by adding to γ_p all the punctured neighbourhoods of p, to form

(5.26) $\delta_p = \gamma_p \cup \{Z \subseteq X : \exists Y \in \gamma_p(Z = Y - \{p\})\}$.

Then in fact

(5.27) $\delta_p = \{Z : \exists S \in \mathcal{D}(p \in S$ and $S - \{p\} \subseteq Z)\}$

as may be seen from (5.25) and (5.26). For each open $R \in \mathcal{D}$, put

(5.28) $j_\delta(R) = \{p : R \in \delta_p\}$

as usual. Then

(5.29) $j_\delta(R) \in \mathcal{D}$,   i.e. $j_\delta : \mathcal{D} \to \mathcal{D}$.

Proof. Let $p \in j_\delta(R)$. We show that p is $\mathcal{D}$-interior to $j_\delta(R)$. By (5.27) there exists $S \in \mathcal{D}$ with $p \in S$ and $S - \{p\} \subseteq R$. Then if $q \in S$, either (i) $q = p$, and so $q \in j_\delta(R)$, or (ii) $q \neq p$, and so $q \in S - \{p\} \subseteq R$, giving $R \in \delta_q$ (since $R \in \mathcal{D}$), and again $q \in j_\delta(R)$. Thus we have $p \in S \subseteq j_\delta(R)$ and $S \in \mathcal{D}$, i.e. p lies in a $\mathcal{D}$-open subset of $j_\delta(R)$ as required.   □

(5.30) j_δ *is inflationary.*

Proof. If $R \in \mathcal{D}$ and $p \in R$, then $R - \{p\} \subseteq R$, so $R \in \delta_p$, hence $p \in j_\delta(R)$. Thus $R \subseteq j_\delta(R)$.   □

(5.31) $j_\delta(R) \subseteq S \cup (-S \cup R)^\circ$ *for all* $R, S \in \mathcal{D}$.

Proof. Let $p \in j_\delta(R)$, so that $S' - \{p\} \subseteq R$, for some S' with $p \in S' \in \mathcal{D}$. Then for any $S \in \mathcal{D}$, if $p \notin S$, $\{p\} \subseteq -S$, so $S' = \{p\} \cup (S' - \{p\}) \subseteq -S \cup R$. But as $p \in S'$, this implies $p \in (-S \cup R)^\circ$ as required.   □

The operator j_δ is not in all cases idempotent. To see this, observe that if $\mathcal{D} = \mathcal{P}^+$ is an order topology then the definition of δ reduces to

(5.32) $\delta_p = \{Z : [p) - \{p\} \subseteq Z\}$.

Then on the order topology of $(\omega, \leq)$ we see that $j_\delta([n + 1)) = [n)$, so that, for example, $j_\delta(j_\delta([2))) - j_\delta([2)) = \{0\} \neq \emptyset$. We shall see later (at the end of Section 6.6) that in a sense the properties (5.30) and (5.31) characterise all algebras of the type $(\mathcal{D}, j_\delta)$. Notice that δ_p is a filter, and so j_δ is always multiplicative.

6.6   Axiomatics and Completeness

In discussing syntax of Ψ we shall follow (and assume) the account of SEGERBERG [84], although with some variations in notation and terminology.

A *logic* is a set $L \subseteq \Psi$ that is closed under *Detachment* (Modus Ponens), i.e. it satisfies

(6.1) $(A \in L$ and $A \to B \in L)$ implies $B \in L$.

Often we write $\vdash_L A$ (A is an *L-theorem*) in place of $A \in L$. We denote by I the logic known as *Heyting's intuitionistic logic*, defined in [84, p. 37].

An extension of I will be called a *normal logic* if it contains all instances of the schema

(3.8) $\nabla(A \rightarrow B) \rightarrow (\nabla A \rightarrow \nabla B)$

and is closed under the rule of *Localisation*:

(6.2) $\vdash_L A$ implies $\vdash_L \nabla A$,

IK denotes the smallest (intersection) of all normal logics. Its theorems are those sentences that are derivable in a finite number of steps from I-axioms and/or (3.8) by the rules of Detachment and Localisation.

A class $\mathcal{C}$ of structures (frames, spaces, algebras ...) is said to *determine* a logic L if for all $A \in \Psi$,

(6.3) $\vdash_L A$ iff $\forall \mathcal{B} \in \mathcal{C}(\mathcal{B} \models A)$.

The "only if" part of (6.3) is expressed by saying that L is *sound* for $\mathcal{C}$, its converse by L being *complete* for $\mathcal{C}$. We shall see that

(6.4) *IK is determined by the class of all frames, as well as by the class of all filter-spaces.*

The smallest logic that extends IK and contains all instances of the schemata

(6.5) $A \rightarrow \nabla A$,
(6.6) $\nabla\nabla A \rightarrow \nabla A$

will be called (as the reader has no doubt guessed) $\mathcal{J}$. Notice that the Localisation rule is implied in $\mathcal{J}$ by (6.5) and so $\mathcal{J}$ is a normal logic. We shall prove

(6.7) *$\mathcal{J}$ is determined by each of (i) the class of $\mathcal{J}$-frames, (ii) the class of $\mathcal{J}$-spaces, and (iii) the class of local algebras.*

In each of the results of (6.4) and (6.7) the "soundness" part is straight-forward. It is simply a matter of checking that the relevant axioms are valid on the structures referred to and that the relevant rules preserve this property. Those cases not already discussed will for the most part be left to the reader.

Most of the axioms we use are of a "positive" (negationless) character, and much of the basic theory to follow is a replication of that for classical modal logic.

An alternative axiomatisation for IK is obtained by adding to I the schemata

(3.9) $\nabla A \wedge \nabla B \rightarrow \nabla(A \wedge B)$,
(6.8) $\nabla\top$,

together with the rule

(6.9) $\vdash_L A \to B$ implies $\vdash_L \nabla A \to \nabla B$.

Taken together, these correspond syntactically to the semantic condition that γ be filter-system. Details are left to the reader.

There are other ways to axiomatise $\mathcal{J}$. Perhaps the simplest is to add just the axioms (6.6), (6.8), and

(6.10) $(A \to B) \to (\nabla A \to \nabla B)$

to I. (6.10) is easily derived from (3.8) and (6.5). To show that this new basis derives the one originally given for $\mathcal{J}$ we have to obtain (3.8) and (6.5) from it. But since

$$\vdash_I A \to (\top \to A),$$

from (6.10) we get

$$A \to (\nabla\top \to \nabla A)$$

which is interderivable over I with

$$\nabla\top \to (A \to \nabla A).$$

From this we detach (6.8) to get (6.5). Next, from (6.10) we derive in I

$$\nabla A \to ((A \to B) \to \nabla B).$$

But as an instance of (6.10) we have

$$((A \to B) \to \nabla B) \to (\nabla(A \to B) \to \nabla\nabla B).$$

Applying Detachment to these last two, and then using (6.6), we derive

$$\nabla A \to (\nabla(A \to B) \to \nabla B),$$

from which (3.8) follows.

It is interesting to note that in *classical* normal logics, i.e. those having $\vdash_L A \vee {\sim}A$, the sentence $\nabla\nabla A \to \nabla A$ is derivable from the rest of the basis for $\mathcal{J}$. On a classical (discrete) frame, where $[p) = \{p\}$, the "increasing" condition becomes $\mu(p) \subseteq \{p\}$, i.e. $\mu(p)$ is either $\{p\}$ or $\emptyset$. Indeed the logic that extends IK by $A \vee {\sim}A$ and $A \to \nabla A$ can be shown to be determined by the two-element discrete frame having $P = 2 = \{0,1\}$, $\mu(0) = \{0\}$, and $\mu(1) = \emptyset$. On this frame $\nabla\nabla A \to \nabla A$ is valid. To see that it is not valid on *all* increasing frames, and hence not derivable from $A \to \nabla A$ over IK itself, we take the example $\mathcal{P} = (2, \sqsubseteq, \mu)$, where $0 \sqsubseteq 1$ but not $1 \sqsubseteq 0$, $\mu(0) = \{1\}$ and $\mu(1) = \emptyset$. This frame is not pseudo-dense. The situation syntactically is that since

$$\vdash_I {\sim}\nabla A \to (\nabla A \to A),$$

in the presence of (6.5) we derive

$$\sim\nabla A \to \nabla(\nabla A \to A),$$

hence by (3.8)

$$\sim\!\nabla A \to (\nabla\nabla A \to \nabla A),$$

and so

$$\nabla\nabla A \to (\sim\!\nabla A \to \nabla A).$$

But in classical logic (not in I) we have the theorem

$$(\sim\!\nabla A \to \nabla A) \to \nabla A$$

which would allow us then to derive (6.6). (The syntactic derivations of this section were developed with the help of K. E. PLEDGER.)

If $\Sigma \subseteq \Psi$ and $A \in \Psi$, then [84, p. 28] A is L-*derivable from* Σ, written $\Sigma \vdash_L A$, if, and only if,

$$\vdash_L B_1 \to (B_2 \to (\ldots \to (B_n \to A)\ldots)$$

holds for some $n \geq 0$ and some $B_1, \ldots, B_n \in \Sigma$. Allowing $n = 0$ includes the possibility that $\vdash_L A$ (i.e. $\emptyset \vdash_L A$). A set $p \subseteq \Psi$ is L-*full* if it contains L, is closed under detachment, is L-*consistent* (i.e. not $p \vdash_L A$, for some A), and *prime* (i.e. $A \vee B \in p$ only if $A \in p$ or $B \in p$). We put

(6.11) $P_L = \{p \subseteq \Psi : p \text{ is } L\text{-full}\}$,
(6.12) $|A|_L = \{p \in P_L : A \in p\}$,
(6.13) $|\Sigma|_L = \{p \in P_L : \Sigma \subseteq p\} = \bigcap\{|A|_L : A \in \Sigma\}$.

In order to obtain our completeness theorem we need a generalisation of the usual statement of LINDENBAUM's Lemma, which itself employs a generalised notion of derivability. If $\Gamma \subseteq \Psi$ we put $\Sigma \vdash_L \Gamma$ iff

(6.14) $\Sigma \vdash_L A_1 \vee \cdots \vee A_m$

holds for some $A_1, \ldots, A_m \in \Gamma$, with $m \geq 1$ (so that $\Sigma \vdash_L A$ as above is the same as $\Sigma \vdash_L \{A\}$). Then we have

(6.15) (LINDENBAUM's Lemma) $\Sigma \vdash_L \Gamma$ *iff* $|\Sigma|_L \subseteq \bigcup\{|A|_L : A \in \Gamma\}$.

Proof. (We make free use of L-full-set properties as developed in [84].) If $\Sigma \vdash_L \Gamma$, so that (6.14) holds for some $A_1, \ldots, A_m \in \Gamma$, then if $p \in |\Sigma|_L$, $\Sigma \subseteq p$, so that $p \vdash_L A_1 \vee \cdots \vee A_m$, hence as p is full, $A_1 \vee \cdots \vee A_m \in p$. But as p is prime this implies that for some $i \leq m$, $p \in |A_i|_L \subseteq \bigcup_{A \in \Gamma}|A|_L$. Conversely, suppose that not $\Sigma \vdash_L \Gamma$, and let $B_0, \ldots, B_j, \ldots$ be an enumeration of Ψ. Put $p_0 = \Sigma$,

$$p_{j+1} = \begin{cases} p_j \cup \{B_j\} & \text{if not } p_j \cup \{B_j\} \vdash_L \Gamma \\ p_j & \text{otherwise,} \end{cases}$$

and $p = \bigcup_{j<\omega} p_j$. By a straightforward adaptation of the proof of Lemma 2.2 of [84] it is shown that $p \in P_L$. Since $\Sigma = p_0 \subseteq p$, $p \in |\Sigma|_L$. Moreover if $A \in \Gamma$ then $A \notin p$, or else $p \vdash_L A$, hence $p_j \vdash_L A$ and so $p_j \vdash_L \Gamma$ for some j. Thus $p \notin \bigcup\{|A|_L : A \in \Gamma\}$. $\qquad\square$

We now confine our attention to *normal* logics L. The *canonical L-frame* is the structure

$$\mathcal{P}_L = (P_L, \sqsubseteq_L, \prec_L),$$

where $p \sqsubseteq_L q$ iff $p \subseteq q$ (set inclusion) and $p \prec_L q$ iff $\{A : \nabla A \in p\} \subseteq q$. The condition (3.4) $p \subseteq q \prec_L r$ only if $p \prec_L r$ is obvious, so $\mathcal{P}_L$ is indeed a frame. Note that it has hereditary monads, where, putting $\nabla_p = \{A : \nabla A \in p\}$, the monad of $p \in P_L$ is

(6.16) $\mu_L(p) = \{q : \nabla_p \subseteq q\} = |\nabla_p|_L$.

(6.17) $\nabla_p \vdash_L B$ *iff* $\nabla B \in p$.

Proof. (As for classical logic, cf. [59].) If $\nabla B \in p$, then $B \in \nabla_p$, hence $\nabla_p \vdash_L B$. Conversely, suppose that

(6.18) $\vdash_L A_1 \rightarrow (\ldots \rightarrow (A_n \rightarrow B)\ldots)$

for some $n \geq 0$, with $A_i \in \nabla_p$, for all $i \leq n$. If in fact $\vdash_L B$, then $\vdash_L \top \rightarrow B$. But $\top \in \nabla_p$ (as $\vdash_L \nabla\top$), so we may always presume $n \geq 1$. By Localisation (6.2), and then n applications of (3.8), (6.18) yields

(6.19) $\vdash_L \nabla A_1 \rightarrow (\ldots \rightarrow (\nabla A_n \rightarrow \nabla B)\ldots)$.

But $A_i \in \nabla_p$, for $i \leq n$, hence $\nabla A_i \in p$. By closure of p under Detachment we get $\nabla B \in p$ as required. $\qquad\square$

The *canonical L-model* based on $\mathcal{P}_L$ is $\mathcal{M}_L = (\mathcal{P}_L, V_L)$, where

$$V_L(\pi) = |\pi|_L = \{p \in P_L : \pi \in p\}, \text{ for all } \pi \in \Psi_0.$$

(6.20) *For all* $A \in \Psi$, $\mathcal{M}_L \models_p A$ *iff* $A \in p$, i.e. $\mathcal{M}_L(A) = |A|_L$.

Proof. By induction, as in Lemma 2.3 of [84]. We treat only the case $A = \nabla B$, under the inductive hypothesis that $\mathcal{M}_L(B) = |B|_L$. We have

$$
\begin{array}{llll}
\mathcal{M}_L \models_p \nabla B & \text{iff} & \mu_L(p) \subseteq \mathcal{M}_L(B) & (3.5) \\
& \text{iff} & |\nabla_p|_L \subseteq |B|_L & (6.16), \text{hypothesis} \\
& \text{iff} & \nabla_p \vdash_L B & (6.15) \\
& \text{iff} & \nabla B \in p & (6.17).
\end{array}
$$

$\qquad\square$

As a corollary we get:

(6.21) $\mathcal{P}_L \models A$ *only if* $\vdash_L A$.

Proof. If $\mathcal{P}_L \models A$, then $\mathcal{M}_L(A) = P_L$, i.e. $|A|_L = |P|_L = |\emptyset|_L$, and so (6.15) $\emptyset \vdash_L A$ as required. $\qquad\square$

This corollary is the key to many completeness theorems. For, to show that any $\mathcal{C}$-valid sentence is an L-theorem, for some class $\mathcal{C}$ of frames, we now know it suffices to show that $\mathcal{P}_L \in \mathcal{C}$. Taking $\mathcal{C}$ as the class of all frames, we have $\mathcal{P}_{IK} \in \mathcal{C}$, giving the completeness theorem for frames of (6.4) (for the filter-space version, observe that $\mathcal{S}^{\mu L}$ (cf. Subsection 6.4.2) is a filter-space and use (4.14)).

(6.22) $\mathcal{P}_{\mathcal{J}}$ is a $\mathcal{J}$-frame.

Proof. Let $p \prec_{\mathcal{J}} q$, and take $A \in p$. Since $\vdash_{\mathcal{J}} A \to \nabla A$, $A \to \nabla A \in p$. By Detachment then, $\nabla A \in p$, and so $A \in q$. Thus $p \subseteq q$, and $\mathcal{P}_{\mathcal{J}}$ is *increasing*. Next we show that $\mathcal{P}_{\mathcal{J}}$ is *dense*. Suppose $p \prec_{\mathcal{J}} q$ and let $\Gamma = \{\nabla B : B \notin q\}$. Then if $\nabla_p \vdash_{\mathcal{J}} \Gamma$ we have

(6.23) $\nabla_p \vdash_{\mathcal{J}} \nabla B_1 \vee \cdots \vee \nabla B_m$

for some m, with $\nabla B_i \in \Gamma$ for all $1 \leq i \leq m$.

But in IK we can derive

(3.10) $\nabla A \vee \nabla B \to \nabla(A \vee B)$

(since it is valid on all frames—a syntactic proof is easy). Applying (3.10) to (6.23) we get

(6.24) $\nabla_p \vdash_{\mathcal{J}} \nabla(B_1 \vee \cdots \vee B_m)$,

and so by (6.17),

(6.25) $\nabla\nabla(B_1 \vee \cdots \vee B_m) \in p$.

The schema $\nabla\nabla B \to \nabla B$ may now be invoked to get

(6.26) $\nabla(B_1 \vee \cdots \vee B_m) \in p$.

But $p \prec_{\mathcal{J}} q$, so $(B_1 \vee \cdots \vee B_m) \in q$ and hence $B_i \in q$ for some $i \leq m$. But this contradicts the fact that $\nabla B_i \in \Gamma$.

The upshot of this argument is that it cannot be that $\nabla_p \vdash_{\mathcal{J}} \Gamma$, and so by LINDENBAUM's Lemma (6.15) there is some $r \in |\nabla_p|_{\mathcal{J}}$, i.e. $p \prec_{\mathcal{J}} r$, that has $r \cap \Gamma = \emptyset$. Then if $\nabla B \in r$, $\nabla B \notin \Gamma$, and so $B \in q$. Hence $r \prec_{\mathcal{J}} q$. □

The proof of (6.22) actually establishes something a little stronger, viz.

(6.27) for any normal logic L, if $\vdash_L A \to \nabla A$ then $\mathcal{P}_L$ is increasing, while if $\vdash_L \nabla\nabla A \to \nabla A$ then $\mathcal{P}_L$ is dense.

To establish the full content of (6.7) we begin with the readily established soundness result that $\mathcal{J}$-theorems are valid on all local algebras. But local algebra validity implies $\mathcal{J}$-space validity (5.14), while the latter implies frame validity (4.16). But by (6.22) and (6.21), $\mathcal{J}$-frame validity in turn implies theoremhood in $\mathcal{J}$, and so all four notions coincide.

Notice that we have in fact constructed a "universal" determining model of each type for $\mathcal{J}$, since for all $A \in \Psi$ we have

(6.28) $\vdash_{\mathcal{J}} A$ iff $\mathcal{P}_{\mathcal{J}} \models A$ iff $\mathcal{P}_{\mathcal{J}}^{+} \models A$ iff $\mathcal{S}^{\mu_{\mathcal{J}}} \models A$.

We turn now to completeness theorems for logics generated by some of the other schemata discussed in Section 6.3.

(6.29) *If* $\vdash_L \nabla A \rightarrow \sim\sim A$, *then in* $\mathcal{P}_L$, $\mu_L(p)$ *is cofinal with* $[p)$ (cf. (3.40)).

Proof. Suppose $q \in [p)$, i.e. $p \subseteq q$. If $\nabla_p \cup q \vdash_L \bot$, then there exist $B_1, \ldots, B_n \in q$ with $\nabla_p \vdash_L B_1 \rightarrow (B_2 \rightarrow (\ldots \rightarrow (B_n \rightarrow \bot)\ldots))$. But then $\nabla_p \vdash_L B \rightarrow \bot$, i.e. $\nabla_p \vdash_L \sim B$, where $B = B_1 \wedge \ldots \wedge B_n \in q$ (if $\nabla_p \vdash_L \bot$, we may take $B = \top \in q$). Now applying result (6.17) we have $\nabla\sim B \in p$. But $\vdash_L \nabla\sim B \rightarrow \sim\sim\sim B$, while $\vdash_I \sim\sim\sim B \rightarrow \sim B$, and so we deduce that $\sim B \in p \subseteq q$. However $B \in q$, and so this contradicts the L-consistency of q. Thus it is not the case that $\nabla_p \cup q \vdash_L \bot$, so (6.15) there exists $r \in \mathcal{P}_L$ with $\nabla_p \cup q \subseteq r$, so that $q \sqsubseteq_L r$ and $r \in \mu_L(p)$ as required. $\qquad\square$

(6.30) *If* $\vdash_L \sim\sim A \rightarrow \nabla A$, *then* $\mathcal{P}_L$ *is increasing and satisfies* $p \prec_L q \subseteq r$ *only if* $q = r$ (cf. (3.41), (3.42)).

Proof. Since $\vdash_L A \rightarrow \sim\sim A$, it follows that $\vdash_L A \rightarrow \nabla A$. so $\mathcal{P}_L$ is increasing (6.27). Next, since $\vdash_I \sim\sim(A \vee \sim A)$, we have $\vdash_L \nabla(A \vee \sim A)$. But then $\nabla(A \vee \sim A) \in p$, all $p \in \mathcal{P}_L$. So, if $p \prec_L q$ we have $A \vee \sim A \in q$ for all $A \in \Psi$. Then if $q \subseteq r$ and $A \notin q$, we have $\sim A \in q \subseteq r$, so $A \notin r$ (or else r would be L-inconsistent), hence $q = r$. $\qquad\square$

From (6.29) and (6.30) it follows that the logic $IK + (\nabla A \equiv \sim\sim A)$ is determined by the class of frames satisfying the properties described therein (soundness is given by (3.39)–(3.41): why are such frames dense?). From the proof of (6.30) it follows that $\sim\sim A \rightarrow \nabla A$ is equivalent deductively (and semantically) over IK to $(A \rightarrow \nabla A) \wedge (\nabla(A \vee \sim A))$. From (6.29) we see that $\nabla A \rightarrow \sim\sim A$ is equivalent to $\nabla\sim A \rightarrow \sim A$ (which reduces to $\nabla A \rightarrow A$ only when $\vdash_L \sim\sim A \rightarrow A$, i.e. when L is classical).

We leave it as an exercise for the reader to prove

(6.31) *if* L *contains all instances of the schema*

(3.29) $\nabla A \rightarrow (B \vee (B \rightarrow A))$,

then $\mathcal{P}_L$ *satisfies* (3.28), *i.e.* $p \sqsubset q$ *only if* $p \prec q$.

Now let N denote the logic $IK + (3.29) + (A \rightarrow \nabla A)$. Then $\mathcal{P}_N$ determines N, since it is increasing and satisfies (3.28). However $\mathcal{P}_N$ does not satisfy the condition

(3.26) $p \prec q$ iff $p \sqsubset q$,

for which N is sound, since $\mathcal{P}_N$ has points with $p \prec p$. To see this, observe that $\Gamma = \{\nabla A \rightarrow A : A \in \Psi\}$ is N-consistent, having a model in the one-element N-frame $\{0\}$, where $0 \prec 0$. Any N-full extension p of Γ will have $p \prec p$. To obtain a determining frame for N that satisfies (3.26) we modify $\mathcal{P}_N$ to replace reflexive points by a denumerable sequence strictly ordered by $\prec$. The simplest way to do this is to put

$$\mathcal{P}_N^* = (P_N \times \omega, \sqsubseteq, \prec)$$

where $(p, n) \sqsubseteq (q, m)$ iff (i) $p \sqsubset_N q$, or (ii) $p = q$ and $n = m$, or (iii) $p = q$ and $p \prec_N q$ and $n < m$; while $(p, n) \prec (q, m)$ iff (i) $p \prec_N q$ and $p \neq q$, or (ii) $p = q$ and $p \prec_N q$ and $n < m$. Then $\mathcal{P}_N^*$ satisfies (3.26) and is an N-frame. If $f : P_N \times \omega \to P_N$ is the projection map $f(p, n) = p$, we have:

(6.32) for any model $\mathcal{M} = (\mathcal{P}_N, V)$ based on $\mathcal{P}_N$, put $\mathcal{M}^* = (\mathcal{P}_N^*, V^*)$, where $V^*(\pi) = f^{-1}(V(\pi))$; then for all $A \in \Psi$ we have $\mathcal{M}^* \models_x A$ iff $\mathcal{M} \models_{f(x)} A$.

It follows that

(6.33) $\mathcal{P}_N^* \models A$ implies $\mathcal{P}_N \models A$ (and so $\vdash_N A$),

giving our desired completeness theorem. (The construction of $\mathcal{P}_N^*$ is essentially the "bulldozer" technique of SEGERBERG [86].)

Representing Algebras.

The canonical frame construction can be translated into the language of lattices to give a representation theory for algebras of the type $\mathfrak{L} = (\mathcal{H}, j)$ (cf. Section 6.5) extending that for HAs sketched in Section 6.2. Define $\mathfrak{L}_+ = (P_\mathcal{H}, \subseteq, \mu_j)$, where $P_\mathcal{H}$ is the set of prime filters of $\mathcal{H}$, and

$$q \in \mu_j(p) \quad \text{iff} \quad \{x : j(x) \in p\} \subseteq q.$$

Then the function $\varphi : x \mapsto \{p : x \in p\}$ is an HA-monomorphism of $\mathfrak{L}$ into the set algebra $\mathfrak{L}_+^{\mu_j}$ defined prior to (5.2). In order that φ be a j-operator homomorphism, which amounts to requiring that

$$j(x) \in p \quad \text{iff} \quad \mu_j(p) \subseteq \varphi(x),$$

it is necessary and sufficient that $\mathfrak{L}$ be an IK-algebra. This in turn amounts to requiring that j be multiplicative and satisfy $j(1) = 1$. In the event that $\mathfrak{L}$ is a local algebra, $\mathfrak{L}_+$ will be a $\mathcal{J}$-frame.

Finally we consider algebras on which j is inflationary and satisfies multiplicativity and

(6.34) $j(x) \sqsubseteq y \sqcup (y \Rightarrow x)$.

By (5.30) and (5.31) these include all algebras of the type $(\mathcal{D}, j_\delta)$, where δ is the punctured-neighbourhood system associated with the topology $\mathcal{D}$. Given any such algebra $\mathfrak{L}$, $\mathfrak{L}_+$ will be an N-frame and so we can apply the $\mathcal{P}_N^*$ construction, as above, to $\mathfrak{L}_+$ to get a frame of the type $\mathfrak{L}_0 = (P_{\mathcal{H}} \times \omega, \sqsubseteq, \sqsubset)$. Composing φ with the inverse to the projection $\mathfrak{L}_0 \to \mathfrak{L}_+$ gives a monomorphism into the algebra of hereditary subsets of $\mathfrak{L}_0$. But by (5.32), the latter is simply the algebra $(\mathcal{D}, j_\delta)$, where $\mathcal{D}$ is the order topology on $\mathfrak{L}_0$. Thus every N-algebra can be represented as a subalgebra of the algebra of open sets determined by the punctured-neighbourhood system of a topology.

6.7   Sites Over Elementary Topoi

We come now to the interpretation of the modal language Ψ in elementary topoi with topologies in the sense of LAWVERE [55]. This will provide us with a new concept of validity that characterises the system $\mathcal{J}$, and yields a formal explication of LAWVERE's description of Grothendieck topologies as modal operators. We presume in this (and only this) section that the reader is familiar with category theory, and with the basics of topos theory as expounded for instance in LAWVERE [56], FREYD [19], KOCK and WRAITH [49], MAC LANE [61], or GOLDBLATT [26].

Let $\mathcal{E}$ be an elementary topos with subobject classifier $1 \xrightarrow{true} \Omega$. Ω is the "object" of truth-values for $\mathcal{E}$, and the basic property of the arrow $true$ is that for each $\mathcal{E}$-monic $a \overset{f}{\rightarrowtail} d$ there is exactly one arrow $\chi_f : d \to \Omega$ (the *character* of f) that makes the diagram

$$
\begin{array}{ccc}
a & \xrightarrow{\ f\ } & d \\
{\scriptstyle !}\big\downarrow & & \big\downarrow{\scriptstyle \chi_f} \\
1 & \xrightarrow{true} & \Omega
\end{array}
$$

a pullback. This establishes a bijection between subobjects of d and elements of the set $\mathcal{E}(d, \Omega)$ of $\mathcal{E}$-arrows of the form $d \to \Omega$. In particular the unique arrow $0 \to 1$ is monic in $\mathcal{E}$, and its character is known as *false*: $1 \to \Omega$.

Let $\Omega \times \Omega \xrightarrow{\cap} \Omega$, $\Omega \times \Omega \xrightarrow{\cup} \Omega$, $\Omega \times \Omega \xrightarrow{\Rightarrow} \Omega$, be the "truth-arrows" of $\mathcal{E}$ corresponding to the connectives $\wedge$, $\vee$, $\to$ (for definitions cf. the above references). These arrows act as operators on $\mathcal{E}(d, \Omega)$, defining binary operations by

(7.1) $h \cap k = \cap \circ (h, k)$,

(7.2) $h \cup k = \cup \circ (h, k)$,

(7.3) $h \Rightarrow k = \Rightarrow \circ (h, k)$,

where (h, k) is the product arrow of h and k:

The structure $(\mathcal{E}(d, \Omega), \cap, \cup, \Rightarrow, false_d)$ thus defined proves to be a Heyting algebra, where the least element $false_d$ is given by

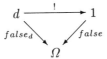

A *topology* on $\mathcal{E}$ is an arrow $\Omega \xrightarrow{j} \Omega$ for which the diagrams

(7.4)

$$\Omega \xrightarrow{(1_\Omega, j)} \Omega \times \Omega$$

with 1_Ω and $\cap$ to Ω

(7.5)

$$\Omega \xrightarrow{j} \Omega$$

with j and j to Ω

(7.6)

$$\begin{array}{ccc} \Omega \times \Omega & \xrightarrow{\cap} & \Omega \\ {\scriptstyle j \times j}\downarrow & & \downarrow{\scriptstyle j} \\ \Omega \times \Omega & \xrightarrow{\cap} & \Omega \end{array}$$

all commute. As shown by FREYD [19], commutativity of (7.4) is equivalent in any topos to that of

(7.7)

$$1 \xrightarrow{true} \Omega$$

with $true$ and j to Ω

A pair of the form $\mathcal{E}_j = (\mathcal{E}, j)$ will be called an *elementary site*. The prime example has $\mathcal{E}$ as the category of presheaves over a topological space $(X, \mathcal{D})$, with j determined by the operator $j_\mathcal{D}$ of (4.18).

The map $h \mapsto j \circ h$ is a local operator on $\mathcal{E}_j(d, \Omega)$. To see, for example, why it is inflationary observe that

$$\begin{aligned}
h \cap (j \circ h) &= \cap \circ (h, j \circ h) & (7.1)\\
&= \cap \circ (1_\Omega, j) \circ h\\
&= 1_\Omega \circ h & (7.4)\\
&= h,
\end{aligned}$$

so that $h \sqsubseteq j \circ h$. The idempotent and multiplicative properties of $h \mapsto j \circ h$ are given by (7.5) and (7.6).

An $\mathcal{E}_j$-*valuation* is a function $V : \Psi_0 \to \mathcal{E}(1, \Omega)$ assigning to each sentence letter π a truth-value ("element" of Ω) in $\mathcal{E}$, i.e. an arrow $V(\pi) : 1 \to \Omega$. This is lifted canonically to a function $V : \Psi \to \mathcal{E}(1, \Omega)$ by putting

(7.8) $V(\bot) = \textit{false}$,

(7.9) $V(A \wedge B) = V(A) \cap V(B)$,

(7.10) $V(A \vee B) = V(A) \cup V(B)$,

(7.11) $V(A \to B) = V(A) \Rightarrow V(B)$,

(7.12) $V(\nabla A) = j \circ V(A)$:

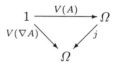

A is $\mathcal{E}_j$-*valid*, $\mathcal{E}_j \models A$, iff $V(A)=\textit{true}$ for every $\mathcal{E}_j$-valuation.

Clearly an $\mathcal{E}_j$-valuation is precisely the same thing as a valuation on the local algebra $\mathcal{E}_j(1, \Omega)$, and it follows from the foregoing that doing topos-theoretic semantics in $\mathcal{E}_j$ is equivalent to local algebra semantics in $\mathcal{E}_j(1, \Omega)$. Thus

(7.13) $\mathcal{E}_j \models A$ iff $\mathcal{E}_j(1, \Omega) \models A$, for all $A \in \Psi$.

From this, by (6.7) we have

(7.14) $\vdash_{\mathcal{J}} A$ only if $\mathcal{E}_j \models A$, for all elementary sites $\mathcal{E}_j$.

The main result of this section is the converse to (7.14), which is derived using a special kind of topos constructed from a frame.

If $\mathcal{P} = (P, \sqsubseteq)$ is a poset, then $\mathcal{P}$ may be construed as a category, with at most one arrow $p \to q$, for any $p, q \in \mathcal{P}$, this arrow existing just in case $p \sqsubseteq q$. Let $Set^{\mathcal{P}}$ be the category of set-valued functors defined on $\mathcal{P}$. An object in this category is a functor $F : \mathcal{P} \to Set$, assigning each $p \in P$ a set F_p and each arrow $p \to q$ (when $p \sqsubseteq q$) a set function $F_{pq} : F_p \to F_q$, such that

(7.15) F_{pp} is the identity function on F_p,

and the following diagram commutes whenever $p \sqsubseteq q \sqsubseteq r$:

(7.16)

$$F_p \xrightarrow{\ F_{pq}\ } F_q$$

with F_{pr} and F_{qr} arrows going down to F_r.

An arrow from functor F to functor G in Set^P is a *natural transformation* $F \xrightarrow{\sigma} G$, i.e. a family

(7.17) $\{F_p \xrightarrow{\sigma_p} G_p : p \in P\}$

of functions (*components* of σ) indexed by P, having the form shown in (7.17), and such that the following diagram commutes whenever $p \sqsubseteq q$:

(7.18)

$$\begin{array}{ccc} F_p & \xrightarrow{\ \sigma_p\ } & G_p \\ {\scriptstyle F_{pq}}\downarrow & & \downarrow{\scriptstyle G_{pq}} \\ F_q & \xrightarrow{\ \sigma_q\ } & G_q \end{array}$$

Informally F may be thought of as a "set" varying (growing) over the "stages" $p \in P$, with the F_{pq}'s as "transition" maps between stages. The similarity with the notion of Kripke-style model for elementary intuitionistic logic is evident.

The relevance of these notions is that Set^P is a topos. Its object of truth-values $\Omega : P \to Set$ has

(7.19) $\Omega_p = [p)^+ =$ the set of $\sqsubseteq$-hereditary subsets of $([p), \sqsubseteq)$.

For each $S \in P^+$ and $p \in P$, let

(7.20) $S_p = [p) \cap S \in [p)^+$.

Then the functions $S \mapsto S_p$ provide the transitions of Ω. That is, if $p \sqsubseteq q$ then

(7.21) $\Omega_{pq}(S) = S_q$, for all $S \in [p)^+$.

The terminal object $1 : P \to Set$ of Set^P has constant values $1_p = \{0\}$, for all $p \in P$. The arrow *true*: $1 \to \Omega$ has p-th component $true_p : \{0\} \to [p)^+$ given by

(7.22) $true_p(0) = [p) =$ the largest element of $[p)^+$.

The arrow *false*: $1 \to \Omega$ is given by

(7.23) $false_p(0) = \emptyset$, for all p.

The truth-arrows $\cap$, $\cup$, $\Rightarrow$, in Set^P have as components the corresponding HA operations on the HAs Ω_p. Thus $\cap_p$ is the operation $(S, R) \mapsto S \cap R$, $\cup_p$ is $(S, R) \mapsto S \cup R$, and $\Rightarrow_p$ is $(S, R) \mapsto (S \Rrightarrow R)$ (2.13) on $[p)^+$.

Let us suppose further that $\mathcal{P}$ is a $\mathcal{J}$-frame $(P, \sqsubseteq, \mu)$. Then for each p we define, for $S \in [p)^+$,

(7.24) $j_p(S) = (j_\mu(S))_p = \{r : p \sqsubseteq r \text{ and } \mu(r) \subseteq S\}$ [cf. (7.20)].

Then $j_p(S)$ is hereditary in $[p)^+$. We leave it to the reader to verify that

(7.25) $j_p : \Omega_p \to \Omega_p$ is a local operator on the Heyting algebra Ω_p.

The right-hand side of equation (7.24) is in fact defined for any $S \in \mathcal{P}^+$. The general situation is

(7.26) $(j_\mu(S))_p = j_p(S_p)$.

Proof.

$$
\begin{aligned}
j_p(S_p) &= j_p([p) \cap S) & (7.20) \\
&= [p) \cap j_\mu([p) \cap S) & (7.24), (7.20) \\
&= [p) \cap j_\mu([p)) \cap j_\mu(S) & (5.6) \\
&= [p) \cap j_\mu(S) & (5.4) \\
&= (j_\mu(S))_p & (7.20).
\end{aligned}
$$

$\square$

From this, noting that $p \sqsubseteq q$ only if $[q) \sqsubseteq [p)$, we can prove that

(7.27) *if* $p \sqsubseteq q$ *then* $\Omega_{pq}(j_p(S)) = j_q(\Omega_{pq}(S))$, *for* $S \in \Omega_p$.

Proof.

$$
\begin{aligned}
\Omega_{pq}(j_p(S)) &= [q) \cap j_p(S) & (7.21) \\
&= [q) \cap [p) \cap j_\mu(S) & (7.24) \\
&= [q) \cap j_\mu(S) & \text{(above note)} \\
&= j_q(S_q) & (7.26) \\
&= j_q(\Omega_{pq}(S)) & (7.21).
\end{aligned}
$$

$\square$

(7.27) asserts that the diagram

$$
\begin{array}{ccc}
\Omega_p & \xrightarrow{\ j_p\ } & \Omega_p \\
{\scriptstyle \Omega_{pq}}\Big\downarrow & & \Big\downarrow{\scriptstyle \Omega_{pq}} \\
\Omega_q & \xrightarrow{\ j_q\ } & \Omega_q
\end{array}
$$

commutes whenever $p \sqsubseteq q$, and so the family $\{j_p : p \in P\}$ form the components of a natural transformation $j_{\mathcal{P}} : \Omega \to \Omega$, i.e. an arrow in Set^P. From (7.25) it can be shown that the diagrams (7.4)–(7.6) commute for $j_{\mathcal{P}}$ (the components of a composite of natural transformations are the composites of their corresponding components). Thus $j_{\mathcal{P}}$ is a topology on Set^P, and so

(7.28) $\mathcal{E}_{\mathcal{P}} = (Set^P, j_{\mathcal{P}})$ *is an elementary site.*

It is interesting to note that the condition that $\mathcal{P}$ be increasing (hence $j_\mathcal{P}$ inflationary) is needed not only to prove its categorial version (7.4), but also in the proof of (7.26) and hence (7.27). In other words it is needed to show that $j_\mathcal{P}$ *is an arrow at all* in $Set^\mathcal{P}$. Thus the axiom $A \to \nabla A$, of little interest in classical modal logic, plays a crucial role in the present theory.

The key to our completeness proof is that for any $\mathcal{J}$-frame $\mathcal{P}$, we have

(7.29) $\mathcal{E}_\mathcal{P} \models A$ *iff* $\mathcal{P} \models A$, *for all* $A \in \Psi$.

Applying this to the canonical $\mathcal{J}$-frame $\mathcal{P}_\mathcal{J}$ gives

(7.30) $\mathcal{E}_{\mathcal{P}_\mathcal{J}} \models A$ iff $\mathcal{P}_\mathcal{J} \models A$ iff $\vdash_\mathcal{J} A$ [cf. (6.28)],

so that the *canonical site* $\mathcal{E}_{\mathcal{P}_\mathcal{J}}$ determines $\mathcal{J}$, and moreover any sentence valid on all sites will be in particular $\mathcal{E}_{\mathcal{P}_\mathcal{J}}$-valid, hence a $\mathcal{J}$-theorem.

As to the proof of (7.29), we already know (7.13) that

(7.31) $\mathcal{E}_\mathcal{P} \models A$ iff $\mathcal{E}_\mathcal{P}(1, \Omega) \models A$,

while by (5.2) we have

(7.32) $\mathcal{P} \models A$ iff $\mathcal{L}^\mu = (\mathcal{P}^+, j_\mu) \models A$.

Since isomorphic local algebras validate the same sentences, our desideratum (7.29) follows from (7.31), (7.32) and

(7.33) *there exists a local algebra isomorphism* $\mathcal{L}^\mu \cong \mathcal{E}_\mathcal{P}(1, \Omega)$.

Proof. Given $S \in \mathcal{P}^+$, define $1 \xrightarrow{\sigma^S} \Omega$ to be the natural transformation whose p-th component $\sigma_p^S : 0 \to \Omega_p$ has

(7.34) $\sigma_p^S(0) = S_p$,

(that σ^S is natural follows since $(S_p)_q = S_q$ when $p \sqsubseteq q$). Conversely, given $\sigma : 1 \to \Omega$ define $S_\sigma \in \mathcal{P}^+$ by

(7.35) $S_\sigma = \bigcup\{\sigma_p(0) : p \in P\}$.

Then the functions $S \mapsto \sigma^S$ and $\sigma \mapsto S_\sigma$, are mutually inverse and provide the asserted isomorphism. The details of this for the Heyting algebra part

$$\mathcal{P}^+ \cong Set^\mathcal{P}(1, \Omega)$$

are fully presented in GOLDBLATT [26]. For the present we show only that $S \mapsto \sigma^S$ preserves the local operators, i.e.

(7.36) $j_\mathcal{P} \circ \sigma^S = \sigma^{j_\mu(S)}$, *for all* $S \in \mathcal{P}^+$.

$\square$

Proof. The p-th component of $j_\mathcal{P} \circ \sigma^S$ is $j_p \circ \sigma_p^S$. But

$$
\begin{aligned}
j_\mathcal{P} \circ \sigma_p^S(0) &= j_p(S_p) && (7.34)\\
&= (j_\mu(S))_p && (7.26)\\
&= \sigma_p^{j_\mu(S)}(0) && (7.34)
\end{aligned}
$$

so that $j_\mathcal{P} \circ \sigma^S$ and $\sigma^{j_\mu(S)}$ have identical components. This completes the proof of (7.29) and hence of (7.30). □

With regard to the quotation of LAWVERE given in Subsection 6.3.8, we note that for any topos $\mathcal{E}$, the pair

$$\mathcal{E}_{\neg\neg} = (\mathcal{E}, \neg \circ \neg),$$

where $\neg \circ \neg$ is the composite of the negation truth arrow with itself, is an elementary site (cf. e.g., FREYD [19]). In $Set^\mathcal{P}$ the natural transformation $\neg$ has component $\neg_p : \Omega_p \to \Omega_p$ given by $\neg_p(S) = (S \mapsto \emptyset)_p$. We leave it to the reader to contemplate the details of

(7.37) if $\mathcal{P}$ is the canonical frame for the logic $IC = IK + (\nabla A \equiv \sim\sim A)$, then for all A, $\vdash_{IC} A$ iff $(Set^\mathcal{P})_{\neg\neg} \models A$;

(7.38) $\models_{IC} A$ iff $(\mathcal{E}_{\neg\neg} \models A$ for all elementary topoi $\mathcal{E}$).

6.8   Finite Models and Decidability

In this final section we establish that the set of $\mathcal{J}$-theorems is recursive by showing $\mathcal{J}$ to have the *finite model property*. This means that each non-theorem A of $\mathcal{J}$ is falsifiable on a finite $\mathcal{J}$-frame whose maximum size is determined effectively by A. Since the concept of $\mathcal{J}$-frame is decidable for finite structures, this gives a decision procedure for $\mathcal{J}$-theoremhood. One simply has to enumerate all finite $\mathcal{J}$-frames up to a certain prescribed size and test A for validity on each of them. The method of *filtration* of models that we use will be applied also to some further completeness theorems, as well as to the decidability of the logic IC.

Let $\mathcal{M}_L$ be the canonical model of a normal logic L, and $\tau \subseteq \Psi$ a set of sentences closed under subsentences. Define an equivalence relation $\approx_\tau$ on P_L by

(8.1) $p \approx_\tau q$ iff $p \cap \tau = q \cap \tau$
$\qquad\qquad$ iff $\forall A \in \tau(\mathcal{M}_L \models_p A$ iff $\mathcal{M}_L \models_q A)$.

Let

(8.2) $|p| = \{q : p \approx_\tau q\}$,

(8.3) $P_\tau = \{|p| : p \in P_L\}$.

Defining a further relation on P_L by

(8.4) $p\tau q$ iff $p \cap \tau \subseteq q \cap \tau$

(so that $p \approx_\tau q$ iff $p\tau q\tau p$), a well defined partial-ordering on P_τ is given by

(8.5) $|p| \sqsubseteq |q|$ iff $p\tau q$.

A valuation on $(P_\tau, \sqsubseteq)$ is given by

(8.6) $V_\tau(\pi) = \{|p| : \pi \in p \cap \tau\}$.

Then if $\prec$ is any relation on P_τ such that the following conditions obtain,

(8.7) $\mathcal{P}_\tau = (P_\tau, \sqsubseteq, \prec)$ is a frame (i.e. satisfies (3.4)),

(8.8) $|p| \prec |q|$ implies $\{B : \nabla B \in p \cap \tau\} \subseteq q$,

(8.9) $p \prec_L q$ implies $|p| \prec |q|$,

we say that the model $\mathcal{M}_\tau = (\mathcal{P}_\tau, V_\tau)$ is a *filtration of* $\mathcal{M}_L$ *through* τ.

(8.10) **Filtration Theorem.**

    For any $A \in \tau$ and $p \in P_L$, $\mathcal{M}_L \models_p A$ iff $\mathcal{M}_\tau \models_{|p|} A$.

Proof. For any non-modal $A \in \Phi$ the proof is given by SEGERBERG [84, Theorem 4.1]. For the inductive case $A = \nabla B$, assuming the result for B, we proceed as follows (exactly as for classical modal logic). First, suppose that $\mathcal{M}_\tau \models_{|p|} \nabla B$. Then if $p \prec_L q$, $|p| \prec |q|$ by (8.9), so $\mathcal{M}_\tau \models_{|q|} B$. The inductive hypothesis then gives $\mathcal{M}_L \models_q B$. Hence $\mathcal{M}_L \models_p \nabla B$. Conversely, if the latter condition obtains, we have $\nabla B \in p$ (6.20). But τ is closed under subsentences, so $B \in \nabla_p \cap \tau$. Then if $|p| \prec |q|$, (8.8) yields $B \in q$, hence by (6.20) and the inductive hypothesis $\mathcal{M}_\tau \models_{|q|} B$. This shows that $\mathcal{M}_\tau \models_{|p|} \nabla B$. $\square$

We shall call τ *logically L-finite* if there exists a finite subset τ_0 of τ such that

(8.11) $\forall A \in \tau \, \exists A_0 \in \tau_0 : \vdash_L (A \equiv A_0)$.

τ_0 will be called an *L-base* for τ. Since

(8.12) $\vdash_L A \equiv A_0$ iff $|A|_L = |A_0|_L$

(this follows from LINDENBAUM's Lemma), we have that if τ has L-base τ_0,

(8.13) $p \approx_\tau q$ iff $p \cap \tau_0 = q \cap \tau_0$,

so that $|p| \mapsto p \cap \tau_0$ gives a well-defined injection of P_τ into the power set of τ_0. Thus $\mathcal{M}_\tau$ will be finite with at most 2^n elements, where n is the cardinality of τ_0.

We observe next that filtrations of $\mathcal{M}_L$ through τ always exist, for we may define

(8.14) $|p| \prec |q|$ iff $\{B : \nabla B \in p \cap \tau\} \subseteq q$

to get a relation satisfying (8.7)–(8.9). Since any other relation satisfying (8.7)–(8.9) is (by (8.8)) contained in the one given by (8.14), the model resulting from (8.14) is called the *largest* filtration of $\mathcal{M}_L$ through τ.

(8.15) **Finite Model Property for** IK.

$\vdash_{IK} A$ *iff* $\mathcal{P} \models A$ *for every frame* $\mathcal{P}$ *of cardinality* $\leq 2^n$, *where* n *is the number of subsentences of* A.

Proof. Soundness is clear. Conversely, if not $\vdash_{IK} A$ then for some $p \in P_{IK}$, not $\mathcal{M}_{IK} \models_p A$ (6.20). Let τ be the set of subsentences of A, and $\mathcal{M}_\tau$ the largest filtration of $\mathcal{M}_{IK}$ through τ. Then A is false in $\mathcal{M}_\tau$ at $|p|$ (8.10), and the finite frame $\mathcal{P}_\tau$ has at most 2^n elements since τ is an IK-base for itself.   □

As well as giving decidability of IK, (8.15) has the corollary

(8.16) IK *is determined by the class of finite frames.*

For the decidability of $\mathcal{J}$ we need to consider a subrelation of (8.14), defined by[1]

(8.17) $|p| \prec' |q|$ iff $\exists t \in \mathcal{P}_L \, \exists q' \in |q| \, (p\tau t \text{ and } t \prec_L q')$.

(8.18) $\mathcal{P}'_\tau = (P_\tau, \sqsubseteq, \prec')$ *is a frame, and* $\mathcal{M}'_\tau = (\mathcal{P}'_\tau, V_\tau)$ *is a filtration of* $\mathcal{M}_L$ *through* τ.

Proof. To show $\mathcal{P}'_\tau$ satisfies (3.4), suppose $|p| \sqsubseteq |q| \prec' |r|$. Then $p \cap \tau \subseteq q \cap \tau \subseteq t \cap \tau$, for some $t \in P_L$, with $t \prec_L r'$ for some $r' \in |r|$ ((8.5) and (8.17)). But then $p\tau t$ and $t \prec_L r'$, so that $|p| \prec' |r'| = |r|$. Hence (8.7) holds. For (8.8), if $p\tau t \prec_L q'$, and $\nabla B \in p \cap \tau$, then $\nabla B \in t \cap \tau$, and so $B \in q' \cap \tau = q \cap \tau$. Finally, for (8.9), if $p \prec_L q$, then $p\tau p \prec_L q$, making $|p| \prec' |q|$.   □

(8.19) *If* L *contains the logic* $\mathcal{J}$, *then* $\mathcal{P}'_\tau$ *is a* $\mathcal{J}$-*frame.*

Proof. If $|p| \prec' |q|$, there exists t with $p\tau t \prec_L q'$ for some $q' \in |q|$. By (6.27), $\mathcal{P}_L$ is increasing, and so $t \subseteq q'$. But then we see that $p\tau t\tau q'$, hence $p\tau q'$, so $|p| \sqsubseteq |q'| = |q|$. Thus $\mathcal{P}'_\tau$ is increasing. Secondly, we show that $\mathcal{P}'_\tau$ is dense, using the fact (6.27) that $\mathcal{P}_L$ is dense. For, if $p\tau t \prec_L q$, then there exists $s \in P_L$, with $t \prec_L s \prec_L q$. But $p\tau t \prec_L s$ gives $|p| \prec' |s|$, while $s\tau s \prec_L q$ gives $|s| \prec' |q|$.   □

The upshot of (8.19) is that any non-theorem of $\mathcal{J}$ is falsifiable on a finite $\mathcal{J}$-frame, namely a filtration of $\mathcal{M}_\mathcal{J}$ of the type $\mathcal{P}'_\tau$. Thus the Finite Model Property Theorem holds exactly as stated in (8.15) with $\mathcal{J}$ in place of IK. We also have now established that $\mathcal{J}$ is determined by the class of all finite $\mathcal{J}$-frames.

---

[1] Definition (8.17) and its application have been modified from the original version of this article, since the latter was incorrect.

It was pointed out in Section 6.3 that the finite frames for the logic $IC = IK + (\nabla A \equiv \sim\sim A)$ (which contains $\mathcal{J}$) are precisely those satisfying

(3.45) $p \prec q$ iff q is a $\sqsubseteq$-maximal member of $[p)$.

Our filtration definition can be used now to show that IC is determined by its finite frames.

(8.20) *If $\vdash_L \nabla A \equiv \sim\sim A$, and τ is closed under negation (i.e. $B \in \tau$ only if $\sim B \in \tau$), then $\mathcal{P}'_\tau$ satisfies (3.45) and so is an IC-frame if it is finite.*

Proof. From (6.29) and (6.30) we know that $\mathcal{P}_L$ is increasing, has $\mu_L(p)$ cofinal with $[p)$, and the members of $\mu_L(p)$ are maximal in $[p)$. By (8.19) we also have $\mathcal{P}'_\tau$ increasing. Now suppose in $\mathcal{P}'_\tau$ that $|p| \prec' |q|$, with $p\tau t \prec_L q$ for some t. Then if $|q| \sqsubseteq |r|$, we have $q \cap \tau \subseteq r \cap \tau$. But from the proof of (6.30) we have that $\nabla(B \vee \sim B) \in t$, and so $B \vee \sim B \in q$, for all $B \in \Psi$. Then if $B \in r \cap \tau$ we must have $B \in q \cap \tau$, or else $\sim B \in q$, and since τ is closed under negation this would make $\sim B \in q \cap \tau \subseteq r$. The latter is incompatible with $B \in r$. Thus $r \cap \tau \subseteq q \cap \tau$, making $|q| = |r|$, and so $|q|$ is $\sqsubseteq$-maximal in $\{|r| : |p| \sqsubseteq |r|\}$.

Conversely, suppose this last condition holds, i.e. that $|p| \sqsubseteq |q|$, and

(8.21) $|q| \sqsubseteq |r|$ implies $|q| = |r|$.

Now in $\mathcal{P}_L$, $\mu_L(q)$ is cofinal with $[q)$, and since $q \in [q)$ there must be some $r \in [q)$, i.e. $q \subseteq r$, with $q \prec_L r$. But then $q \cap \tau \subseteq r \cap \tau$, i.e. $|q| \sqsubseteq |r|$, so by (8.21) $|q| = |r|$. However, since $|p| \sqsubseteq |q|$, we have $p\tau q \prec_L r$, so $|p| \prec' |r| = |q|$. This completes the derivation of (3.45) and the proof of (8.20).   □

In order to apply (8.20), we take τ to be the closure under negation of the set of subsentences of a given non-theorem A of IC. Denoting this set of subsentences by τ_A, it follows from the fact that $\vdash_I (\sim\sim\sim B \equiv \sim B)$ that τ is logically IC-finite, and that

$$\tau_0 = \tau_A \cup \{\sim B : B \in \tau_A\} \cup \{\sim\sim B : B \in \tau_A\}$$

is an IC-base for τ. The filtration $\mathcal{P}'_\tau$ of $\mathcal{M}_{IC}$ through τ will be a finite IC-frame with $\mathcal{M}'_\tau$ a falsifying model on it for A. Thus the Finite Model Property holds for IC with the upper bound in (8.15) modified to 2^{3n}.

Finally, we return once more to the limit point condition

(3.26) $p \prec q$ iff $p \sqsubset q$.

Let FN denote the logic obtained by adjoining to IK the schemata

(3.29) $\nabla A \to (B \vee (B \to A))$,

(3.27) $(\nabla A \to A) \to \nabla A$.

Notice that since $\vdash_I A \to (\nabla A \to A)$, we have $\vdash_{FN} A \to \nabla A$, so FN contains the logic N of Section 6.6: the latter, as shown there, being determined by the frame condition (3.26). We are going to prove that FN is determined by the class of finite frames satisfying this condition (soundness was noted in Section 6.3). To do this we reverse our approach, so that instead of showing that a filtration has the relevant frame condition, we prove that the condition implies the Filtration Theorem.

(8.22) *Let L be a normal logic containing FN and τ a set of sentences closed under subsentences and under the implication connective. Then in the model $\mathcal{M} = (P_\tau, \sqsubseteq, \sqsubset)$ we have for all $p \in P_L$ and $A \in \tau$, $\mathcal{M}_L \models_p A$ iff $\mathcal{M} \models_{|p|} A$.*

Proof. The only new part is the inductive step $A = \nabla B$, for which we use the obvious fact (cf. (8.4), (8.5)) that

(8.23) $|p| \sqsubset |q|$ iff $p \cap \tau \subsetneq q \cap \tau$.

Now if $\mathcal{M}_L \models_p \nabla B$, then $\nabla B \in p$ (6.20). Then if $|p| \sqsubset |q|$ there exists (8.23) some $C \in q \cap \tau$ with $C \notin p$. Since

$$(\nabla B \to (C \vee (C \to B))) \in p \qquad (3.29),$$

we then get $(C \to B) \in p$. But τ is closed under $\to$, so $(C \to B) \in p \cap \tau \subset q \cap \tau$. Using $C \in q$ we then get $B \in q$, whence by (6.20) and the induction hypothesis, $\mathcal{M} \models_{|q|} B$. This shows $\mathcal{M} \models_{|p|} \nabla B$. On the other hand, suppose that not $\mathcal{M}_L \models_p \nabla B$, i.e. $\nabla B \notin p$. By the FN-axiom (3.27) it follows that $(\nabla B \to B) \notin p$. But then by properties of $\mathcal{M}_L$, and the semantic clause for $\to$ (2.5), there exists $q \in P_L$ with $p \subseteq q$, $\nabla B \in q$, and $B \notin q$. By induction hypothesis, $\mathcal{M} \models_{|q|} B$ fails. But $\nabla B \notin p$, and so $p \cap \tau \subset q \cap \tau$, giving $|p| \sqsubset |q|$. Thus $\mathcal{M} \models_{|p|} \nabla B$ fails. This completes the proof of (8.22). $\square$

Taking τ now to be the closure of the set of subsentences of A under $\to$, then if not $\vdash_{FN} A$ the model $\mathcal{M}$ constructed from $\mathcal{M}_{FN}$ as in (8.22) will falsify A at some point and, by definition, will be based on an FN-frame. But it is known from the work of DIEGO [9] that there exists a primitive recursive function f such that if σ is a set of sentences of finite cardinality n, there are at most a finite number $f(n)$ of sentences constructible from σ by the implication connective that are deductively non-equivalent over I. Thus the τ just described is logically finite, hence $\mathcal{M}$ is finite. We see then that the Finite Model Property holds for FN, with an upper-bound on the models of the form $2^{f(n)}$.

7

The Semantics of Hoare's Iteration Rule

ABSTRACT. Hoare's Iteration Rule is a principle of reasoning that is used to derive correctness assertions about the effects of implementing a **while**-command. We show that the propositional modal logic of this type of command is axiomatised by Hoare's Rule in conjunction with two additional axioms. The proof also establishes decidability of the logic. The paper concludes with a discussion of the relationship between the logic of "**while**" and Segerberg's axiomatisation of propositional dynamic logic.

Introduction

The *modal logic of programs* proposed by PRATT [71] associates with each command α a modal connective $[\alpha]$ that is read "after α terminates...". Thus the symbolism

$$A \to [\alpha]B$$

expresses the *partial correctness assertion* "if A is true (now), then after α terminates B will be true". In these terms, the *Iteration Rule* introduced by HOARE [45] for reasoning about **while**-commands takes the following form:

$$\text{if} \quad \vdash e \wedge A \to [\alpha]A,$$
$$\text{then} \quad \vdash A \to [\textbf{while } e \textbf{ do } \alpha](A \wedge \neg e).$$

The validity of this inference rule is based on the fact that a performance of (**while** e **do** α) consists of a finite sequence of executions of α, leading to a state in which e is false, with each execution starting in a state in which e is true. The premiss of the rule asserts that the sentence A is an *invariant* of each step in such a sequence, i.e. if it is true at the start of the step, then it is still true when the step ends. From this the rule infers that if A is true at the outset, then when the whole sequence is finished we will have A still true, with e false.

The Iteration Rule has been used to establish the correctness of many specifications of algorithms (cf. ALAGIC and ARBIB [1] for an introduction to this methodology). Examples have also been given (cf. WAND [105]) of correctness assertions that it is incapable of deriving. A potential explanation of this phenomenon is that we really need an *infinitary* inference rule to obtain certain assertions about **while**-commands, since there are infinitely many possibilities for the length of a finite sequence, and we need one premiss for each such possibility.

Another source of "incompleteness" of Hoare's Rule is that while it allows us to draw conclusions about what happens *if* a **while**-command terminates, it does not allow us to establish *that* it terminates. Thus, for instance, we could use the Rule to infer that the command fails to terminate, by deriving

[**while** e **do** α]**false**,

(where **false** is some constantly false, or contradictory, assertion), but we cannot, as we shall see, use it to derive the sentence

(1) $\neg e \to \neg$[**while** e **do** α]**false**,

which expresses the valid principle that a **while**-command terminates if its test expression is false.

In this article a completeness theorem for a propositional logic of programs is presented which establishes that the meaning of (**while** e **do** α) is exactly characterised by the Iteration Rule in consort with (1) and

(2) $e \to ($[**while** e **do** α]$A \to [\alpha]$[**while** e **do** α]$A)$.

Hoare's Rule itself corresponds to the semantic principle that every execution of (**while** e **do** α) consists of a sequence of the type described above. (1) and (2) are needed for the other side of the coin, viz. that every such sequence constitutes an execution of the command.

These results were first announced in GOLDBLATT [38]. In a subsequent monograph (GOLDBLATT [23]) the author has developed a completeness theorem for the program logic over a general first-order language, using an infinitary analogue of Hoare's Rule. The first stage of the proof is a completeness theorem for a propositional logic, using the same infinitary rule. However, whereas this rule is unavoidable in general in the presence of elementary quantification, at the propositional level the set of valid formulae is decidable and can be given a finitary axiomatisation. The burden of this article is to establish that fact. This will be done in the context of a simplification: we overlook the distinction drawn in [23] between *external* and *internal* logic, i.e. between the logical operations performed by the programmer in reasoning about program behaviour, and those performed by the computer in evaluating test expressions. Internal logic is the logic of the Boolean expression e in (**while** e **do** α). It involves a "sequential" interpretation of connectives, and a *three*-valued semantics to accomodate the possibility that the computer may leave an expression undefined – e.g. when its evaluation fails to terminate. On the other hand, external logic, which is a version of "logic without existence assumptions" at the first-order level, concerns modal formulae that express assertions about programs, and is *two*-valued – such assertions either being the case or not. However for this paper we will simplify matters by using a two-valued semantics throughout. Readers are invited to satisfy themselves that the results developed below extend to the logical system of Chapter 2 of [23].

Syntax

Our formal language contains the following syntactic categories:

Boolean variables:	$p \in Bvb$
Program letters:	$\pi \in Prl$
Boolean expressions:	$e \in Bxp$
Commands:	$\alpha \in Cmd$
Formulae:	$A \in Fma$

Bvb and Prl are two disjoint denumerable sets, from which Bxp, Cmd, and Fma are generated by the BNF-style definitions

$$
\begin{aligned}
e \quad &::= \quad p \mid \textbf{false} \mid e_1 \rightarrow e_2 \\
\alpha \quad &::= \quad \pi \mid \textbf{skip} \mid \textbf{abort} \mid \alpha_1 ; \alpha_2 \mid \textbf{if } e \textbf{ then } \alpha_1 \textbf{ else } \alpha_2 \mid \\
&\qquad \alpha_1 \textbf{ or } \alpha_2 \mid \textbf{while } e \textbf{ do } \alpha \\
A \quad &::= \quad p \mid \textbf{false} \mid A_1 \rightarrow A_2 \mid [\alpha]A.
\end{aligned}
$$

Thus $Bxp \subseteq Fma$. From the material implication connective $\rightarrow$, and the propositional constant **false**, the standard Boolean connectives $\neg$, $\wedge$, $\vee$, $\leftrightarrow$ are defined in the usual way. **skip** and **abort** are constants whose meaning will be evident from the formal semantics to follow. $(\alpha_1 ; \alpha_2)$ is the *composite* of α_1 and α_2, executed by doing α_1 and then doing α_2. (**if** e **then** α_1 **else** α_2) is the *conditional* command executed by performing α_1 if e is true, and α_2 otherwise. $(\alpha_1 \textbf{ or } \alpha_2)$ is a non-deterministic command executed by arbitrarily choosing to execute α_1 or α_2.

Semantics

A *model* is a structure $\mathcal{M} = (S, V, R(\cdot))$, where

(i) S is a non-empty set (of "states");

(ii) V is a *valuation* that assigns to each $p \in Bvb$ a subset $V(p)$ of S;

(iii) $R(\cdot)$ is an operator that assigns to each $\alpha \in Cmd$ a binary relation $R(\alpha)$ on S.

The property "A is *true (holds) at s in* $\mathcal{M}$", symbolised $\mathcal{M} \models_s A$, is defined by induction on the number of symbols in A, as follows (the prefix $\mathcal{M}$ may be dropped if it is clear which model is intended).

$$
\begin{array}{lll}
\mathcal{M} \models_s p & \text{iff} & s \in V(p) \\
\mathcal{M} \not\models_s \textbf{false} & & (\text{i.e. not } \mathcal{M} \models_s \textbf{false}) \\
\mathcal{M} \models_s A_1 \rightarrow A_2 & \text{iff} & \mathcal{M} \models_s A_1 \text{ implies } \mathcal{M} \models_s A_2 \\
\mathcal{M} \models_s [\alpha]A & \text{iff} & \text{for all } t \in S, sR(\alpha)t \text{ implies } \mathcal{M} \models_t A
\end{array}
$$

(notice that the set $\{s : \mathcal{M} \models_s e\}$ is determined, for any $e \in Bxp$, by V and does not depend at all on $R(\cdot)$).

We say that A is *true in* $\mathcal{M}$, denoted $\mathcal{M} \models A$, if A is true at every $s \in S$ in $\mathcal{M}$.

Our attention will be focused on models in which the properties of $R(\alpha)$ reflect the intuitive meaning of the command α . To describe these properties we introduce some notation about binary relations R on S.

$$
\begin{array}{llll}
\textit{Restrictions}: & A \upharpoonright R & = & \{(s,t) : sRt \text{ and } \models_s A\} \\
& R \upharpoonright A & = & \{(s,t) : sRt \text{ and } \models_t A\} \\
\textit{Composition}: & P \circ R & = & \{(s,t) : \text{ for some } u, sPuRt\} \\
\textit{Equality}: & E_S & = & \{(s,s) : s \in S\} \\
\textit{Iteration}: & R^0 & = & E_S \\
& R^{n+1} & = & R^n \circ R \\
\textit{Closure}: & R^* & = & \bigcup_{n<\omega} R^n.
\end{array}
$$

Now in any model there are standard set-theoretical operations that can be applied to the subcommands of α and their interpretations to assign to α a relation on S that corresponds to the intended way that α is to be performed. This *standard meaning* of α will be denoted $\mathcal{M}(\alpha)$, and is defined as follows.

$$
\begin{array}{lll}
\mathcal{M}(\textbf{skip}) & = & E_s, \text{ i.e. } s\mathcal{M}(\textbf{skip})t \text{ iff } s = t \ ; \\
\mathcal{M}(\textbf{abort}) & = & \emptyset, \text{ i.e. not } s\mathcal{M}(\textbf{abort})t \text{ for any } s,t; \\
\mathcal{M}(\alpha_1;\alpha_2) & = & R(\alpha_1) \circ R(\alpha_2) \\
\mathcal{M}(\textbf{if } e \textbf{ then } \alpha_1 \textbf{ else } \alpha_2) & = & (e \upharpoonright R(\alpha_1)) \cup (\neg e \upharpoonright R(\alpha_2)), \\
\mathcal{M}(\alpha_1 \textbf{ or } \alpha_2) & = & R(\alpha_1) \cup R(\alpha_2) \\
\mathcal{M}(\textbf{while } e \textbf{ do } \alpha) & = & (e \upharpoonright R(\alpha))^* \upharpoonright \neg e, \text{ i.e.}
\end{array}
$$

$s\mathcal{M}(\textbf{while } e \textbf{ do } \alpha)t$ if, and only if, for some $n < \omega$, and some $s_0, \ldots, s_n \in S$, we have $s_0 = s$, $s_n = t$, and $\mathcal{M} \not\models_t e$, with $s_i R(\alpha) s_{i+1}$ and $\mathcal{M} \models_{s_i} e$ whenever $0 \leq i < n$.

(For program letters we may take $\mathcal{M}(\pi) = R(\pi)$).

A model is *standard for α* if $\mathcal{M}(\alpha) = R(\alpha)$, and *standard*, simpliciter, if it is standard for all $\alpha \in Cmd$. Given S, V, and $R(\pi)$ for all $\pi \in Prl$, we can systematically construct a standard model: proceeding by induction on the length of α we *define* $R(\alpha)$ to be $\mathcal{M}(\alpha)$ as given by the above equations. In other words, once the $R(\pi)$'s are given the standard-model structure is uniquely determined.

(3) **Theorem.** *Let $\mathcal{M}$ be a model in which*

$$R(\textbf{while } e \textbf{ do } \alpha) \subseteq \mathcal{M}(\textbf{while } e \textbf{ do } \alpha).$$

Then $\mathcal{M}$ validates Hoare's Rule, i.e.

$$\mathcal{M} \models e \wedge A \rightarrow [\alpha]A \quad only \; if$$
$$\mathcal{M} \models A \rightarrow [\textbf{while } e \textbf{ do } \alpha](A \wedge \neg e).$$

Proof. Let A be true at s in $\mathcal{M}$. To show that $[\textbf{while } e \textbf{ do } \alpha](A \wedge \neg e)$ must then also be true at s we must show that if $sR(\textbf{while } e \textbf{ do } \alpha)t$, then $A \wedge \neg e$ holds at t. But given such a t, the hypothesis on $\mathcal{M}$ implies that there exists a sequence $s = s_0, \ldots, s_n = t$, for some $n \geq 0$, that has the properties described in the definition of $\mathcal{M}(\textbf{while } e \textbf{ do } \alpha)$. But if $e \wedge A \rightarrow [\alpha]A$ is true in $\mathcal{M}$, then whenever $rR(\alpha)u$, with A and e true at r, we have A true at u. But then as A is true at s_0, it follows by induction on i that A is true at s_i for all $i \leq n$. In particular A is true at $s_n = t$ as desired. Moreover, as e fails to hold at t, $\neg e$ is true there, and hence so is $A \wedge \neg e$. $\qquad \qquad \square$

Proof Theory

A *logic* is any set L of formulae that satisfies:

(i) L contains all instances of the schemata

     A1: $A \rightarrow (B \rightarrow A)$

     A2: $(A \rightarrow (B \rightarrow C)) \rightarrow ((A \rightarrow B) \rightarrow (A \rightarrow C))$

     A3: $\neg\neg A \rightarrow A$;

(ii) L is closed under *Detachment*, i.e.

$$A, (A \rightarrow B) \in L \text{ only if } B \in L;$$

(iii) L contains all instances of

     A4: $[\alpha](A \rightarrow B) \rightarrow ([\alpha]A \rightarrow [\alpha]B);$

(iv) L is closed under the α - *Termination Rule*, for all α, i.e.

$$A \in L \text{ only if } [\alpha]A \in L.$$

As is well known, (1) and (2) provide an adequate basis for the classical Propositional Calculus (PC), so that every instance in Fma of a PC-tautology belongs to L.

If X is a subset of Fma, we say that A is *deducible from X in L*, in symbols $X \vdash_L A$, if there is a finite sequence $A_1, \ldots, A_n$ of formulae such that $A_n = A$, and for all $i \leq n$ either $A_i \in X$ or $A_i \in L$, or there are $j, k < i$ such that A_k is $A_j \to A_i$ (so that A_i is deducible from A_j and A_k by Detachment). As a special case of this relation, we put $\vdash_L A$ if $\emptyset \vdash_L A$, and observe that this obtains iff $A \in L$.

A set X is *L-consistent* if $X \nvdash_L$ **false**, and *L-maximal* if it is L-consistent and contains one of A and $\neg A$, for each $A \in Fma$ (this is equivalent to requiring that X not be a subset of any other L-consistent set).

The presence of PC in L suffices to establish the *Deduction Theorem* for L:

$$X \cup \{A\} \vdash_L B \quad \text{iff} \quad X \vdash_L (A \to B),$$

and this is used to prove, still only using PC, the result known as *Lindenbaum's Lemma*, viz.

Every L-consistent set has an L-maximal extension.

From this follows

(4) $X \vdash_L A$ *iff A belongs to every L-maximal extension of X;*

(5) $\vdash_L A$ *iff A belongs to every L-maximal set.*

The essential role of A4 and the α-Termination Rule in the proof theory of L is to yield

(6) $X \vdash_L [\alpha]A$ *iff A belongs to every L-maximal extension of $X(\alpha) = \{B : [\alpha]B \in X\}$.*

The Canonical Model

Let $\mathcal{M}_L = (S_L, V_L, R_L(\cdot))$, where

(i) S_L is the set of all L-maximal subsets of Fma;

(ii) $V_L(p) = \{s \in S_L : p \in s\}$;

(iii) $sR_L(\alpha)t$ iff $s(\alpha) \subseteq t$ iff $\{B : [\alpha]B \in s\} \subseteq t$.

The fundamental property of this model is that for any $A \in Fma$, and any $s \in S_L$,

(7) $\mathcal{M}_L \models_s A$ iff $A \in s$.

This is proven by induction on the length of A, with (4) and (6) being invoked to show that

$$[\alpha]A \in s \quad \text{iff} \quad sR_L(\alpha)t \text{ implies } A \in t.$$

From (5) and (7) we obtain that in general,

$$\vdash_L A \quad \text{iff} \quad \mathcal{M}_L \models A,$$

i.e. the formulae true in $\mathcal{M}_L$ are precisely the L-theorems. Hence $\mathcal{M}_L$ is known as the *canonical* model for L.

$\mathcal{M}_L$ as a Standard Model

The theory just outlined is by now standard material in the study of propositional modal logics, and the reader will find a full account of it, with proofs, in e.g. LEMMON [59] or CHELLAS [8]. The usefulness of the canonical model resides in the fact that in order to prove that L is determined by a certain class $\mathcal{C}$ of models, i.e. that

$$\vdash_L A \quad \text{iff} \quad \text{for all } \mathcal{M} \in \mathcal{C}, \ \mathcal{M} \models A,$$

it suffices to show

(i)   each member of $\mathcal{C}$ is an L-model;   and

(ii) $\mathcal{M}_L \in \mathcal{C}$.

Now consider the following axiom schemata:

A5: $[\text{skip}]A \leftrightarrow A$

A6: $[\text{abort}]\text{false}$

A7: $[\alpha_1; \alpha_2]A \leftrightarrow [\alpha_1][\alpha_2]A$

A8: $[\text{if } e \text{ then } \alpha_1 \text{ else } \alpha_2]A \leftrightarrow (e \rightarrow [\alpha_1]A) \wedge (\neg e \rightarrow [\alpha_2]A)$

A9: $[\alpha_1 \text{ or } \alpha_2]A \leftrightarrow [\alpha_1]A \wedge [\alpha_2]A.$

It is readily seen that A5–A9 are true in all standard models. To show then that these axioms *characterise* the standard-model conditions for the commands they refer to, it suffices to show that if L contains all instances of the axiom in question, then $\mathcal{M}_L$ satisfies the corresponding condition. We leave this as an exercise for the reader. The only case that is not straightforward is to show that $\mathcal{M}_L$ satisfies

$$R_L(\alpha_1; \alpha_2) \subseteq R_L(\alpha_1) \circ R_L(\alpha_2)$$

in the presence of A7. A proof of this fact may be found in [88, §4] or [32, Theorem 10.3].

The situation for **while**-commands is however rather different. If we confine ourselves to finitary proof theory, then there is no way to make $\mathcal{M}_L$ standard for all **while**-commands. To see this, take a particular Boolean variable p and program letter π and consider the set

$$X = \{[\pi]^n p : n < \omega\} \cup \{\neg[\textbf{while } p \textbf{ do } \pi]\textbf{false}\},$$

where

$$[\pi]^0 p = p, \quad \text{and}$$
$$[\pi]^{n+1} p = [\pi][\pi]^n p.$$

Now in any model we have

$$\models_s [\pi]^n p \quad \text{iff} \quad sR(\pi)^n t \text{ implies } \models_t p,$$

so that in a *standard* model X cannot be satisfied, i.e. there is no point s at which all members of X are simultaneously true. For, if $[\pi]^n p$ holds at s for all n, then (**while** p **do** π) will not terminate if started in s. But every finite subset of X can be satisfied in some standard model. Adapting the example of SEGERBERG [88, §4], for each $k < \omega$ let $\mathcal{M}_k = (\omega, V_k, R(\cdot))$ be a standard model based on ω in which $R(\pi)$ is the graph of the successor function, and $V_k(p) = \{0, \dots, k\}$. Then every member of $\{[\pi]^n p : n \leq k\} \cup \{\neg[\textbf{ while } p \textbf{ do } \pi] \textbf{ false}\}$ holds at 0 in $\mathcal{M}_k$.

Now if L is a logic generated by adding to A1–A9 any number of axioms and any number of *finitary* inference rules (i.e. ones taking finitely many formulae as premisses), then a set will be L-consistent whenever each of its finite subsets is. If all L-theorems are true in standard models, then for the above construction every finite subset of X is satisfied in some L-model and so must be L-consistent. Hence X itself will be L-consistent and so by Lindenbaum's Lemma will have an L-maximal extension, say s. But then by (7), X is simultaneously satisfied at s in $\mathcal{M}_L$, and so it follows that $\mathcal{M}_L$ cannot be standard for the command (**while** p **do** π) (even though $\mathcal{M}_L$ may well validate the Iteration Rule).

Filtrations

We wish to show that under certain conditions a logic is determined by its standard models, i.e.

$$\vdash_L A \quad \text{iff} \quad A \text{ is true in all standard } L\text{-models.}$$

Now if $\nvdash_L A$, then we know that $\mathcal{M}_L$ is always a falsifying model for A, but not in general a standard one. To remedy this defect we will use the method of *filtrations* to "collapse" $\mathcal{M}_L$ to a standard falsifying model for A. This will produce a different falsifying model for each A, and moreover a finite one whose size is effectively determined by the length of A.

Filtrations are constructed as follows. Let Z be a set of formulae that is closed under subformulae. Then Z determines an equivalence relation on S_L by putting

$$s \sim t \quad \text{iff} \quad s \cap Z = t \cap Z.$$

Let $|s| = \{t : s \sim t\}$ be the $\sim$-equivalence class of s, and put

$$S/Z = \{|s| : s \in S_L\}.$$

The assignment of $s \cap Z$ to $|s|$ is a well-defined injection of S/Z into the power-set of Z. Thus if Z is finite, with say n members, then S/Z has at most 2^n members and is finite.

A valuation V_Z is well-defined on S/Z by putting

$$V_Z(p) = \begin{cases} \{|s| : p \in s\} & \text{if } p \in Z, \\ \emptyset & \text{otherwise.} \end{cases}$$

(Actually, the definition of $V_Z(p)$ when $p \notin Z$ is immaterial.) We shall be concerned with truth of members of Z in models of the form

$$\mathcal{M} = (S/Z, V_Z, R(\cdot)),$$

constructed by considering various definitions of $R(\alpha)$ on S/Z (hence the definition of $R(\alpha)$ will only be significant for those α's that occur in members of Z). For any model of this form, the following result is evident.

(8) **Theorem.** *If $e \in Bxp$, and every Boolean variable in e belongs to Z, then for any $s \in S_L$,*

$$\begin{aligned} e \in s & \quad \text{iff} \quad \mathcal{M} \models_{|s|} e, \quad i.e. \\ \mathcal{M}_L \models_s e & \quad \text{iff} \quad \mathcal{M} \models_{|s|} e. \end{aligned}$$

$\square$

The next result is crucial to our analysis of the Iteration Rule below.

(9) **Theorem.** *If Z is finite, then for any subset T of S/Z there is a formula A_T such that for all $s \in S_L$,*

$$A_T \in s \quad \text{iff} \quad |s| \in T.$$

Proof. For each $t \in S_L$, let A_t be the conjunction of

$$(t \cap Z) \cup \{\neg A : A \in Z - t\}$$

(which is finite as Z is). The definition of A_t depends only on $|t|$, in that $A_t = A_s$ iff $|t| = |s|$, and indeed

$$A_t \in s \quad \text{iff} \quad |s| = |t|.$$

Now if $T = \emptyset$, let A_T be **false** to obtain our desired conclusion. Otherwise, since S/Z is finite we may take T to be $\{|t_1|, \ldots, |t_m|\}$ for some $m \geq 1$, and some $t_1, \ldots, t_m \in S_L$. Let A_T then be

$$A_{t_1} \vee \ldots \vee A_{t_m},$$

so that

$$A_T \in s \quad \text{iff} \quad |s| = |t_1| \text{ or } \ldots \text{ or } |s| = |t_m|$$

as desired.   □

Now let α be a command that occurs in Z. A relation $R(\alpha)$ on S/Z will be called a *filtration of $R_L(\alpha)$ through Z* if the following two conditions hold.

I(α): $sR_L(\alpha)t$ implies $|s|R(\alpha)|t|$;

II(α): $|s|R(\alpha)|t|$ implies $\{A : [\alpha]A \in s \cap Z\} \subseteq t$.

Such relations always exist. The *smallest* is given by

$$|s|R(\alpha)|t| \quad \text{iff} \quad s'R_L(\alpha)t' \text{ for some } s' \sim s \text{ and } t' \sim t,$$

while the *largest* has

$$|s|R(\alpha)|t| \quad \text{iff} \quad \{A : [\alpha]A \in s \cap Z\} \subseteq t.$$

A model of the form $\mathcal{M} = (S/Z, V_Z, R(\cdot))$ will be called a *filtration of $\mathcal{M}_L$ through Z* if $R(\alpha)$ is a filtration of $R_L(\alpha)$ through Z for all α's that occur in Z. The proof of the next result may be found, e.g., in [59, §3], or [8, §3.5]. It extends Theorem (8) to include modalised formulae.

(10) **Theorem.** *If $\mathcal{M}$ is a filtration of $\mathcal{M}_L$ through Z, then for any $A \in Z$, and any $s \in S_L$,*

$$\mathcal{M}_L \models_s A \quad \text{iff} \quad \mathcal{M} \models_{|s|} A.$$

Hence for all $A \in Z$,

$$\mathcal{M}_L \models A \quad \text{iff} \quad \mathcal{M} \models A.$$

□

This result provides us with finite falsifying models: if $\nvdash_L A$, then $\mathcal{M}_L \nvDash A$, and so A is falsified by any filtration of $\mathcal{M}_L$ through any (finite) Z that contains A (e.g. Z could be the set of subformulae of A). In order to show that the filtration satisfies some desired property we can then make use of the properties of $\mathcal{M}_L$, as determined by the properties of L itself. We will do this in the next result, which displays the essential role of Hoare's Rule in the present theory of models.

(11) **Theorem.** *Let L be a logic that is closed under Hoare's Iteration Rule for the command* (**while** e **do** α), *i.e.*

$$\vdash_L e \wedge A \to [\alpha]A \quad \text{only if} \quad \vdash_L A \to [\textbf{while } e \textbf{ do } \alpha](A \wedge \neg e).$$

Suppose that Z is finite, closed under subformulae, and contains all Boolean variables appearing in e. Then if $\mathcal{M}$ is any model of the form $(S/Z, V_Z, R(\cdot))$ that satisfies the condition I(α), we have

$$sR_L(\textbf{while } e \textbf{ do } \alpha)t \quad \text{only if} \quad |s|\mathcal{M}(\textbf{while } e \textbf{ do } \alpha)|t|,$$

for all $s, t \in S_L$.

Proof. Take a particular $s \in S_L$ and define a subset T of S/Z by putting

$$x \in T \quad \text{iff} \quad \text{for some } n \geq 0, \; |s|(e \restriction R(\alpha))^n x.$$

Then we have, for any t,

$$|s|\mathcal{M}(\textbf{while } e \textbf{ do } \alpha)|t| \quad \text{iff} \quad |t| \in T \text{ and } \mathcal{M} \models_{|t|} \neg e.$$

Hence by Theorem (8),

(i) $\qquad |s|\mathcal{M}(\textbf{while } e \textbf{ do } \alpha)|t| \quad \text{iff} \quad |t| \in T \text{ and } \neg e \in t.$

Now by Theorem (9) there is a formula A such that in general

$$A \in t \quad \text{iff} \quad |t| \in T.$$

We can now show that the formula $(e \wedge A \to [\alpha]A)$ is in every L-maximal set, and hence is an L-theorem. To see this, we have to show that if $(e \wedge A) \in u \in S_L$, then $[\alpha]A \in u$. But if $(e \wedge A) \in u$, then $e \in u$ and $A \in u$, and so $|u| \in T$, i.e. $|s|(e \restriction R(\alpha))^n|u|$ for some n. Thus if $uR_L(\alpha)t$, for any t, then by I(α) we have $|u|R(\alpha)|t|$, and so by Theorem (8) $|u|(e \restriction R(\alpha))|t|$. This means that $|s|(e \restriction R(\alpha))^{n+1}|t|$, so that $|t| \in T$, and thus $A \in t$. It follows that $[\alpha]A \in u$ as desired.

By the assumed closure condition on L, we conclude that the formula $A \to [\textbf{while } e \textbf{ do } \alpha](A \wedge \neg e)$ is an L-theorem, and hence belongs to s. But $A \in s$, since $|s| \in T$ (take $n = 0$), and thus we have $[\textbf{while } e \textbf{ do } \alpha](A \wedge \neg e) \in s$. This allows us to complete the Theorem, since for any t, if $sR_L(\textbf{while } e \textbf{ do } \alpha)t$ we can now infer that $(A \wedge \neg e) \in t$, giving $A \in t$, whence $|t| \in T$, and also $\neg e \in t$. By (i) this implies that $|s|\mathcal{M}(\textbf{while } e \textbf{ do } \alpha)|t|$ as required. $\qquad \square$

Theorem (11) can be used to obtain a simple semantic characterisation of the Iteration Rule: if H is the smallest logic that is closed under the Rule, then the theorems of H are precisely those formulae that are true in all models that satisfy

(12) $\qquad R(\textbf{while } e \textbf{ do } \alpha) \subseteq \mathcal{M}(\textbf{while } e \textbf{ do } \alpha).$

Theorem (3) established that in any such model the set of true formulae is a logic that is closed under the Rule, and hence contains H. In other words, any model satisfying (12) is an H-model. To complete the characterisation we have to show that if $\nvdash_H A$, then A is falsified by some model satisfying (12). For this model we take a filtration of $\mathcal{M}_H$ through the set Z_A of subformulae of A in which $R(\alpha)$ is the smallest filtration of $R_L(\alpha)$ through Z_A if α occurs in A, and $\emptyset$ if not. The definition of the smallest filtration given earlier, together with Theorem (11), ensure that this model satisfies (12), as the reader may confirm.

Now the conclusion of Theorem (11) states that the relation

$$\mathcal{M}(\textbf{while } e \textbf{ do } \alpha)$$

satisfies the first of the two conditions necessary to make it a filtration of $R_L(\textbf{while } e \textbf{ do } \alpha)$. Criteria for the second condition are given by the following result.

(13) **Theorem.** *Let L be a logic that contains all instances of the schemata*

A10: $e \to ([\textbf{while } e \textbf{ do } \alpha]A \to [\,\alpha\,][\textbf{while } e \textbf{ do } \alpha]A)$

A11: $\neg e \to ([\textbf{while } e \textbf{ do } \alpha]A \to A)$.

Suppose that Z satisfies the closure condition

C1: $[\textbf{while } e \textbf{ do } \alpha]A \in Z$ *only if* $[\,\alpha\,][\textbf{while } e \textbf{ do } \alpha]A \in Z$.

Then if $\mathcal{M}$ is any model of the form $(S/Z, V_Z, R(\cdot))$ that satisfies the condition $\text{II}(\alpha)$, we have

$|s|\mathcal{M}(\textbf{while } e \textbf{ do } \alpha)|t|$ *only if* $\{A : [\textbf{while } e \textbf{ do } \alpha]A \in s \cap Z\} \subseteq t$.

Proof. Let $|s|\mathcal{M}(\textbf{while } e \textbf{ do } \alpha)|t|$. Then for some $n \geq 0$ there exist points $|s_0|, \ldots, |s_n|$, with $s_0 = s$, $s_n = t$, $\mathcal{M} \nvDash_{|t|} e$, and $|s_i|R(\alpha)|s_{i+1}|$ and $\mathcal{M} \vDash_{|s_i|} e$ whenever $0 \leq i < n$.

Now let $[\textbf{while } e \textbf{ do } \alpha]A \in s \cap Z$. We wish to show that $A \in t$. First we prove by induction on i that $[\textbf{while } e \textbf{ do } \alpha]A \in s_i$ whenever $0 \leq i \leq n$.

The case $i = 0$ holds by assumption. Next assume the result for some $i < n$. But $e \in s_i$ by Theorem (8), and so as all instances of A10 belongs to s_i we get $[\,\alpha\,][\textbf{while } e \textbf{ do } \alpha]A \in s_i$. By C1 this last formula is also in Z, and hence by $\text{II}(\alpha)$, since $|s_i|R(\alpha)|s_{i+1}|$ we get $[\textbf{while } e \textbf{ do } \alpha]A \in s_{i+1}$ as desired. In particular we can conclude that $[\textbf{while } e \textbf{ do } \alpha]A \in s_n = t$. But by Theorem (8) once more, $\neg e \in t$, and so we can apply A11 to obtain $A \in t$.   □

The two Theorems 11 and 13 combine as follows.

(14) **Theorem.** *Let L be a logic that is closed under the Iteration Rule and contains A10 and A11. Suppose that Z is finite, satisfies C1, is closed under subformulae, and contains all Boolean variables that occur in e. Then in any model of the form $\mathcal{M} = (S/Z, V_Z, R(\cdot))$, if $R(\alpha)$ is a filtration of $R_L(\alpha)$ through Z, then $\mathcal{M}(\textbf{while } e \textbf{ do } \alpha)$ is a filtration of $R_L(\textbf{while } e \textbf{ do } \alpha)$ through Z.*   □

In order to obtain a full axiomatisation of the formulae true in all standard models, we need analogues of C1 for the other types of commands we are dealing with. If $\mathcal{M}_L$ is standard for **skip** and **abort** (when L contains A5 and A6) then $\mathcal{M}(\textbf{skip})$ and $\mathcal{M}(\textbf{abort})$ will be filtrations of $R_L(\textbf{skip})$ and $R_L(\textbf{abort})$, respectively, through Z. Moreover, if $\mathcal{M}_L$ is standard for β, where β is any of $(\alpha_1; \alpha_2)$, $(\alpha_1 \textbf{ or } \alpha_2)$, (if e then α_1 else α_2), then the property $I(\beta)$ will hold with $\mathcal{M}(\beta)$ in place of $R(\beta)$, provided that $I(\alpha_1)$ and $I(\alpha_2)$ hold. For the condition $II(\beta)$ to hold however, Z must satisfy further closure conditions, viz.

C2: $[\alpha_1; \alpha_2]A \in Z$ only if $[\alpha_1][\alpha_2]A \in Z$;

C3: $[\alpha_1 \text{ or } \alpha_2]A \in Z$ only if $[\alpha_1]A, [\alpha_2]A \in Z$;

C4: $[\text{if } e \text{ then } \alpha_1 \text{ else } \alpha_2]A \in Z$ only if $[\alpha_1]A, [\alpha_2]A \in Z$.

(15) **Theorem.** *Suppose that L contains the schemata A7, A8, A9, and Z satisfies C2, C3, C4. Then in any model $\mathcal{M} = (S/Z, V_Z, R(\cdot))$, if $R(\alpha_1)$ and $R(\alpha_2)$ are filtrations of $R_L(\alpha_1)$ and $R_L(\alpha_2)$ through Z, respectively, then $\mathcal{M}(\beta)$ is a filtration of $R_L(\beta)$ through Z, where β is any of $(\alpha_1; \alpha_2)$, $(\alpha_1 \text{ or } \alpha_2)$, or $(\text{if } e \text{ then } \alpha_1 \text{ else } \alpha_2)$ with all Boolean variables of e occurring in Z.* $\square$

To obtain a filtration that is a standard model, we define

$$\mathcal{M}_L/Z = (S/Z, V_Z, R_Z(\cdot)),$$

where

(i) if α is a program letter, then $R_Z(\alpha)$ is the least filtration of $R_L(\alpha)$ through Z if $\alpha \in Z$, and $\emptyset$ otherwise;

(ii) if $\alpha \notin Prl$, $R_Z(\alpha)$ is inductively *defined* to be $\mathcal{M}_L/Z(\alpha)$.

The idea here, as explained earlier, is that the $R_Z(\pi)$'s, once given, generate a uniquely determined standard model based on $(S/Z, V_Z)$, and this is the model we take as $\mathcal{M}_L/Z$. In fact the definition of $R_Z(\pi)$ is immaterial if $\pi \notin Z$, and otherwise it matters only that $R_Z(\pi)$ be *some* filtration of $R_L(\pi)$. For then we have the following result, proved by induction on the lengths of commands.

(16) **Theorem.** *Let L contain all of A5–A10 and be closed under the Iteration Rule. Then if Z satisfies C1–C4 and is finite, the standard model $\mathcal{M}_L/Z$ is a filtration of $\mathcal{M}_L$ through Z.* $\square$

We denote by FPL (Finitary Program Logic) the smallest logic that contains A5–A10 and is closed under the Iteration Rule. Then any standard model is an FPL model. To show that FPL consists precisely of the formulae true in all standard models, we have to show that any non-theorem A of FPL is falsified by some standard model. But if Z is a subset of Fma that contains A, is closed under subformulae and satisfies C1–C4, then *if* Z is finite the standard model $\mathcal{M}_{FPL}/Z$ will be a finite filtration of $\mathcal{M}_{FPL}$ through Z, and hence (by (10)) will falsify A because the canonical model $\mathcal{M}_{FPL}$ does. Our proof will therefore be complete once we have established

(17) **Theorem.** *For any formula A there exists a finite subset Z_A of Fma that contains A, is closed under subformulae, and satisfies C1–C4.*

Proof. Define an ordering $<$ on Fma by declaring $B < C$ to hold iff one of the following four cases obtains.

1. C is of the form [while e do α]D, and B is the formula
   $[\alpha]$[while e do α]D.
2. C has the form $[\alpha; \beta]D$, and B is either $[\alpha][\beta]D$ or $[\beta]D$.
3. C is $[\alpha$ **or** $\beta]D$, and B is either $[\alpha]D$ or $[\beta]D$.
4. C is [**if** e **then** α **else** β]D, and B is either $[\alpha]D$ or $[\beta]D$.

Let $<^*$ be the reflexive transitive closure (ancestral) of $<$, and for each $C \in Fma$, put

$$C^* = \{B : B <^* C\}.$$

Then by cases 1–4, C^* satisfies C1–C4. Hence the set

$$Z^* = \bigcup\{C^* : C \in Z\},$$

for any $Z \subseteq Fma$, is an extension of Z that satisfies C1–C4. But if X is any set of formulae that is closed under subformulae, then if $B < D \in X$, inspection of cases 1–4 shows that $X \cup \{B\}$ is closed under subformulae as well (this is the point at which we need, in case 2, the extra formula $[\beta]D$ that is not required by C2). From this we can establish that in general if Z is closed under subformulae then so too is $Z \cup C^*$ for any $C \in Z^*$, and hence so too is Z^*.

Now let us take our given formula A, and put Z/A to be the *finite* set of all subformulae of A. Then Z/A is closed under subformulae, and so from all that we have said it follows that by letting

$$Z_A = (Z/A)^*$$

we establish Z_A as a set that contains A, is closed under subformulae, and satisfies C1–C4. It remains then only to show that Z_A is finite.

In fact we shall show that C^* is finite for any C. Since Z/A is finite, it will then follow that Z_A is the union of finitely many finite sets, giving our desired conclusion. The basis of the proof is that the relation $<$ is well-founded on C^*. For, in general if $B < D$, then B and D are of the form $[\alpha]E$ and $[\beta]F$, respectively, with α a proper subcommand of β and hence a subexpression of length *strictly less* than that of B. It follows that there can be no infinitely descending $<$-chains. Moreover, each formula has only finitely many *immediate* predecessors under $<$ (indeed at most two). In other words, under the relation $<$, C^* is a finitely-branching tree in which every path is finite. Hence, by König's Lemma, there is an upper bound to the possible lengths of the paths, and the tree is finite. □

Decidability of FPL

Rather than simply appeal to the all-powerful König's Lemma to complete the proof of Theorem (17), we continue the analysis to observe that the proof itself indicates that the finite set Z_A can be effectively generated, given A, and its number of elements, denoted n_A, thereby effectively calculated. First of all, the set Z/A of subformulae can be generated by direct inspection of A. Then each C^* can be generated: if C is of the form $[\alpha]B$, then the size of C^* is determined by the nature and complexity of α, and the four cases making up the definition of $<$ provide a set of rules for generating C^* as a finite tree. Otherwise, C^* is just $\{C\}$. In this way we obtain n_A, and hence the upper bound 2^{n_A} on the size of the model $\mathcal{M}_{FPL}/Z_A$ that has A true iff $\mathcal{M}_{FPL}$ does.

We can now say that $\vdash_{FPL} A$ iff A is true in all models with at most 2^{n_A} elements. But then our analysis yields an algorithm for deciding theoremhood in FPL, for the procedure of generating all of the standard models up to a prescribed finite size and testing the truth of a given formula in each of them is an effective one.

An Alternative Axiom

Although A11 is the natural axiom to use for the proof of Theorem (13), in the presence of the Iteration Rule it can be weakened to

A12: $\neg e \to \neg[\textbf{while } e \textbf{ do } \alpha]\textbf{false}$,

which is the special case of A11 in which A is **false**. We have

(18) **Theorem.** *If L is closed under the Iteration Rule and contains all instances of A12, then L contains all instances of A11.*

Proof. We first show that

(i) $[\textbf{while } e \textbf{ do } \alpha]\neg e$,   and
(ii) $\neg e \to (A \to [\textbf{while } e \textbf{ do } \alpha]A)$

are always L-theorems in the presence of the Iteration Rule. For (i), if A is any tautology (e.g. **false** $\to$ **false**) then the formula

$$e \wedge A \to [\alpha]A$$

is true in all models, hence in $\mathcal{M}_L$, and so is an L-theorem. Applying the Iteration Rule, we conclude that

$$\vdash_L A \to [\textbf{while } e \textbf{ do } \alpha](A \wedge \neg e).$$

Since A is true in $\mathcal{M}_L$, so too then is $[\textbf{while } e \textbf{ do } \alpha](A \wedge \neg e)$, and hence $[\textbf{while } e \textbf{ do } \alpha]\neg e$, making the latter an L-theorem.

For (ii), given A we take B to be $(\neg e \wedge A)$, so that

$$e \wedge B \rightarrow [\alpha]B$$

is an instance of a tautology, hence an L-theorem. Thus we get

$$\vdash_L \neg e \wedge A \rightarrow [\textbf{while } e \textbf{ do } \alpha](B \wedge \neg e).$$

From this it follows easily that (ii) is true in $\mathcal{M}_L$, and so is an L-theorem.

To show that A11 is an L-theorem it suffices to show that it is true in $\mathcal{M}_L$, for which we can now use the fact that A12 and any instance of (ii) are true in $\mathcal{M}_L$. The argument from here on is very general: it works in any model. For, suppose that $\neg e$ is true at s. We wish to show that $([\textbf{while } e \textbf{ do } \alpha]A \rightarrow A)$ is also true at s. So, let $[\textbf{while } e \textbf{ do } \alpha]A$ hold at s. Since A12 is true, there must exist a t with $sR(\textbf{while } e \textbf{ do } \alpha)t$. Hence we have A true at t. Now if A were not true at s, applying the instance of (ii) that has $\neg A$ in place of A, we would get $[\textbf{while } e \textbf{ do } \alpha]\neg A$ true at s, hence $\neg A$ true at t – a contradiction. Thus A must be true at s, as needed to establish the truth of A11.   □

The reader is invited to develop a more "proof-theoretic" derivation of A11 from A12 and the Iteration Rule.

An Infinitary Rule

Given $e \in Bxp$, $\alpha \in Cmd$, and $A \in Fma$, a sequence of formulae $A_n(e, \alpha)$, for all $n < \omega$, is defined by putting

$$A_0(e, \alpha) \;=\; (\neg e \rightarrow A)$$
$$A_{n+1}(e, \alpha) \;=\; (e \rightarrow [\alpha]A_n(e, \alpha)).$$

Then in any model it is the case that

$$\models_s A_n(e, \alpha) \quad \text{iff} \quad (s(e \upharpoonright R(\alpha))^n t \text{ and } \models_t \neg e) \text{ implies } \models_t A.$$

Hence in any *standard* model we get

$$\models_s [\textbf{while } e \textbf{ do } \alpha]A \quad \text{iff} \quad \text{for all } n, \models_s A_n(e, \alpha).$$

It follows that if $\mathcal{M}_L$ were standard, then for any L-maximal set s we would have

(19) $[\textbf{while } e \textbf{ do } \alpha]A \in s$ iff $\{A_n(e, \alpha) : n < \omega\} \subseteq s$.

There are instances where s does not satisfy (19), as may be seen by adapting the counter-example to the standardness of $\mathcal{M}_L$ given earlier. However by confining ourselves to those members of S_L that do satisfy (19), we can obtain a kind of canonical model for L that is standard. For this we need to know that L is closed under various infinitary rules that have premisses of the type $A_n(e, \alpha)$. To present these rules in a systematic way we employ the device of *admissible forms*, which are expressions Φ generated by the recursive definition

$$\Phi ::= \mathbf{w} \mid A \to \Phi \mid [\alpha]\Phi.$$

Each such Φ has a unique (and innermost) occurrence of the letter $\mathbf{w}$. Replacing this occurrence by a formula B turns Φ into a member of Fma, which we denote $\Phi(B)$. Then the infinitary rule schema we have in mind is

(20)   If $\vdash \Phi(A_n(e, \alpha))$ for all $n < \omega$, then $\vdash \Phi([\textbf{while } e \textbf{ do } \alpha]A)$.

In Chapter 2 of [23], a canonical model construction is given for logics that are closed under (20). But this rule preserves the property of truth in standard models, and so it follows from the characterisation of this article that FPL is itself closed under (20). In other words, as far as generating the set of theorems is concerned, the infinitary (20) reduces to the finitary Iteration Rule (the reader may enjoy the challenge of, conversely, deriving the Iteration Rule from a suitable instance of (20)). On the other hand, the difference between the two rules emerges at the level of the deducibility relation "$X \vdash_L A$". If we allowed the use of (20) in deriving A from X, then more formulae would become derivable from certain sets X than would be derivable with only the Iteration Rule. Then we would have a logical system with the same theorems (i.e. formulae derivable from $\emptyset$) as FPL, but with fewer consistent sets of formulae. For instance, the set X used earlier to show that $\mathcal{M}_L$ was not standard would no longer be consistent.

To put these observations in full perspective, we should note that the commands studied in this paper are generated by primitive "program letters" π, rather than actual commands, such as assignments. Thus we have been concerned with the logical structure, or form, of commands, rather than the logic of actual commands. Similarly, the members of Fma represent, not actual assertions about actual programs, but only the propositional "shape" of assertions. Actual programs have variables in them that take values in various data types (numbers, strings, truth-values). To study assertions about such programs we need to move to the level of first-order languages with quantifiers for individual variables. At this level the infinitary proof theory is unavoidable: not only does the set of theorems become undecidable – it is not even effectively enumerable (cf. [23] for details).

Segerberg's Axioms For Dynamic Logic

The language of propositional dynamic logic PDL (cf. FISCHER AND LADNER [16]) differs from the language of this paper in that the class of commands is specified by

$$\alpha ::= \pi \mid \textbf{abort} \mid \alpha_1 ; \alpha_2 \mid \alpha_1 \textbf{ or } \alpha_2 \mid A? \mid \alpha^*.$$

$A?$ is the command "test A", with the standard model condition

$$R(A?) = \{(s, s) : \models_s A\},$$

and the characteristic axiom

$$[A?]B \leftrightarrow (A \to B).$$

In our present language, $e?$ can be defined as

$$(\textbf{if } e \textbf{ then skip else abort}).$$

α^* is the non-deterministic command "do α some finite number of times" and has the standard model condition

$$R(\alpha^*) = R(\alpha)^*.$$

An axiomatisation of the PDL-formulae true in all standard models was first announced by Segerberg [87], and the proof found independently by Parikh [70], Dov Gabbay (unpublished) and Segerberg [88]. The construct α^* is handled by two axioms:

(21) $[\alpha^*]A \to A \wedge [\alpha][\alpha^*]A,$ and

(22) $A \wedge [\alpha^*](A \to [\alpha]A) \to [\alpha^*]A.$

(21) is an analogue for α^* of the combination of our A10 and A11, as may be more clearly seen from the fact that the conjunction of A10 and A11 is equivalent, by PC, to

$$[\textbf{while } e \textbf{ do } \alpha]A \to (\neg e \to A) \wedge (e \to [\alpha][\textbf{while } e \textbf{ do } \alpha]A).$$

Axiom (22) plays a similar role to the Iteration Rule, and is itself replaceable by an analogous rule, viz.

(23)   if $\vdash A \to [\alpha]A$ then $\vdash A \to [\alpha^*]A.$

Any logic that contains (21) will contain (22) iff it is closed under (23). To derive (23) from (22) is straightforward via the α^*-Termination Rule. For the converse, replace A in (23) by the whole of the antecedent of (22) and apply principles that hold for all logics.

In PDL, the commands $(\textbf{if } e \textbf{ then } \alpha \textbf{ else } \beta)$ and $(\textbf{while } e \textbf{ do } \alpha)$ are defined, respectively, as

$$((e?; \alpha) \textbf{ or } (\neg e?; \beta)), \quad \text{and} \quad ((e?; \alpha)^*; \neg e?).$$

Using these definitions, our axiomatisation of FPL can be derived in PDL.

8

An Abstract Setting for Henkin Proofs

ABSTRACT. A general result is proved about the existence of maximally consistent theories satisfying prescribed closure conditions. The principle is then used to give streamlined proofs of completeness and omitting-types theorems, in which inductive Henkin-style constructions are replaced by a demonstration that a certain theory "respects" a certain class of inference rules.

By a *Henkin proof* is meant an application of the technique introduced by Henkin [43] for constructing maximally consistent theories that satisfy certain prescribed closure conditions. The method is to build up the desired theory by induction along an enumeration of some relevant class of formulae, with choices being made at each inductive step to include certain formulae, in such a way that when the induction is finished the theory has the properties desired. The character of this procedure is neatly captured in a phrase of Sacks [78, p. 30], who attributes its importance to the fact that it "takes account of decisions made at intermediate stages of the construction".

In this article the Henkin method is used to derive a general principle about the existence of maximal theories closed under abstract "inference rules". This principle may then be used to give alternative proofs of standard completeness and omitting-types theorems, proofs in which the Henkin method is replaced by a demonstration that a certain theory "respects" a certain class of inference rules. This alternative approach is illustrated by a re-working of the completeness and omitting-types theorems for first-order logic and countable fragments of $L_{\infty\omega}$, as well as for the completeness of modal predicate logic with the Barcan formula, and modal propositional logic with infinitary inference rules.

8.1   The Abstract Henkin Principle

Consider a formal language that includes a constant false sentence $\bot$, and a negation connective $\neg$. Let Φ be any class of formulae of this language such that $\bot \in \Phi$ and Φ is closed under $\neg$.

Let $\vdash$ be a subset of $2^\Phi \times \Phi$, i.e. a binary relation from the powerset of Φ to Φ. For $\Gamma \subseteq \Phi$ and $\varphi \in \Phi$, write $\Gamma \vdash \varphi$ if (Γ, φ) belongs to $\vdash$, and $\Gamma \nvdash \varphi$ otherwise. The relation $\vdash$ is called a *deducibility relation on* Φ if it satisfies

D1: If $\Gamma \vdash \varphi$ and $\Gamma \subseteq \Delta$, then $\Delta \vdash \varphi$;

D2: If $\varphi \in \Gamma$, then $\Gamma \vdash \varphi$;

D3: If $\Gamma \vdash \varphi$ and $\Gamma \cup \{\varphi\} \vdash \bot$, then $\Gamma \vdash \bot$;

D4: $\Gamma \cup \{\neg\varphi\} \vdash \bot$ iff $\Gamma \vdash \varphi$.

A subset Γ of Φ is called $\vdash$-*consistent* if $\Gamma \nvdash \bot$, and *finitely* $\vdash$-*consistent* if each finite subset of Γ is $\vdash$-consistent in this sense. Γ is *maximally* $\vdash$-*consistent* if it is $\vdash$-consistent but has no $\vdash$-consistent proper extension in Φ. Replacing "$\vdash$-consistent" by "finitely $\vdash$-consistent" in this last definition yields the notion of Γ being *maximally finitely* $\vdash$-*consistent*.

Now from D1 it follows that any $\vdash$-consistent set is finitely $\vdash$-consistent. The relation $\vdash$ is called *finitary* if, conversely, it satisfies

D5: Every finitely $\vdash$-consistent set is $\vdash$-consistent, i.e. if $\Gamma \vdash \bot$, then for some finite $\Gamma_0 \subseteq \Gamma$, $\Gamma_0 \vdash \bot$.

If $\mathcal{C}$ is a collection of finitely $\vdash$-consistent subsets of Φ that is linearly ordered by set inclusion, i.e. $\Gamma \subseteq \Delta$ or $\Delta \subseteq \Gamma$ for all $\Gamma, \Delta \in \mathcal{C}$, then the union $\bigcup \mathcal{C}$ of $\mathcal{C}$ is finitely $\vdash$-consistent. This follows immediately from the fact that any finite subset of $\bigcup \mathcal{C}$ is a subset of some $\Gamma \in \mathcal{C}$. Thus if

$$P = \{\Delta \subseteq \Phi : \Gamma \subseteq \Delta \ \& \ \Delta \text{ is finitely } \vdash\text{-consistent}\},$$

then under the partial ordering of set inclusion P fulfills the hypothesis of Zorn's Lemma. From the latter we deduce

LINDENBAUM'S LEMMA. *Every finitely $\vdash$-consistent subset of Φ has a maximally finitely $\vdash$-consistent extension in Φ.*   □

(Note that this result uses no properties of $\vdash$ other than the definitions of the concepts referred to in the statement of the Lemma.)

An ordered pair (Π, χ) with $\Pi \subseteq \Phi$ and $\chi \in \Phi$ will be called an *inference* in Φ. As motivation, the reader may care to think of Π as a set of "premises" and χ as a "conclusion", but the notion of inference is quite abstract and applies to any such pair. A set Γ will be said to *respect* the inference (Π, χ) when

$$(\Gamma \vdash \varphi, \text{ all } \varphi \in \Pi) \text{ implies } \Gamma \vdash \chi.$$

Γ is *closed under* (Π, χ) if

$$\Pi \subseteq \Gamma \text{ implies } \chi \in \Gamma.$$

Γ respects (is closed under) a *set* $\mathcal{I}$ of inferences if it respects (is closed under) each member of $\mathcal{I}$.

The cardinality of a set X will be denoted card$X$. If κ is a cardinal number, then X is κ-*finite* if card$X < \kappa$. A κ-*finite extension* of X is a set of the form $X \cup Y$ with Y κ-finite. In other words a κ-finite extension of X is a set obtained by adding *fewer than* κ elements to X.

Theorem 8.1.1
Let $\vdash$ be a finitary deducibility relation on Φ. If $\mathcal{I}$ is a set of inferences in Φ of cardinality κ, and Γ is a $\vdash$-consistent subset of Φ such that

$$\text{every } \kappa\text{-finite extension of } \Gamma \text{ respects } \mathcal{I},$$

then Γ has a maximally $\vdash$-consistent extension in Φ that is closed under $\mathcal{I}$.

This theorem will be established by first separating out that part of its content that does not involve Lindenbaum's Lemma. To do this requires a further concept: a set $\Gamma \subseteq \Phi$ will be said to *decide* (Π, χ) if

$$\text{either } \chi \in \Gamma, \text{ or for some } \varphi \in \Pi, \ \neg\varphi \in \Gamma.$$

Γ decides a set of inferences if it decides each member of the set.

The following result holds for any deducibility relation.

Lemma 8.1.2

(1) If Γ decides (Π, χ) and $\Gamma \subseteq \Delta$, then Δ decides (Π, χ).
(2) If Γ decides (Π, χ), then Γ respects (Π, χ).
(3) If Γ is finitely $\vdash$-consistent, and Γ decides (Π, χ), then Γ is closed under (Π, χ).
(4) If Γ is $\vdash$-consistent, and Γ respects (Π, χ), then for some $\psi \in \Phi$, $\Gamma \cup \{\psi\}$ is $\vdash$-consistent and decides (Π, χ).

Proof.

(1) Immediate.
(2) Suppose $\Gamma \vdash \varphi$, all $\varphi \in \Pi$. Then if $\Gamma \nvdash \chi$, by D2 $\chi \notin \Gamma$, so if Γ decides (Π, χ) then $\neg\psi \in \Gamma$ for some $\psi \in \Pi$. But by assumption $\Gamma \vdash \psi$, and so by D4 $\Gamma \cup \{\neg\psi\} \vdash \bot$, i.e. $\Gamma \vdash \bot$. But then by D1, $\Gamma \cup \{\neg\chi\} \vdash \bot$, and so by D4 again, $\Gamma \vdash \chi$. Hence $\Gamma \vdash \chi$.

(3) Suppose Γ decides (Π, χ), and $\Pi \subseteq \Gamma$. Then if $\chi \notin \Gamma$, $\neg\psi \in \Gamma$ for some $\psi \in \Pi \subseteq \Gamma$. Now by D2, $\{\psi\} \vdash \psi$, and so by D4, $\{\psi, \neg\psi\} \vdash \bot$. But $\{\psi, \neg\psi\} \subseteq \Gamma$, so then Γ is not finitely $\vdash$-consistent.

(4) If $\Gamma \cup \{\neg\varphi\}$ is $\vdash$-consistent for some $\varphi \in \Pi$, then the result follows with $\psi = \neg\varphi$. Otherwise, for all $\varphi \in \Pi$, $\Gamma \cup \{\neg\varphi\} \vdash \bot$, and so by D4, $\Gamma \vdash \varphi$. But Γ respects (Π, χ), hence $\Gamma \vdash \chi$. Since $\Gamma \nvdash \bot$, D3 then implies that $\Gamma \cup \{\chi\} \nvdash \bot$, so the result follows with $\psi = \chi$.

$\qquad\qquad\qquad\qquad\qquad\qquad\qquad\qquad\qquad\qquad\qquad\qquad\qquad$ $\square$

ABSTRACT HENKIN PRINCIPLE.   *Let $\vdash$ be a finitary deducibility relation on Φ. If $\mathcal{I}$ is a set of inferences in Φ of cardinality κ, and Γ is a $\vdash$-consistent subset of Φ such that*

$$(*) \text{ every } \kappa\text{-finite extension of } \Gamma \text{ respects } \mathcal{I},$$

then Γ has a $\vdash$-consistent extension Δ that decides $\mathcal{I}$.

Note that by applying Lindenbaum's Lemma to the $\vdash$-consistent extension Δ given by the conclusion of this result, an extension of Γ is obtained that is maximally $\vdash$-consistent (since $\vdash$ is finitary), decides $\mathcal{I}$ by Lemma 8.1.2(1), and hence is closed under $\mathcal{I}$ by Lemma 8.1.2(3). This argument proves Theorem 8.1.1.

To prove the Abstract Henkin Principle, let $\{(\Pi_\alpha, \chi_\alpha) : \alpha < \kappa\}$ be an indexing of the members of $\mathcal{I}$ by the ordinals less than κ. A sequence $\{\Delta_\alpha : \alpha < \kappa\}$ of extensions of Γ is then defined such that

(i) Δ_α is $\vdash$-consistent;

(ii) $\Delta_\gamma \subseteq \Delta_\alpha$ whenever $\gamma < \alpha$;

(iii) $\mathrm{card}(\Delta_\alpha - \Gamma) \leq \alpha$, hence Δ_α is a κ-finite extension of Γ;

and such that $\Delta_{\alpha+1}$ decides $(\Pi_\alpha, \chi_\alpha)$. The definition proceeds by transfinite induction on α.

Case 1: If $\alpha = 0$, put $\Delta_\alpha = \Gamma$, so that Δ_α is $\vdash$-consistent by assumption, and $\mathrm{card}(\Delta_\alpha - \Gamma) = 0 = \alpha$.

Case 2: Suppose $\alpha = \beta + 1$, and assume inductively that Δ_β has been defined such that (i)–(iii) hold with β in place of α. Then as Δ_β is a κ-finite extension of Γ, the hypothesis $(*)$ on Γ implies that Δ_β respects (Π_β, χ_β). Hence by Lemma 8.1.2(4), there is a $\psi \in \Phi$ such that $\Delta_\beta \cup \{\psi\}$ is $\vdash$-consistent and decides (Π_β, χ_β). Put $\Delta_\alpha = \Delta_\beta \cup \{\psi\}$, so that (i) holds for α. Since $\Delta_\beta \subseteq \Delta_\alpha$, and $\gamma < \alpha$ iff $\gamma \leq \beta$, (ii) follows readily. For (iii), since $\Delta_\alpha - \Gamma \subseteq (\Delta_\beta - \Gamma) \cup \{\psi\}$, $\mathrm{card}(\Delta_\alpha - \Gamma) \leq \mathrm{card}(\Delta_\beta - \Gamma) + 1 \leq \beta + 1 = \alpha$.

Case 3: Suppose α is a limit ordinal and that for all $\beta < \alpha$, Δ_β has been defined to satisfy (i)–(iii). Put

$$\Delta_\alpha = \bigcup_{\beta<\alpha}\Delta_\beta.$$

Then (ii) is immediate for α. For (i), observe that Δ_α is the union of a chain of $\vdash$-consistent, hence finitely $\vdash$-consistent, sets Δ_β, and so Δ_α is finitely $\vdash$-consistent as in the proof of Lindenbaum's Lemma. But $\vdash$ is finitary, so Δ_α is then $\vdash$-consistent. For (iii), observe that

$$(\Delta_\alpha - \Gamma) = \bigcup_{\beta<\alpha}(\Delta_\beta - \Gamma),$$

and note that by the inductive hypothesis, if $\beta < \alpha$ then $\mathrm{card}(\Delta_\beta - \Gamma) \le \beta < \alpha$. Thus $(\Delta_\alpha - \Gamma)$ is the union of a collection of at most $\mathrm{card}\,\alpha$ sets, each of which has at most $\mathrm{card}\,\alpha$ members. Hence $\mathrm{card}(\Delta_\alpha - \Gamma) \le \mathrm{card}\,\alpha \le \alpha$.

This completes the definition of Δ_α for all $\alpha < \kappa$. Now put

$$\Delta = \bigcup_{\alpha<\kappa}\Delta_\alpha.$$

Then by the argument of Case 3, Δ is a $\vdash$-consistent extension of Γ. Moreover, for each $\beta < \alpha$, $\Delta_{\beta+1}$ decides (Π_β, χ_β) by Case 2, and so Δ decides (Π_β, χ_β) by Lemma 8.1.2(1). $\square$

8.2   The Countable Case

In the proof of the Abstract Henkin Principle, the assumption that $\vdash$ is finitary is used only in Case 3, and in the final formation of Δ, to show that the union of an increasing sequence of $\vdash$-consistent sets is $\vdash$-consistent. But if κ is countable, then Case 3 does not arise. Case 2 is iterated countably many times, and then Δ is constructed as the union of the Δ_α's. Then if $\vdash$ is not finitary, Δ may not be $\vdash$-consistent. However, it will at least be *finitely* $\vdash$-consistent, and this gives the following result.

COUNTABLE HENKIN PRINCIPLE. *Let $\vdash$ be any deducibility relation on Φ. If $\mathcal{I}$ is a countable set of inferences in Φ, and Γ is a $\vdash$-consistent subset of Φ such that*

$$(*)\quad \Gamma \cup \Sigma \text{ respects } \mathcal{I} \text{ for all finite } \Sigma \subseteq \Phi,$$

then Γ has a finitely $\vdash$-consistent extension that decides $\mathcal{I}$. $\square$

(By Lindenbaum's Lemma, the extension of Γ deciding $\mathcal{I}$ in this result can be taken to be *maximally* finitely $\vdash$-consistent.)

The Countable Henkin Principle will be used below to prove an omitting-types theorem for countable first-order languages, and the completeness theorem for countable fragments of $\mathsf{L}_{\infty\omega}$. The analysis given here provides one way of "putting one's finger" on the role of countability restrictions in such applications.

If the ambient formal language has a conjunction connective, allowing the formation of the conjunction $\bigwedge \Sigma$ of any *finite* subset Σ of Φ, then a natural constraint on $\vdash$ would be to require that for all $\Gamma \subseteq \Phi$, and all $\varphi \in \Phi$,

$$\Gamma \cup \Sigma \vdash \varphi \text{ iff } \Gamma \cup \{\bigwedge \Sigma\} \vdash \varphi.$$

A deducibility relation satisfying this condition will be called *conjunctive*. Thus for a conjunctive deducibility relation, the hypothesis (∗) in the Countable Henkin Principle can be weakened to

$$\Gamma \cup \{\psi\} \text{ respects } \mathcal{I} \text{ for all } \psi \in \Phi.$$

Applications

8.3   Completeness for First-Order Logic

Let L be a set of relation, function, and individual-constant symbols, and Γ a set of sentences in the first-order language of L that is consistent under the standard deducibility relation of first-order logic.

The *Completeness Theorem* asserts that Γ has a model. To prove this, a new language $K = L \cup C$ is formed by adding to L a set C of new individual constants of cardinality κ, where κ is the maximum of cardL and $\aleph_0$. The usual construction of a model for Γ involves two phases.

Phase 1: Γ is extended by the "Henkin method" to a maximally consistent set Γ^* of K-sentences such that for each K-formula $\varphi(x)$ with at most one variable (x) free,

(a) if $\exists x \varphi \in \Gamma^*$, then $\varphi(c) \in \Gamma^*$ for some $c \in C$.

Phase 2: A model $\mathfrak{A}^*$ is defined, based on the quotient set $C/\sim$, where $\sim$ is the equivalence relation

$$c \sim d \text{ iff } (c = d) \in \Gamma^*.$$

For each K-formula $\psi(x_1, \ldots, x_n)$, this model satisfies

(b) $\mathfrak{A}^* \models \psi[c_1/\sim, \ldots, c_n/\sim]$ iff $\psi(c_1, \ldots, c_n) \in \Gamma^*$.

In particular, $\mathfrak{A}^* \models \sigma$ iff $\sigma \in \Gamma^*$, where σ is any K-sentence, so as $\Gamma \subseteq \Gamma^*$, $\mathfrak{A}^* \models \Gamma$.

The Abstract Henkin Principle of this article may be used to give a succinct development of Phase 1. For this, let Φ be the set of all first-order sentences of K, and $\vdash$ the standard (finitary) first-order deducibility relation on Φ. Then Γ is $\vdash$-consistent. The key property of $\vdash$ that will be used is

(c) if $\Delta \vdash \varphi(c)$, and the constant c does not occur in Δ or $\varphi(x)$, then $\Delta \vdash \forall x \varphi(x)$.

Now the closure condition (a) on Γ^* in Phase 1 is equivalent to:

if $\varphi(c) \in \Gamma^*$ for all $c \in C$, then $\forall x \varphi(x) \in \Gamma^*$,

i.e. to the closure of Γ^* under the inference

$$\varphi_C = (\{\varphi(c) : c \in C\}, \forall x \varphi(x)).$$

Let $\mathcal{I}$ be the set of inferences φ_C for all first-order K-formulae φ with one free variable. The number of such formulae is κ, since $\text{card} K = \kappa$. Hence $\text{card} \mathcal{I} = \kappa$. Thus to prove the existence of Γ^* it suffices to show that if Δ is a κ-finite subset of Φ, then

$$\Gamma \cup \Delta \text{ respects } \mathcal{I}.$$

But if $\text{card} \Delta < \kappa$, then for any φ, $\text{card}(\Delta \cup \{\varphi\}) < \kappa$, since κ is infinite. Hence fewer that κ members of C appear in $\Delta \cup \{\varphi\}$. But *none* of these constants appear in Γ. Thus if

$$\Gamma \cup \Delta \vdash \varphi(c) \text{ for all } c \in C,$$

then

$$\Gamma \cup \Delta \vdash \varphi(c) \text{ for some c not occurring in } \Gamma \cup \Delta \cup \{\varphi\},$$

and so by (c),

$$\Gamma \cup \Delta \vdash \forall x \varphi(x). \qquad \square$$

8.4 Omitting Types

Let L be a *countable* language, and G the set of all first-order L-formulae all of whose free variables are among $x_1, \ldots, x_n$. A consistent subset Σ of G is called an *n-type* if it has no proper consistent extension in G, or, equivalently, if for each $\psi \in G$, exactly one of $\psi, \neg \psi$ belongs to Σ. An L-structure $\mathfrak{A}$ realises an *n-type* Σ if there are individuals $a_1, \ldots, a_n$ in $\mathfrak{A}$ such that

$$\mathfrak{A} \models \varphi[a_1, \ldots, a_n] \qquad \text{for all } \varphi \in \Sigma.$$

$\mathfrak{A}$ *omits* Σ if it does not realise Σ.

If Γ is a consistent set of L-sentences, then an n-type Σ is *principal over* Γ if there is some $\varphi \in \Sigma$ such that

$$\Gamma \vdash \varphi \to \psi \qquad \text{for all } \psi \in \Sigma.$$

The basic omitting-types theorem asserts that if Σ is not principal over Γ, then Γ has a model that omits Σ. The proof is a refinement of the proof of the completeness theorem sketched above, and the required model is the structure $\mathfrak{A}^*$ given there.

To simplify the exposition, let Σ be a 1-type. Since each individual of $\mathfrak{A}^*$ is of the form $c/\sim$ for some $c \in C$, to ensure that $\mathfrak{A}^*$ does not realise Σ it suffices, by clause (b) of the description of $\mathfrak{A}^*$ to show that for each $c \in C$ there is some formula $\varphi(x_1) \in \Sigma$ such that $\varphi(c) \notin \Gamma^*$. Since $\perp \notin \Gamma^*$, this amounts to requiring, for each $c \in C$, that Γ^* be closed under the inference

$$\Sigma_c = (\{\varphi(c) : \varphi \in \Sigma\}, \perp).$$

Lemma 8.4.1 *For any* K-*sentence* σ, $\Gamma \cup \{\sigma\}$ *respects* Σ_c.

Proof. σ may contain members of C other than c. To simplify the notation again, let σ contain just one C-constant, d, other than c.

Suppose that

$$\Gamma \cup \{\sigma(c, d)\} \vdash \varphi(c),$$

and hence

(d) $\Gamma \vdash \sigma(c, d) \rightarrow \varphi(c)$,   for all $\varphi(x_1) \in \Sigma$.

Then as c and d do not occur in Γ, it follows that

$$\Gamma \vdash \exists x_2 \sigma(x_1, x_2) \rightarrow \varphi(x_1), \text{ for all } \varphi(x_1) \in \Sigma.$$

But $\exists x_2 \sigma(x_1, x_2) \in G$, and so $\exists x_2 \sigma(x_1, x_2) \notin \Sigma$, or else Σ would be principal over Γ. Since Σ is a 1-type, it follows that $\neg \exists x_2 \sigma(x_1, x_2) \in \Sigma$, and so by (d),

$$\Gamma \vdash \sigma(c, d) \rightarrow \neg \exists x_2 \sigma(c, x_2).$$

But

$$\Gamma \vdash \sigma(c, d) \rightarrow \exists x_2 \sigma(c, x_2),$$

by a basic axiom of quantification logic, and so $\Gamma \vdash \sigma(c, d) \rightarrow \perp$, hence

$$\Gamma \cup \{\sigma(c, d)\} \vdash \perp.$$

$\square$

Now as C is countable, there are countably many rules of the form Σ_c. Since the standard deducibility relation of first-order logic is conjunctive, the lemma just proved applies to the Countable Henkin Principle and yields, with Lindenbaum's Lemma, a maximally $\vdash$-consistent extension Γ^* of Γ that is closed under Σ_c for all $c \in C$. But K is countable, since L is countable, and so there are countably many inferences of the form φ_C, for φ a K-formula with at most one free variable. Hence if the latter inferences are added to the Σ_c's, there are still only countably many inferences involved altogether, and so Γ^* can be taken to be closed under each φ_C as before.

In fact the whole argument can begin with a countable number of types, not just one. Each type will contribute a countable number of inferences of the form Σ_C, and so, as a countable union of countable sets is countable, this will still involve only countably many inferences altogether. Thus with no extra work, other than these observations about the sizes of sets of inferences, it may be concluded that any countable collection of non-principal types is simultaneously omitted by some model of Γ.

8.5 Completeness of Infinitary Logic

The infinitary logic $L_{\infty\omega}$ generated by a language L has a proper class of individual variables, and a proper class of formulae obtained by allowing, in addition to $\neg\varphi$ and $\forall v\varphi$, formation of the conjunction $\bigwedge\Psi$ of any set Ψ of formulae (disjunction being definable by $\bigwedge$ and $\neg$ as usual).

The deducibility relation for infinitary logic has, in addition to the defining properties of deducibility for first-order logic, the axiom schema

$$\bigwedge\Psi \to \varphi \quad \text{if } \varphi \in \Psi,$$

and the rule of deduction
if

$$\Gamma \vdash \psi \to \varphi \quad \text{for all } \varphi \in \Psi,$$

then

$$\Gamma \vdash \psi \to \bigwedge\Psi.$$

Each formula involved in the following discussion will be assumed to have only a finite number of free variables. This restriction is justified by the fact that it includes all subformulae of infinitary *sentences*.

A *fragment* of $L_{\infty\omega}$ is a set L_A of $L_{\infty\omega}$-formulae that includes all first-order L-formulae and is closed under $\neg, \forall$, *finite* conjunctions, subformulae, and substitution for variables of terms each of whose variables appears in L_A (cf. [3, p. 84]).

A "weak" completeness theorem [3, Section III.4] asserts that if L_A is a *countable* fragment of $L_{\infty\omega}$ and Γ is a set of L_A-sentences that is consistent, then Γ has a model. To prove this, let C be a denumerable set of new constants, $K = L \cup C$, and K_A the set of all formulae obtained from formulae $\varphi \in L_A$ by replacing *finitely* many free variables by constants $c \in C$. Then K_A is countable, and is the smallest fragment of $K_{\infty\omega}$ which contains L_A. A crucial point to note is that each member of K_A contains only finitely many constants from C.

Now let Φ be the (countable) set of sentences in K_A, and $\vdash$ the restriction of the $K_{\infty\omega}$-deducibility relation to Φ. To obtain a Γ-model,

Γ is to be extended to a subset Γ^* of Φ for which the definition of the model $\mathfrak{A}^*$ can be carried through as for first-order logic, and for which the condition

(b) $\mathfrak{A}^* \models \psi[c_1/\sim, \dots, c_n/\sim]$ iff $\psi(c_1, \dots c_n) \in \Gamma^*$

can be established for each formula $\psi(x_1, \dots, x_n)$ *that belongs to* K_A. Then $\mathfrak{A}^*$ will be a Γ-model, as $\Gamma \subseteq \Gamma^*$.

In order for (b) to hold for all K_A-formulae it is sufficient (and necessary) that the following hold.

(i) Γ^* is maximally *finitely* $\vdash$-consistent: this is sufficient to ensure that $\mathfrak{A}^*$ is well-defined; $\neg\varphi \in \Gamma^*$ iff $\varphi \notin \Gamma^*$; $\varphi \to \psi \in \Gamma^*$ iff $\varphi \in \Gamma^*$ implies $\psi \in \Gamma^*$; if $\bigwedge \Psi \in \Gamma^*$ then $\varphi \in \Gamma^*$ for all $\varphi \in \Gamma^*$; and if $\forall x \varphi \in \Gamma^*$ then $\varphi(c) \in \Gamma^*$ for all $c \in \mathsf{C}$.

(ii) Γ^* is closed under the inference φ_{C} for each K_A-formula φ with at most one free variable.

(iii) If $\bigwedge \Psi \in \mathsf{K}_A$, and $\Psi \subseteq \Gamma^*$, then $\bigwedge \Psi \in \Gamma^*$, i.e. if $\bigwedge \Psi \in \mathsf{K}_A$, then Γ^* is closed under the inference $(\Psi, \bigwedge \Psi)$.

Since K_A is countable, there are countably many inferences involved in fulfilling (ii) and (iii). Hence by the Countable Henkin Principle, and Lindenbaum's Lemma, it suffices to show that for all $\sigma \in \Phi$, $\Gamma \cup \{\sigma\}$ respects each such inference. The proof that $\Gamma \cup \{\sigma\}$ respects φ_{C} is just as for first-order logic, since, as noted above, σ has only finitely many constants from C, while Γ has no such constants.

For an inference of the form $(\Psi, \bigwedge \Psi)$, observe that if

$$\Gamma \cup \{\sigma\} \vdash \varphi \qquad \text{for all } \varphi \in \Psi,$$

then

$$\Gamma \vdash \sigma \to \varphi \qquad \text{for all } \varphi \in \Psi,$$

so

$$\Gamma \vdash \sigma \to \bigwedge \Psi,$$

hence

$$\Gamma \cup \{\sigma\} \vdash \bigwedge \Psi.$$

□

It is left as an exercise for the reader to formulate and derive an omitting-types theorem for countable fragments of $\mathsf{L}_{\infty\omega}$.

8.6   Completeness for the Barcan Formula

In *modal* first-order logic, formulae are generated from a language L by means of the modal connective $\square$ in addition to the usual connectives and quantifiers of classical first-order logic. One notion of model for this logic is a structure of the form

$$\mathfrak{A} = \langle W, R, \{\mathfrak{A}_w : w \in W\}\rangle,$$

where W is a non-empty set, R is a binary relation on W, and $\{\mathfrak{A}_w : w \in W\}$ is a collection of classical first-order L-structures that are all based on the same underlying set A, and all give the same interpretation in A to any constant from L (cf. [94]). The relation

$$\mathfrak{A} \models_w \varphi[v]$$

of *satisfaction of formula φ at w in $\mathfrak{A}$ by valuation $v = \langle a_0, \ldots, a_n, \ldots \rangle$* is defined by induction on the formation of φ. The key conditions are

$$\mathfrak{A} \models_w \varphi[v] \qquad \text{iff} \quad \mathfrak{A}_w \models \varphi[v], \qquad \text{if } \varphi \text{ is atomic;}$$
$$\mathfrak{A} \models_w \Box\varphi[v] \qquad \text{iff} \quad \text{for all } z \text{ such that } wRz, \ \mathfrak{A} \models_z \varphi[v];$$
$$\mathfrak{A} \models_w \forall x_n\varphi[v] \quad \text{iff} \quad \text{for all } a \in A, \ \mathfrak{A} \models_w \varphi[v(n/a)],$$

where the sequence $v(n/a)$ is identical to v except in having a as its n-th term. The classical truth-functional connectives are treated as usual.

A formula φ is *true in* $\mathfrak{A}$ if $\mathfrak{A} \models_w \varphi[v]$ holds for all $w \in W$ and all valuations v. The *Barcan formula*

$$BF: \qquad \forall x \Box \varphi \to \Box \forall x \varphi$$

then turns out to be true in all models. For countable languages it was shown in [94] that BF can be used as a schema to give an axiomatisation of the set of sentences made true by this semantics. We will now show how to formulate this completeness proof with the help of the Countable Henkin Principle.

Let $\vdash$ be the finitary deducibility relation that results from extending the standard proof theory of first-order logic with identity by the additional modal axioms BF and

$$K: \quad \Box(\varphi \to \psi) \to (\Box\varphi \to \Box\psi),$$
$$Id: \quad x \neq y \to \Box x \neq y,$$

and the rule of Necessitation:

$$\text{from } \varphi \text{ derive } \Box\varphi.$$

Then the schema

$$Id^+: \quad x = y \to \Box x = y,$$

is $\vdash$-derivable.

Let Γ be a $\vdash$-consistent set of sentences in the modal language generated by L. In order to construct a model (as above) that satisfies Γ at some point, we proceed just as in Section 8.3 to form a new language $\mathsf{K} = \mathsf{L} \cup \mathsf{C}$, with C a new set of constants, and extend Γ to a maximally consistent set Γ^* of modal K-sentences that is closed under the rules φ_C for all modal K-formulae φ with one free variable. Let Dg^* be the set

$$\{(c = d) : c, d \in K \ \& \ (c = d) \in \Gamma^*\} \cup \{(c \neq d) : c, d \in K \ \& \ (c = d) \notin \Gamma^*\}$$

of all equations and inequalities between K-constants that are true of the structure $\mathfrak{A}^*$ described in Phase 2 of Section 8.3. We call Dg^* the *diagram* of Γ^*.

In general, a set Δ of sentences will be said to be ∀-*complete* if it *respects* all of the rules φ_C. Δ is ∀-*closed* if it is closed under these rules. For maximally consistent Δ these two notions coincide, since $\Delta \vdash \varphi$ if, and only if, $\varphi \in \Delta$.

Let W_Γ be the set of maximally consistent ∀-closed sets of modal K-sentences that contain the diagram Dg^*. Define a binary relation R_Γ on W_Γ by putting

$$\Delta R_\Gamma \Theta \quad \text{iff} \quad \{\varphi : \Box\varphi \in \Delta\} \subseteq \Theta.$$

Each $\Delta \in W_\Gamma$ determines a classical K-structure $\mathfrak{A}_\Delta$, defined as for the model $\mathfrak{A}^*$ in Phase 2 of Section 8.3. Since $Dg^* \subseteq \Delta$ we have

$$(c = d) \in \Delta \quad \text{iff} \quad (c = d) \in \Gamma^*$$

for any $c, d \in K$, and this ensures that all structures $\mathfrak{A}_\Delta$ are based on the same set, and give the same interpretation to each constant $c \in K$. Put

$$\mathfrak{A}_\Gamma = \langle W_\Gamma, R_\Gamma, \{\mathfrak{A}_\Delta : \Delta \in W_\Gamma\}\rangle.$$

For each K-formula $\psi(x_1, \ldots, x_n)$, this model satisfies

(†) $\quad \mathfrak{A}_\Gamma \models_\Delta \psi[c_1/\sim, \ldots, c_n/\sim] \quad$ iff $\quad \psi(c_1, \ldots, c_n) \in \Delta.$

From this it follows that $\mathfrak{A}_\Gamma \models_{\Gamma^*} \Gamma$, establishing the desired completeness theorem.

The proof of (†) proceeds by induction, with ∀-closure taking care of the case $\psi = \forall x\varphi$. For the case $\psi = \Box\varphi$, the part that is not straightforward is to show

(‡) $\quad$ If $\Box\varphi \notin \Delta$, then $\varphi \notin \Theta$ for some $\Theta \in W_\Gamma$ with $\Delta R_\Gamma \Theta.$

Lemma 8.6.1 *If a set Σ of modal K-sentences is ∀-complete, then so is $\Sigma \cup \{\sigma\}$ for any modal K-sentence σ.*

Proof. (cf. [44, Lemma, p. 3]) Suppose that φ has only x free, and

$$\Sigma \cup \{\sigma\} \vdash \varphi(c)$$

for all $c \in C$. Then for all such c,

$$\Sigma \vdash \sigma \rightarrow \varphi(c).$$

But $\sigma \rightarrow \varphi(c) = (\sigma \rightarrow \varphi)(c)$, since σ is a sentence, so the ∀-completeness

of Σ then implies
$$\Sigma \vdash \forall x(\sigma \to \varphi).$$
Hence
$$\Sigma \vdash \sigma \to \forall x\varphi$$
as σ does not have x free, and this gives
$$\Sigma \cup \{\sigma\} \vdash \forall x\varphi,$$
establishing that $\Sigma \cup \{\sigma\}$ respects the rule $\varphi\mathsf{C}$.   □

The essential role of the Barcan formula in proving (‡) is contained in the following result.

Lemma 8.6.2 *If a set Σ of modal K-sentences is $\forall$-complete, then so is*
$$\Sigma/\Box = \{\sigma : \Sigma \vdash \Box\sigma\}.$$

Proof. (cf. [94, p. 59]) Using the axiom K and the rule of Necessitation, one shows that for any ψ,
$$\Sigma/\Box \vdash \psi \quad \text{iff} \quad \Sigma \vdash \Box\psi.$$
Thus if $\Sigma/\Box \vdash \varphi(\mathsf{c})$ for all $\mathsf{c} \in \mathsf{C}$, then for all such c we have $\Sigma \vdash \Box\varphi(\mathsf{c})$, so the $\forall$-completeness of Σ implies $\Sigma \vdash \forall x\Box\varphi$. Hence by BF we infer $\Sigma \vdash \Box\forall x\varphi$, and so $\Sigma/\Box \vdash \forall x\varphi$.   □

We are now in a position to prove (‡), and thereby finish the completeness proof. Assuming that L is countable, it follows that K, and hence the set of K-sentences, is denumerable. Thus the Countable Henkin Principle can be applied.

Suppose $\Box\varphi \notin \Delta$. Let
$$\Theta_0 = \Delta/\Box \cup \{\neg\varphi\}.$$
Θ_0 is consistent, or else $\Delta/\Box \vdash \varphi$, implying that $\Box\varphi \in \Delta$. Also $Dg^* \subseteq \Theta_0$, because $Dg^* \subseteq \Delta$ by definition of W_Γ, so if $\sigma \in Dg^*$ then $\sigma \in \Delta$, whence $\Box\sigma \in \Delta$ by the schemata Id and Id^+, giving $\sigma \in \Delta/\Box \subseteq \Theta_0$.

Now Δ is $\forall$-complete, so by Lemma 8.6.2, $\Delta/\Box$ is $\forall$-complete. Then by Lemma 8.6.1, Θ_0 is $\forall$-complete. But then applying Lemma 8.6.2 again, $\Theta_0 \cup \{\sigma\}$ is $\forall$-complete for all K-sentences σ. Thus by the Countable Henkin Principle and Lindenbaum's Lemma, Θ_0 has a maximally consistent extension Θ that decides all inferences $\varphi\mathsf{C}$.

Since $Dg^* \subseteq \Theta_0$ and Θ is $\forall$-closed, we get $\Theta \in W_\Gamma$. Since $\Delta/\Box \subseteq \Theta_0$ we get $\Delta R_\Gamma \Theta$. Finally, since $\neg\varphi \in \Theta_0$, we have $\varphi \notin \Theta$.   □

8.7 Infinitary Rules in Modal Logic

The Ancestral Rule

Consider a propositional language which, in addition to the classical truth-functional connectives, has two modalities $\Box$ and $\boxast$. An *ancestral model* for this language is a structure

$$\mathcal{M} = \langle W, R, V \rangle,$$

with R a binary relation on set W, and V a valuation assigning a subset $V(p)$ of W to each propositional variable p. The relation $\mathcal{M} \models_w A$ of *truth of formula A at w in $\mathcal{M}$* it defined by

$$
\begin{aligned}
&\mathcal{M} \models_w p && \text{iff} && w \in V(p); \\
&\mathcal{M} \nvDash_w \bot && && \text{i.e. not } \mathcal{M} \models_w \bot; \\
&\mathcal{M} \models_w A \to B && \text{iff} && \mathcal{M} \models_w A \text{ implies } \mathcal{M} \models_w B; \\
&\mathcal{M} \models_w \Box A && \text{iff} && \text{for all } z \text{ such that } wRz, \ \mathcal{M} \models_z A; \\
&\mathcal{M} \models_w \boxast A && \text{iff} && \text{for all } z \text{ such that } wR^*z, \ \mathcal{M} \models_z A,
\end{aligned}
$$

where R^* is the ancestral (reflexive transitive closure) of relation R. For a set Γ of formulae we write $\mathcal{M} \models_w \Gamma$ to mean that $\mathcal{M} \models_w A$ holds for all $A \in \Gamma$.

Formula A is *true in $\mathcal{M}$*, $\mathcal{M} \models A$, if $\mathcal{M} \models_w A$ for all $w \in W$. The set

$$\{A : \mathcal{M} \models A \text{ for all ancestral } \mathcal{M}\}.$$

is a normal multimodal logic in the sense of [32, §5], and its properties are well understood. It can be viewed as a fragment of the dynamic logic analysed in §10 of [32], and is axiomatised by adding to a standard basis for classical Propositional Calculus the axioms

$$
\begin{aligned}
K: \quad & \Box(A \to B) \to (\Box A \to \Box B), \\
K^*: \quad & \boxast(A \to B) \to (\boxast A \to \boxast B), \\
Mix: \quad & \boxast A \to A \wedge \Box \boxast A, \\
Ind: \quad & \boxast(A \to \Box A) \to (A \to \boxast A),
\end{aligned}
$$

and the rule of Necessitation for $\Box$ and $\boxast$:

$$\text{from } A \text{ derive } \Box A \text{ and } \boxast A.$$

The deducibility relation $\Gamma \vdash^* A$ is defined to mean that for some n there exist $B_0, \ldots, B_{n-1} \in \Gamma$ such that the formula

$$B_0 \to (B_1 \to (\cdots \to (B_{n-1} \to A) \cdots))$$

is derivable in the axiomatic system just described. When $n = 0$, i.e. when A itself is so derivable, we write $\vdash^* A$. Then it can be proved that

$$\vdash^* A \quad \text{iff} \quad \text{for all ancestral } \mathcal{M}, \ \mathcal{M} \models A,$$

or equivalently,

every $\vdash^*$-consistent formula is satisfiable (true at some point) in an ancestral model.

Ancestral models define a natural *semantic consequence relation* $\models^*$, given by putting

$$\Gamma \models^{\mathcal{M}} A \quad \text{iff} \quad \text{for all } w \in W, \; \mathcal{M} \models_w \Gamma \text{ implies } \mathcal{M} \models_w A;$$

and then

$$\Gamma \models^* A \quad \text{iff} \quad \text{for all ancestral models } \mathcal{M}, \; \Gamma \models^{\mathcal{M}} A.$$

However, whereas the notion "$\vdash^* A$" proof-theoretically characterises truth in ancestral models, the full deducibility relation $\Gamma \vdash^* A$ is not equivalent to $\Gamma \models^* A$. While $\vdash^*$ is finitary, $\models^*$ is not, and in general $\Gamma \vdash^* A$ implies $\Gamma \models^* A$, but not conversely. To see this, consider formulae of the form $\square^n A$, defined inductively by

$$\begin{aligned} \square^0 A \quad &= A, \\ \square^{n+1} A \quad &= \square\square^n A. \end{aligned}$$

The *Ancestral Rule* is the set of all inferences of the form

$$(\{\square^n A : n \geq 0\}, \boxasterisk A)$$

(there are denumerably many such inferences, given denumerably many formulae A).

This Rule is preserved by $\models^*$, in the sense that

$$\{\square^n A : n \geq 0\} \models^{\mathcal{M}} \boxasterisk A$$

for all ancestral $\mathcal{M}$. But we do not have $\{\square^n A : n \geq 0\} \vdash^* \boxasterisk A$. For example, if p is a propositional variable, and

$$\Gamma^p = \{\square^n p : n \geq 0\} \cup \{\neg\boxasterisk p\},$$

then Γ^p is not satisfiable in any ancestral model, but each finite subset of Γ^p is so satisfiable. Indeed if $\mathcal{M}$ is an ancestral model of the form $\langle \omega, R, V \rangle$, with R the successor relation on ω and $V(p) = \{0, \ldots, m\}$, then

$$\mathcal{M} \models_0 \{\square^n p : n \leq m\} \cup \{\neg\boxasterisk p\}.$$

Thus $\Gamma^p \models^* \perp$, but any finite $\Gamma_0 \subseteq \Gamma^p$ has $\Gamma_0 \nvDash^* \perp$, and so $\Gamma_0 \nvdash^* \perp$. Hence every finite subset of Γ^p is $\vdash^*$-consistent. Since $\vdash^*$ is finitary, it follows that Γ^p itself is $\vdash^*$-consistent, so $\Gamma^p \nvdash^* \perp$, and

$$\{\square^n p : n \geq 0\} \nvdash^* \boxasterisk p$$

(notice that this is essentially the same example as that used in the previous chapter to show that the canonical model for Finitary Program Logic is not standard).

The most common approach to completeness theorems in modal logic is to build models whose points are maximally consistent sets of formulae. If "consistent" here means $\vdash^*$-consistent, then this will not produce ancestral models, since there will be points, such as those containing Γ^p, which are not closed under the Ancestral Rule. The relation $\vdash^*$ is weaker than $\models^*$, so to proof-theoretically characterise the latter, $\vdash^*$ will have to be strengthened in some way, at least by adjoining the Ancestral Rule. The effect will be to reduce the number of consistent sets, eliminating such sets as Γ^p.

A General Approach

We will now take up the matter of adjoining infinitary rules to modal logics in a general context, and return later to applying our results to the Ancestral Rule.

Let Φ be the set of formulae of a countable propositional language that includes the classical truth-functional connectives and a modal connective $\square$. Consider the following properties of a relation $\vdash$ from 2^Φ to Φ:

PC: $\Gamma \vdash A$ if A is a tautological consequence of Γ.

CT: If $\Gamma \vdash B$ for all $B \in \Delta$, and $\Delta \vdash A$, then $\Gamma \vdash A$.

DT: $\Gamma \cup \{A\} \vdash B$ implies $\Gamma \vdash A \rightarrow B$.

IR: $\Gamma \vdash B$ implies $(A \rightarrow \Gamma) \vdash A \rightarrow B$,
      where
$$(A \rightarrow \Gamma) = \{A \rightarrow C : C \in \Gamma\}.$$

BR: $\Gamma \vdash A$ implies $\square\Gamma \vdash \square A$,
      where
$$\square\Gamma = \{\square B : B \in \Gamma\}.$$

Here PC stands for "Propositional Calculus", CT for "Cut Rule", DT for "Deduction Theorem", IR for "Implication Rule", and BR for "Box Rule".

Lemma 8.7.1 *If $\vdash$ satisfies PC, CT, and DT, then it satisfies IR, and*
$$\Gamma \vdash A \rightarrow B \quad implies \quad \Gamma \cup \{A\} \vdash B.$$

Proof. Suppose $\Gamma \vdash B$. Since PC gives
$$(A \rightarrow \Gamma) \cup \{A\} \vdash C, \quad \text{for all } C \in \Gamma,$$
it then follows by CT that $(A \rightarrow \Gamma) \cup \{A\} \vdash B$. Hence by DT,
$$(A \rightarrow \Gamma) \vdash A \rightarrow B,$$
establishing IR.

Next, suppose $\Gamma \vdash A \to B$. But by PC, $\Gamma \cup \{A\} \vdash C$ for all $C \in \Gamma$, so this yields $\Gamma \cup \{A\} \vdash A \to B$ by CT. Since $\{A \to B, A\} \vdash B$ and $\Gamma \cup \{A\} \vdash A$ (both by PC), CT then gives $\Gamma \cup \{A\} \vdash B$. □

We will write $\vdash A$ to mean that $\emptyset \vdash A$, where $\emptyset$ is the empty set of formulae. Since $\Box \emptyset = \emptyset$, it follows that when $\vdash$ satisfies the Box Rule it must also satisfy Necessitation:

$$\text{if } \vdash A \text{ then } \vdash \Box A.$$

Now fix a relation $\vdash$ satisfying PC, CT, DT, and BR (hence IR). Then $\vdash$ is a deducibility relation, since conditions D1–D4 of Section 8.1 can be derived from PC, CT, and DT, as the reader may verify.

Let $\mathcal{I}$ be a *countable* subset of $2^\Phi \times \Phi$, i.e. a countable set of inferences, that is included in $\vdash$:

$$(\Gamma, A) \in \mathcal{I} \text{ implies } \Gamma \vdash A.$$

Let $\mathcal{I}_\omega$ be the smallest extension of $\mathcal{I}$ in $2^\Phi \times \Phi$ that satisfies IR and BR, in the sense that

$$(\Gamma, A) \in \mathcal{I}_\omega \text{ implies } (\Box\Gamma, \Box A), (B \to \Gamma, B \to A) \in \mathcal{I}_\omega.$$

Then $\mathcal{I}_\omega$ is countable, because Φ is countable and so there are only countably many instances of IR and BR that need be added to $\mathcal{I}$ to obtain $\mathcal{I}_\omega$. Also, since $\vdash$ satisfies IR and BR, it extends $\mathcal{I}_\omega$:

$$(\Gamma, A) \in \mathcal{I}_\omega \text{ implies } \Gamma \vdash A.$$

A set Δ of formulae will be called $(\mathcal{I}, \vdash)$-*saturated* if

- Δ is maximally finitely $\vdash$-consistent, and
- Δ is closed under $\mathcal{I}_\omega$, i.e.

$$\text{if } (\Gamma, A) \in \mathcal{I}_\omega \text{ and } \Gamma \subseteq \Delta, \text{ then } A \in \Delta.$$

Being maximally finitely $\vdash$-consistent is enough to ensure that membership of Δ reflects the classical truth-functions, i.e.

$$\bot \notin \Delta$$
$$\neg A \in \Delta \quad \text{iff} \quad A \notin D$$
$$A \to B \in \Delta \quad \text{iff} \quad A \in \Delta \text{ implies } B \in \Delta$$

etc., but whether Δ is actually $\vdash$-consistent, rather than finitely $\vdash$-consistent, is not evident. To show this it would be enough to show that Δ was $\vdash$-deductively closed, in the sense that

$$\Delta \vdash A \quad \text{implies} \quad A \in \Delta,$$

for then $\Delta \vdash \bot$ would give $\{\bot\} \subseteq \Delta$, contrary to Δ being finitely $\vdash$-consistent. But if Δ were not deductively closed, then $\Delta \vdash A$ and $A \notin \Delta$ for some A. Hence $\neg A \in \Delta$, and so by D4, $\Delta \vdash \bot$.

Thus the question as to whether a $(\mathcal{I},\vdash)$-saturated Δ is $\vdash$-consistent is equivalent to the question as to whether Δ is $\vdash$-deductively closed. From the finite $\vdash$-consistency of Δ we can conclude that Δ is *finitely $\vdash$-deductively closed* in the sense that

$$\text{if } \Delta_0 \vdash A \text{ for some finite } \Delta_0 \subseteq \Delta, \text{ then } A \in \Delta,$$

hence in particular

$$\vdash A \quad \text{implies} \quad A \in \Delta,$$

but the question of full deductive closure is one that we will have to set aside for now, and resolve later (cf. Corollary 8.7.7 and the discussion following it).

Lemma 8.7.2 (Extension Lemma) *Every $\vdash$-consistent set of formulae has a $(\mathcal{I},\vdash)$-saturated extension.*

Proof. This is a direct application of the Countable Henkin Principle. Let Σ be $\vdash$-consistent. Then for any formula A, we show that

$$\Sigma \cup \{A\} \text{ respects } \mathcal{I}_\omega.$$

For if $(\Gamma, C) \in \mathcal{I}_\omega$ and $\Sigma \cup \{A\} \vdash B$ for all $B \in \Gamma$, then by DT

$$\Sigma \vdash A \to B \text{ for all } B \in \Gamma.$$

But

$$(A \to \Gamma) \vdash A \to C,$$

since $\mathcal{I}_\omega$ is closed under IR and contained in $\vdash$. Hence by CT

$$\Sigma \vdash A \to C,$$

whence by Lemma 8.7.1 $\Sigma \cup \{A\} \vdash C$.

Thus indeed $\Sigma \cup \{A\}$ respects $\mathcal{I}_\omega$. But $\vdash$ is conjunctive (by PC and CT), so by the Countable Henkin Principle and then Lindenbaum's Lemma, Σ has a maximally finitely $\vdash$-consistent extension that is $\mathcal{I}_\omega$-closed. $\qquad\qquad\qquad\qquad\qquad\qquad\qquad\qquad\qquad\qquad\qquad\square$

Lemma 8.7.3 (Box Lemma) *If Γ is $(\mathcal{I},\vdash)$-saturated and $\Box A \notin \Gamma$, then there exists a $(\mathcal{I},\vdash)$-saturated Δ with $A \notin \Delta$ and*

$$\{B : \Box B \in \Gamma\} \subseteq \Delta.$$

Proof. In finitary modal logic, this result is obtained by showing that the set

$$\{B : \Box B \in \Gamma\} \cup \{\neg A\}$$

is $\vdash$-consistent, and then applying a version of the Extension Lemma to obtain the desired Δ (as in the proof of (‡) in Section 8.6). Now if this set were not $\vdash$-consistent, then using the properties D4 and D1 of a deducibility relation as well as BR we could conclude that $\Gamma \vdash \Box A$. If

Γ were $\vdash$-deductively closed, this would then contradict the assumption that $\square A \notin \Gamma$. However we do not know at this stage whether Γ is in fact $\vdash$-deductively closed (cf. the discussion prior to the Extension Lemma).

Thus in the absence of an assumption that $\vdash$ is finitary, the Box Lemma requires a more detailed analysis. The proof we give now is a refinement of the proof of the Countable Henkin Principle, and for this purpose it is convenient to use the dual modality $\Diamond$, given by $\Diamond C = \neg\square\neg C$. Two general facts that will be needed are

(i) If $\square B, \Diamond C \in \Gamma$, then $\Diamond(B \wedge C) \in \Gamma$;
(ii) If $\Diamond C \in \Gamma$, then for any B, one of $\Diamond(C \wedge B), \Diamond(C \wedge \neg B)$ is in Γ.

To prove (i), observe that PC and BR give

$$\{\square B, \square\neg(B \wedge C)\} \vdash \square\neg C,$$

so if $\square B, \Diamond C \in \Gamma$ then $\square\neg C \notin \Gamma$, so finite $\vdash$-deductive closure of Γ implies $\square\neg(B \wedge C) \notin \Gamma$, hence $\neg\square\neg(B \wedge C) \in \Gamma$ as desired.

The proof of (ii) is similar, using the fact that

$$\{\square\neg(C \wedge B), \square\neg(C \wedge \neg B)\} \vdash \square\neg C.$$

Now let

$$\Gamma_\square = \{B : \square B \in \Gamma\}.$$

Then $\Gamma_\square$ is denumerable, since Φ is, and so $\Gamma_\square \cup \mathcal{I}_\omega$ is denumerable. Let

$$\Gamma_\square \cup \mathcal{I}_\omega = \{\sigma_n : n < \omega\}.$$

We construct an increasing sequence

$$\Delta_0 \subseteq \cdots \subseteq \Delta_n \subseteq \cdots\cdots$$

of finite sets such that, for all $n < \omega$,

(1_n) $\Diamond(\bigwedge \Delta_n) \in \Gamma$;
(2_n) If $\sigma_n \in \Gamma_\square$, then $\sigma_n \in \Delta_{n+1}$;
(3_n) If $\sigma_n \in \mathcal{I}_\omega$, then Δ_{n+1} decides σ_n.

First put $\Delta_0 = \{\neg A\}$. Now $\neg\square A \in \Gamma$, since $\square A \notin \Gamma$, and $\Diamond\neg A \in \Gamma$ follows readily from this, giving (1_0).

Next, make the inductive assumption that Δ_n has been defined and (1_n) holds. If $\sigma_n \in \Gamma_\square$, then $\square\sigma_n \in \Gamma$, so as $\Diamond(\bigwedge \Delta_n) \in \Gamma$, result (i) above implies $\Diamond(\sigma_n \wedge (\bigwedge \Delta_n)) \in \Gamma$. Thus putting

$$\Delta_{n+1} = \Delta_n \cup \{\sigma_n\}$$

makes $\Diamond(\bigwedge \Delta_{n+1}) \in \Gamma$, so (1_{n+1}) and (2_n) are fulfilled, and (3_n) holds vacuously.

If however $\sigma_n \in \mathcal{I}_\omega$, let $\sigma_n = (\Sigma, B)$. By (1_n) and result (ii), either $\Diamond((\bigwedge \Delta_n) \wedge B) \in \Gamma$ or $\Diamond((\bigwedge \Delta_n) \wedge \neg B) \in \Gamma$.

If $\Diamond((\bigwedge \Delta_n) \wedge B) \in \Gamma$, then putting $\Delta_{n+1} = \Delta_n \cup \{B\}$ makes (1_{n+1}) true, and as $B \in \Delta_{n+1}$, Δ_{n+1} decides (Σ, B), so (3_n) is true.

Alternatively, when $\Diamond((\bigwedge \Delta_n) \wedge \neg B) \in \Gamma$, we have

$$\Box((\bigwedge\Delta_n) \to B) \notin \Gamma.$$

But

$$(\Box((\bigwedge\Delta_n) \to \Sigma), \Box((\bigwedge\Delta_n) \to B)) \in \mathcal{I}_\omega,$$

since $(\Sigma, B) \in \mathcal{I}_\omega$ and $\mathcal{I}_\omega$ satisfies IR and BR (indeed this is why we needed to introduce $\mathcal{I}_\omega$!). Since the $(\mathcal{I}, \vdash)$-saturated set Γ is closed under $\mathcal{I}_\omega$, there must then exist some $C \in \Sigma$ with $\Box((\bigwedge \Delta_n) \to C) \notin \Gamma$, whence $\Diamond((\bigwedge \Delta_n) \wedge \neg C) \in \Gamma$. Hence putting $\Delta_{n+1} = \Delta_n \cup \{\neg C\}$ makes (1_{n+1}) and (3_n) true.

This completes the definition of the Δ_n's satisfying (1_n)–(3_n). Each Δ_n is $\vdash$-consistent, for if $\Delta_n \vdash \bot$ then by PC and DT, $\vdash \neg(\bigwedge \Delta_n)$, hence using Necessitation we get $\Box\neg(\bigwedge \Delta_n) \in \Gamma$. But in view of (1_n), this contradicts the fact that Γ is finitely $\vdash$-consistent.

It follows that

$$\Delta_\omega = \bigcup_{n<\omega}\Delta_n$$

must be finitely $\vdash$-consistent, and in view of (2_n) have $\Gamma_\Box \subseteq \Delta_\omega$. Moreover by (3_n), Δ_ω decides $\mathcal{I}_\omega$. Hence the desired Δ can be taken to be any maximally finitely $\vdash$-consistent extension of Δ_ω. Since $\neg A \in \Delta_0 \subseteq \Delta$, $A \notin \Delta$. $\mathcal{I}_\omega$ is decided by Δ_ω, and thus by Δ, so Δ is closed under $\mathcal{I}_\omega$. As $\Gamma_\Box \subseteq \Delta_\omega \subseteq \Delta$, this completes the proof of the Box Lemma.   □

Ancestral Logic

We now return to the propositional language of $\Box$ and $\boxasterisk$. Let $\mathcal{I}^a$ be the Ancestral Rule, i.e. the (denumerable) set of all inferences of the form

$$(\{\Box^n A : n \geq 0\}, \boxasterisk A).$$

Define $\vdash^a$ to be the *smallest* relation from 2^Φ to Φ such that

(1_a) $\vdash^a$ satisfies PC, CT, DT, and BR;
(2_a) $(\Gamma, A) \in \mathcal{I}^a$ implies $\Gamma \vdash^a A$;
(3_a) $\{\boxasterisk A\} \vdash^a A \wedge \boxasterisk A$.

The last condition is a version of the axiom *Mix*. Using it, by PC, CT, and BR, we can inductively derive

$$\{\boxasterisk A\} \vdash^a \Box^n A.$$

From this it follows that any $(\mathcal{I}^a, \vdash^a)$-saturated set Δ satisfies

(4_a) $\boxasterisk A \in \Delta$ iff $\{\Box^n A : n \geq 0\} \subseteq \Delta$.

Lemma 8.7.4 (Soundness) *If $\Sigma \vdash^a A$, then $\Sigma \models^* A$.*

Proof. If $\mathcal{M}$ is an ancestral model, then (1_a)–(3_a) hold when $\vdash^a$ is replaced by $\models^{\mathcal{M}}$. Since $\vdash^a$ is defined to be the smallest relation satisfying these conditions, $\Sigma \vdash^a A$ implies $\Sigma \models^{\mathcal{M}} A$. □

Now consider the ancestral model

$$\mathcal{M}^a = \langle W^a, R^a, V^a \rangle,$$

where

- $W^a = \{\Delta \subseteq \Phi : \Delta \text{ is } (\mathcal{I}^a, \vdash^a)\text{-saturated}\}$,
- $\Gamma R^a \Delta$ iff $\{B : \Box B \in \Gamma\} \subseteq \Delta$,
- $V^a(p) = \{\Delta \in W^a : p \in \Delta\}$.

Lemma 8.7.5 (Truth Lemma) *For any formula A, and any $\Gamma \in W^a$,*

$$\mathcal{M}^a \models_\Gamma A \quad iff \quad A \in \Gamma.$$

Proof. The case $A = p$ holds by definition of $\mathcal{M}^a$, the inductive cases of the truth-functional connectives are given by the previously observed membership properties of saturated sets, and the case of $\Box$ follows from the Box Lemma (8.7.3). For the case of ⊞, use the fact that any ancestral model satisfies

$$\mathcal{M} \models_w \boxplus A \quad iff \quad \text{for all } n \geq 0, \; \mathcal{M} \models_w \Box^n A,$$

and apply (4_a). □

Theorem 8.7.6 (Completeness)

$$\Sigma \vdash^a A \quad iff \quad \Sigma \models^* A \quad iff \quad \Sigma \models^{\mathcal{M}^a} A.$$

Proof. It is evident that the implications hold from left to right. But if $\Sigma \nvdash^a A$, then $\Sigma \cup \{\neg A\}$ is $\vdash^a$-consistent (D4), so is contained in some $\Gamma \in W^a$ by the Extension Lemma 8.7.2. Then by Lemma 8.7.5,

$$\mathcal{M}^a \models_\Gamma \Sigma \quad \text{and} \quad \mathcal{M}^a \nvDash_\Gamma A,$$

and so $\Sigma \nvDash^{\mathcal{M}^a} A$. □

Corollary 8.7.7 *If Γ is $(\mathcal{I}^a, \vdash^a) - saturated$, then Γ is $\vdash^a$-closed and maximally $\vdash^a$-consistent.*

Proof. Suppose $\Gamma \vdash^a A$. Then $\Gamma \models^{\mathcal{M}^a} A$ by Soundness (8.7.4), and $\mathcal{M}^a \models_\Gamma \Gamma$ by Lemma 8.7.5, so $\mathcal{M}^a \models_\Gamma A$, hence $A \in \Gamma$.

Consequently if Γ were not $\vdash^a$-consistent, then $\Gamma \vdash^a \bot$, whence $\bot \in \Gamma$, contradicting the fact that Γ is finitely $\vdash^a$-consistent. Thus Γ is $\vdash^a$-consistent, and cannot have any $\vdash^a$-consistent extensions, since it has no finitely $\vdash^a$-consistent extensions. □

A Better Approach ?

At the beginning of the discussion of the proof of the Box Lemma 8.7.3, we observed that the proof would be rather short if we knew that $(\mathcal{I}, \vdash)$-saturated sets were $\vdash$-closed. We have just established that this is indeed true in the $(\mathcal{I}^a, \vdash^a)$ case, but the result was obtained as a corollary to the Truth Lemma 8.7.5, which itself depends on the Box Lemma.

A similar analysis could be carried out for other infinitary modal logics (i.e. other pairs $(\mathcal{I}, \vdash)$). In each case a model would be built out of saturated sets, leading, via a Truth Lemma, to the proof-theoretic result that maximally finitely consistent sets closed under certain inferences $(\mathcal{I})$ are in fact maximally consistent and deductively closed.

We might ask whether it is possible to obtain such a result by purely proof-theoretic means. Can we make a *syntactic* construction that extends any $\vdash$-consistent set to a maximally $\vdash$-consistent $\mathcal{I}$-closed set without making a model-theoretic detour to get there ?

We will answer this question in the next chapter.

Related Principles

The Countable Henkin Principle is intimately related to the Principle of Dependent Choice in set theory and the Rasiowa-Sikorski Lemma for Boolean algebras. A discussion of these connections may be found in [27].

9

A Framework for Infinitary Modal Logic

9.1 Introduction

In propositional modal logic there are certain systems, defined by a notion of semantic consequence $\Gamma \models A$ over some class $\mathcal{C}$ of models, for which the following obtains.

(1) The set
$$\{A: \models A\}$$
of formulae true in all models from $\mathcal{C}$ is characterised by a *finitary* proof relation $\vdash$. We have
$$\vdash A \quad \text{iff} \quad \models A,$$
giving the completeness property:

> every $\vdash$-consistent formula is $\models$-satisfiable (i.e. satisfiable in some $\mathcal{C}$-model).

(2) The relation $\Gamma \models A$ is not compact: there are cases where $\Gamma \models A$ but $\Gamma_0 \nvDash A$ for all finite $\Gamma_0 \subseteq \Gamma$. Then $\Gamma \cup \{\neg A\}$ is not $\models$-satisfiable, but all of its finite subsets are. It follows that the relation $\Gamma \models A$ has no finitary proof theory, and so the relation $\vdash$ of (1) is not *strongly* complete for $\models$: there are $\vdash$-consistent *sets* of sentences that are not $\models$-satisfiable.

Natural examples of this situation arise when there is a pair $\Box, \boxast$ of modal connectives for which the binary relation interpreting $\boxast$ in a Kripke model is the reflexive transitive closure of the relation interpreting $\Box$. We studied this in the Ancestral Logic of the previous chapter. It also applies to Temporal Logic, where $\Box$ means "at the next moment" and $\boxast$ means "from now on" [32, §9], as well as to Program Logic, where

213

□ means "after program α terminates" and ⊞ makes the same assertion about the iteration α^* of α (cf. Chapter 7, or [32, §10]).

For logics of this type, the semantic relation $\models$ can only be characterised proof-theoretically by the use of *infinitary* rules of inference. One approach to this is to construct models whose points are sets of formulae that are maximally *finitely* consistent and closed under infinitary rules. In the analysis we gave for Ancestral Logic it turned out, as a consequence of the resulting completeness theorem, that these sets are actually maximally consistent, not just finitely consistent, and deductively closed, not just finitely deductively closed.

The problem then suggests itself of making a purely proof-theoretic construction of sets of this last type: maximally $\vdash$-consistent sets with specified closure properties, where $\vdash$ is a non-finitary proof relation. This issue will now be taken up and resolved in a way that gives a proof-theoretic explanation as to why certain maximally finitely consistent sets turn out to be fully consistent and deductively closed (cf. Corollary 9.3.6). To achieve this we develop in a general setting the approach first used for Program Logic in [23, Chapter 2], and work out the details for an n-ary modality $\Box(A_1, \ldots, A_n)$, rather than just a unary connective.

Although Section 8.7 is recommended reading for motivation of what follows, the present chapter has been written to be self-contained. This has involved only a small amount of repetition.

9.2   Truth and Deducibility

Models

Let Φ be the (denumerable) set of formulae of a propositional language that has a countable set of atomic variables, and whose connectives include $\bot$, $\rightarrow$, and an n-ary connective $\Box$ (and possibly others). The other standard truth-functional connectives $\neg$, $\wedge$, $\vee$, $\leftrightarrow$ are taken to be defined in terms of $\bot$ and $\rightarrow$, while $\Diamond$, the *dual* connective to $\Box$, is given by

$$\Diamond(A_1, \ldots, A_n) = \neg\Box(\neg A_1, \ldots, \neg A_n).$$

In order to handle formulae containing the connective $\Box$, we may use substitutional notation. This if

$$C = \Box(B_1, \ldots, B_{i-1}, A, B_{i+1}, \ldots, B_n),$$

we denote by $C_A[B]$ the formula

$$\Box(B_1, \ldots, B_{i-1}, B, B_{i+1}, \ldots, B_n),$$

obtained by replacing A by B *in the indicated position* in C.

To define models, we use structures

$$\mathcal{M} = \langle W, R, \models \rangle,$$

where R is an $n+1$-placed relation on the set W, and $\models$ is a *satisfaction relation* between points of W and formulae, i.e. a subset of $W \times \Phi$. We write

$$\mathcal{M} \models_w A$$

whenever the pair (w, A) belongs to the relation $\models$.

A structure $\mathcal{M}$ of this type will be called a *model* if

(m1) $\mathcal{M} \not\models_w \bot$ i.e. not $\mathcal{M} \models_w \bot$;

(m2) $\mathcal{M} \models_w A \to B$ iff $\mathcal{M} \models_w A$ implies $\mathcal{M} \models_w B$;

(m3) $\mathcal{M} \models_w \Box(A_1, \ldots, A_n)$ iff for all $z_1, \ldots, z_n$ such that $R(w, z_1, \ldots, z_n)$ there exists $i \leq n$ with $\mathcal{M} \models_{z_i} A_i$;

and consequently

(m4) $\mathcal{M} \models_w \Diamond(A_1, \ldots, A_n)$ iff there exist $z_1, \ldots, z_n$ such that $R(w, z_1, \ldots, z_n)$ and for all $i \leq n$, $\mathcal{M} \models_{z_i} A_i$.

We put $\mathcal{M} \models_w \Gamma$ if $\mathcal{M} \models_w A$ for all $A \in \Gamma$. Truth in the model $\mathcal{M}$ is defined by

$$\mathcal{M} \models A \quad \text{iff} \quad \text{for all } w \in W, \ \mathcal{M} \models_w A,$$

and semantic consequence is defined by

$$\Gamma \models^{\mathcal{M}} A \quad \text{iff} \quad \text{for all } w \in W, \ \mathcal{M} \models_w \Gamma \text{ implies } \mathcal{M} \models_w A.$$

Observe that if $C = \Box(B_1, \ldots, B_{i-1}, A, B_{i+1}, \ldots, B_n)$, then in any model:

- $\mathcal{M} \models A$ implies $\mathcal{M} \models C$;
- $\mathcal{M} \models C_A[B \to D] \to (C_A[B] \to C_A[D])$.

Note that the more common practice in modal logic is to define a model as a structure $\langle W, R, V \rangle$, with V being a function that assigns a subset $V(p)$ of W to each atomic variable p. Then (m1)–(m3) serve to *define* the relation $\mathcal{M} \models_w A$ inductively, starting from the base

$$\mathcal{M} \models_w p \quad \text{iff} \quad w \in V(p).$$

Here we are taking a different approach because we wish to develop a very general theory that allows our object language to have additional unspecified connectives. (m1)–(m3) are the minimum conditions that $\mathcal{M}$ must satisfy in order to be a model, but there may be other conditions to be imposed, depending on which particular language or logical system we apply the theory to.

For instance, if $\mathcal{I}$ is a subset of $2^{\Phi} \times \Phi$, then a model $\mathcal{M}$ will be called $\mathcal{I}$-*sound* if

$$(\Gamma, A) \in \mathcal{I} \quad \text{implies} \quad \Gamma \models^{\mathcal{M}} A$$

for all $\Gamma \subseteq \Phi$ and $A \in \Phi$.

Typically here $\mathcal{I}$ will be the set of all instances of some *rule of inference*, and so an $\mathcal{I}$-sound model is one whose semantic consequence relation preserves this rule.

Logics

In this chapter, a *logic* will mean a set Λ of formulae that includes all tautologies and is closed under *Detachment*:

$$\text{if } A, (A \rightarrow B) \in \Lambda, \text{ then } B \in \Lambda.$$

Λ is a *normal* logic if it is closed under *Necessitation*:

$$A \in \Lambda \quad \text{implies} \quad \Box(B_1, \ldots, B_{i-1}, A, B_{i+1}, \ldots, B_n) \in \Lambda,$$

and contains all instances of the schema

$$K: \quad C_A[B \rightarrow D] \rightarrow (C_A[B] \rightarrow C_A[D])$$

where $C = \Box(B_1, \ldots, B_{i-1}, A, B_{i+1}, \ldots, B_n)$ (when $\Box$ is a unary modality, this is the standard definition of normality ([32, p. 20])).

From what we noted above about the truth relation $\mathcal{M} \models A$, for any model $\mathcal{M}$ the set

$$\{A : \mathcal{M} \models A\}$$

of formulae true in $\mathcal{M}$ is a normal logic.

If Λ is any logic, then a Λ-*model* is a model $\mathcal{M}$ such that $\mathcal{M} \models \Lambda$, i.e. $\mathcal{M} \models A$ for all $A \in \Lambda$. If $\mathcal{I} \subseteq 2^{\Phi} \times \Phi$, then the set

$$\Lambda\mathcal{I} = \{A : \mathcal{M} \models A \text{ for all } \mathcal{I}\text{-sound } \Lambda\text{-models } \mathcal{M}\}$$

is a normal logic that contains Λ and is closed under $\mathcal{I}$:

$$\text{if } (\Gamma, A) \in \mathcal{I} \text{ and } \Gamma \subseteq \Lambda\mathcal{I}, \text{ then } A \in \Lambda\mathcal{I}.$$

We might ask for the exact relationship between Λ and $\Lambda\mathcal{I}$. Is $\Lambda\mathcal{I}$ the *smallest normal logic extending Λ that is closed under $\mathcal{I}$* ? An answer to that question will be given in Corollary 9.5.3.

A more general question concerns the semantic consequence relation $\models_{\Lambda\mathcal{I}}$ determined by Λ and $\mathcal{I}$, where

$$\Gamma \models_{\Lambda\mathcal{I}} A \quad \text{iff} \quad \Gamma \models^{\mathcal{M}} A \text{ for all } \mathcal{I}\text{-sound } \Lambda\text{-models } \mathcal{M}.$$

The problem is to axiomatise $\models_{\Lambda\mathcal{I}}$: to give a purely proof-theoretic characterisation of this relation. We will see that this can be done in a very satisfactory way for countable $\mathcal{I}$.

Deducibility

If Λ is a logic, then A is $\vdash_\Lambda$-*deducible from* Γ, $\Gamma \vdash_\Lambda A$, if for some n there exist $B_0, \ldots, B_{n-1} \in \Gamma$ such that the formula

$$B_0 \to (B_1 \to (\cdots \to (B_{n-1} \to A) \cdots)$$

belongs to Λ (in the case $n = 0$ this means that $A \in \Lambda$). We write $\vdash_\Lambda A$ when $\emptyset \vdash_\Lambda A$. Hence

$$\vdash_\Lambda A \quad \text{iff} \quad A \in \Lambda.$$

Consider the following properties of a relation $\vdash$ from 2^Φ to Φ:

PC: $\Gamma \vdash A$ if A is a tautological consequence of Γ.

CT: If $\Gamma \vdash B$ for all $B \in \Delta$, and $\Delta \vdash A$, then $\Gamma \vdash A$.

DT: $\Gamma \cup \{A\} \vdash B$ implies $\Gamma \vdash A \to B$.

IR: $\Gamma \vdash B$ implies $(A \to \Gamma) \vdash A \to B$,

     where

$$(A \to \Gamma) = \{A \to C : C \in \Gamma\}.$$

BR: $\Gamma \vdash A$ implies

$$\Box(B_1, \ldots, B_{i-1}, \Gamma, B_{i+1}, \ldots, B_n) \vdash \Box(B_1, \ldots, B_{i-1}, A, B_{i+1}, \ldots, B_n),$$

    where $\Box(B_1, \ldots, B_{i-1}, \Gamma, B_{i+1}, \ldots, B_n)$ is the set

$$\{\Box(B_1, \ldots, B_{i-1}, C, B_{i+1}, \ldots, B_n) : C \in \Gamma\}.$$

Here PC stands for "Propositional Calculus", CT for "Cut Rule", DT for "Deduction Theorem", IR for "Implication Rule", and BR for "Box Rule".

For any model $\mathcal{M}$, the semantic consequence relation $\models^\mathcal{M}$ satisfies all five of these properties. For any logic Λ, the Λ-deducibility relation $\vdash_\Lambda$ satisfies PC, CT, DT, and IR. $\vdash_\Lambda$ satisfies BR if, and only if, Λ is *normal*.

Lemma 9.2.1 *If $\vdash$ satisfies PC, CT, and DT, then it satisfies IR, and*

$$\Gamma \vdash A \to B \quad \text{implies} \quad \Gamma \cup \{A\} \vdash B.$$

Proof. (As for Lemma 8.7.1.) Suppose $\Gamma \vdash B$. Since PC gives

$$(A \to \Gamma) \cup \{A\} \vdash C, \quad \text{for all } C \in \Gamma,$$

it then follows by CT that $(A \to \Gamma) \cup \{A\} \vdash B$. Hence by DT,

$$(A \to \Gamma) \vdash A \to B,$$

establishing IR.

Next, suppose $\Gamma \vdash A \to B$. But by PC, $\Gamma \cup \{A\} \vdash C$ for all $C \in \Gamma$, so this yields $\Gamma \cup \{A\} \vdash A \to B$ by CT. Since $\{A \to B, A\} \vdash B$ and $\Gamma \cup \{A\} \vdash A$ (both by PC), CT then gives $\Gamma \cup \{A\} \vdash B$. $\Box$

Let $\vdash_{\Lambda\mathcal{I}}$ be the smallest relation from 2^{Φ} to Φ such that

- $\vdash_{\Lambda\mathcal{I}}$ extends $\vdash_{\Lambda}$: $\Gamma \vdash_{\Lambda} A$ implies $\Gamma \vdash_{\Lambda\mathcal{I}} A$;
- $\vdash_{\Lambda\mathcal{I}}$ extends $\mathcal{I}$: $(\Gamma, A) \in \mathcal{I}$ implies $\Gamma \vdash_{\Lambda\mathcal{I}} A$;
- $\vdash_{\Lambda\mathcal{I}}$ satisfies CT, DT, and BR.

Then from the above, $\vdash_{\Lambda\mathcal{I}}$ also satisfies PC and IR. Our main result (Theorem 9.5.2) is going to be that when Λ is normal and $\mathcal{I}$ is countable, $\vdash_{\Lambda\mathcal{I}}$ characterises $\models_{\Lambda\mathcal{I}}$:

$$\Gamma \vdash_{\Lambda\mathcal{I}} A \quad \text{iff} \quad \Gamma \models_{\Lambda\mathcal{I}} A.$$

Here are some properties of $\vdash_{\Lambda\mathcal{I}}$ that will be needed below.

Lemma 9.2.2

(1) $A \in \Gamma$ implies $\Gamma \vdash_{\Lambda\mathcal{I}} A$.

(2) (*Monotonicity*) If $\Gamma \vdash_{\Lambda\mathcal{I}} A$ and $\Gamma \subseteq \Delta$, then $\Delta \vdash_{\Lambda\mathcal{I}} A$.

(3) (*Detachment*) If $\Gamma \vdash_{\Lambda\mathcal{I}} A$ and $\Gamma \vdash_{\Lambda\mathcal{I}} A \rightarrow B$, then $\Gamma \vdash_{\Lambda\mathcal{I}} B$.

(4) If $\Gamma \vdash_{\Lambda\mathcal{I}} A$ and $\Gamma \cup \{A\} \vdash_{\Lambda\mathcal{I}} \bot$, then $\Gamma \vdash_{\Lambda\mathcal{I}} \bot$.

(5) If $\Gamma \cup \{\neg A\} \vdash_{\Lambda\mathcal{I}} \bot$, then $\Gamma \vdash_{\Lambda\mathcal{I}} A$.

Proof.

(1) If $A \in \Gamma$ then $\Gamma \vdash_{\Lambda} A$ by the tautology $A \rightarrow A$. But $\vdash_{\Lambda\mathcal{I}}$ extends $\vdash_{\Lambda}$.

(2) If $\Gamma \vdash_{\Lambda\mathcal{I}} A$ and $\Gamma \subseteq \Delta$, then $\Delta \vdash_{\Lambda\mathcal{I}} B$ for all $B \in \Gamma$ by (1), so CT yields $\Delta \vdash_{\Lambda\mathcal{I}} A$.

(3) We have $\{A, A \rightarrow B\} \vdash_{\Lambda\mathcal{I}} B$, since $\vdash_{\Lambda\mathcal{I}}$ extends $\vdash_{\Lambda}$, so the desired result follows by CT.

(4) If $\Gamma \vdash_{\Lambda\mathcal{I}} A$ and $\Gamma \cup \{A\} \vdash_{\Lambda\mathcal{I}} \bot$, then $\Gamma \vdash_{\Lambda\mathcal{I}} A \rightarrow \bot$ by DT, so (3) gives $\Gamma \vdash_{\Lambda\mathcal{I}} \bot$.

(5) If $\Gamma \cup \{\neg A\} \vdash_{\Lambda\mathcal{I}} \bot$, then $\Gamma \vdash_{\Lambda\mathcal{I}} \neg A \rightarrow \bot$ by DT. Then apply the tautology $(\neg A \rightarrow \bot) \rightarrow A$ and CT.

9.3   Theories

Let $\mathcal{I}$ be a subset of $2^{\Phi} \times \Phi$. Define $\mathcal{I}_{\omega}$ to be the smallest extension of $\mathcal{I}$ in $2^{\Phi} \times \Phi$ that satisfies IR and BR, in the sense that if $(\Gamma, A) \in \mathcal{I}_{\omega}$ then $(B \rightarrow \Gamma, B \rightarrow A)$ and

$$(\Box(B_1, \ldots, B_{i-1}, \Gamma, B_{i+1}, \ldots, B_n), \Box(B_1, \ldots, B_{i-1}, A, B_{i+1}, \ldots, B_n))$$

belong to $\mathcal{I}_{\omega}$.

For any model $\mathcal{M}$, the semantic consequence relation $\models^{\mathcal{M}}$ satisfies IR and BR. Hence if $\models^{\mathcal{M}}$ extends $\mathcal{I}$ it must also extend $\mathcal{I}_{\omega}$. In other words, *any $\mathcal{I}$-sound model is $\mathcal{I}_{\omega}$-sound.*

Lemma 9.3.1 *If $\mathcal{I}$ is countable, then so is $\mathcal{I}_\omega$.*

Proof. Let $\mathcal{I}_0 = \mathcal{I}$, and inductively let $\mathcal{I}_{n+1}$ be the result of adding to $\mathcal{I}_n$ all pairs of the form $(B \to \Gamma, B \to A)$ and

$$(\Box(B_1, \ldots, B_{i-1}, \Gamma, B_{i+1}, \ldots, B_n), \Box(B_1, \ldots, B_{i-1}, A, B_{i+1}, \ldots, B_n))$$

for which $(\Gamma, A) \in \mathcal{I}_n$. Then $\mathcal{I}_\omega = \bigcup_{n<\omega} \mathcal{I}_n$.

But if $\mathcal{I}_n$ is countable, then so is $\mathcal{I}_{n+1}$, since Φ is countable and so there are only countably many pairs to be added to $\mathcal{I}_n$ to form $\mathcal{I}_{n+1}$. Thus if $\mathcal{I}_0$ is countable, then $\mathcal{I}_\omega$ is a countable union of countable sets, hence is countable. $\qquad\qquad\qquad\qquad\qquad\qquad\qquad\qquad\qquad\qquad\Box$

From now on let Λ be a fixed *normal* logic, and $\mathcal{I}$ a fixed *countable* subset of $2^\Phi \times \Phi$. A $\Lambda\mathcal{I}$-theory is a set $\Delta \subseteq \Phi$ such that

- $\Lambda \subseteq \Delta$;
- Δ is closed under Detachment:

$$\text{if } A, (A \to B) \in \Delta, \text{ then } B \in \Delta;$$

- Δ is closed under $\mathcal{I}_\omega$:

$$\text{if } \Sigma \subseteq \Delta, \text{ and } (\Sigma, A) \in \mathcal{I}_\omega, \text{ then } A \in \Delta.$$

The intersection of any collection of $\Lambda\mathcal{I}$-theories is a $\Lambda\mathcal{I}$-theory. This implies that there is a smallest $\Lambda\mathcal{I}$-theory, and we will prove (Corollary 9.5.3) that this is just $\Lambda\mathcal{I}$ itself.

Moreover, for any set Γ of formulae there is a smallest $\Lambda\mathcal{I}$-theory that contains Γ. We describe the members of this theory in terms of another relation $\vdash^+_{\Lambda\mathcal{I}}$ from 2^Φ to Φ. Define

$$\Gamma \vdash^+_{\Lambda\mathcal{I}} A \quad \text{iff} \quad A \in \bigcap\{\Delta : \Gamma \subseteq \Delta \text{ and } \Delta \text{ is a } \Lambda\mathcal{I}\text{-theory}\}.$$

Thus $\Gamma \vdash^+_{\Lambda\mathcal{I}} A$ if, and only if, A belongs to every $\Lambda\mathcal{I}$-theory containing Γ. Consequently, $\Lambda\mathcal{I}$-*theories are* $\vdash^+_{\Lambda\mathcal{I}}$-*closed*: it is immediate from the definitions that

$$\text{if } \Gamma \text{ is an } \Lambda\mathcal{I}\text{-theory and } \Gamma \vdash^+_{\Lambda\mathcal{I}} A, \text{ then } A \in \Gamma.$$

This point will be important later, and will be used to prove that $\vdash^+_{\Lambda\mathcal{I}}$ and $\vdash_{\Lambda\mathcal{I}}$ are in fact the same relation.

Theorem 9.3.2 (Soundness) $\Gamma \vdash^+_{\Lambda\mathcal{I}} A$ *implies* $\Gamma \models_{\Lambda\mathcal{I}} A$.

Proof. Suppose $\Gamma \vdash^+_{\Lambda\mathcal{I}} A$. Let $\mathcal{M}$ be any $\mathcal{I}$-sound Λ-model and suppose $\mathcal{M} \models_w \Gamma$. We have to show $\mathcal{M} \models_w A$. Put

$$\Delta = \{B : \mathcal{M} \models_w B\}.$$

It is enough to show that Δ is a $\Lambda\mathcal{I}$-theory containing Γ, for then as $\Gamma \vdash^+_{\Lambda\mathcal{I}} A$ we get $A \in \Delta$ as desired.

It is straightforward from our assumptions that $\Gamma \cup \Lambda \subseteq \Delta$ and Δ is closed under Detachment. But as noted earlier, an $\mathcal{I}$-sound model is $\mathcal{I}_\omega$-sound, and from $\mathcal{I}_\omega$-soundness it follows that Δ is $\mathcal{I}_\omega$-closed.   □

Theorem 9.3.3 $\vdash^+_{\Lambda\mathcal{I}}$ *extends* $\vdash_\Lambda$ *and* $\mathcal{I}$, *and satisfies CT, DT, and BR.*

Proof.

(1) That $\Gamma \vdash_\Lambda A$ implies $\Gamma \vdash^+_{\Lambda\mathcal{I}} A$ follows readily from the fact that $\Lambda\mathcal{I}$-theories contain Λ and are closed under Detachment.

(2) That $(\Gamma, A) \in \mathcal{I}$ implies $\Gamma \vdash^+_{\Lambda\mathcal{I}} A$ follows immediately from the fact that $\Lambda\mathcal{I}$-theories are closed under $\mathcal{I}$.

(3) CT: If $\Gamma \vdash^+_{\Lambda\mathcal{I}} B$ for all $B \in \Delta$, and $\Delta \vdash^+_{\Lambda\mathcal{I}} A$, then any $\Lambda\mathcal{I}$-theory containing Γ will contain Δ, and hence A. Thus $\Gamma \vdash^+_{\Lambda\mathcal{I}} A$.

(4) DT: Suppose $\Gamma \cup \{A\} \vdash^+_{\Lambda\mathcal{I}} B$. Let

$$\Delta = \{C : \Gamma \vdash^+_{\Lambda\mathcal{I}} A \to C\}.$$

We want to prove $B \in \Delta$, so by our supposition it is enough to show that Δ is a $\Lambda\mathcal{I}$-theory containing $\Gamma \cup \{A\}$.

Now since $C \to (A \to C)$ is a tautology, it belongs to Λ, and this leads to $\Gamma \vdash^+_{\Lambda\mathcal{I}} A \to C$, hence $C \in \Delta$, in case that $C \in \Gamma$ or $C \in \Lambda$. Similarly, using the tautology $A \to A$ we get $A \in \Delta$. Thus $\Lambda \cup \Gamma \cup \{A\} \subseteq \Delta$.

Next, to show that Δ is closed under Detachment, suppose C and $C \to D$ are in Δ. Then the tautology

$$(A \to C) \to ((A \to (C \to D)) \to (A \to D))$$

leads to $\Gamma \vdash^+_{\Lambda\mathcal{I}} A \to D$, as desired.

Finally, to show Δ is closed under $\mathcal{I}_\omega$, let $(\Sigma, C) \in \mathcal{I}_\omega$ and suppose $\Sigma \subseteq \Delta$. Then $\Gamma \vdash^+_{\Lambda\mathcal{I}} A \to D$ for all $D \in \Sigma$. Hence any $\Lambda\mathcal{I}$-theory containing Γ will contain $A \to \Sigma$. But

$$(A \to \Sigma, A \to C) \in \mathcal{I}_\omega,$$

since $\mathcal{I}_\omega$ was defined to satisfy IR, so every $\Lambda\mathcal{I}$-theory containing Γ will contain $A \to C$, being $\mathcal{I}_\omega$-closed. Thus $C \in \Delta$.

This completes the proof that Δ is a $\Lambda\mathcal{I}$-theory containing $\Gamma \cup \{A\}$, and hence the proof that $\vdash^+_{\Lambda\mathcal{I}}$ satisfies DT.

(5) BR: Suppose $\Gamma \vdash^+_{\Lambda\mathcal{I}} A$. To show

$$\Box(B_1, \ldots, B_{i-1}, \Gamma, B_{i+1}, \ldots, B_n) \vdash^+_{\Lambda\mathcal{I}} C,$$

where $C = \Box(B_1, \ldots, B_{i-1}, A, B_{i+1}, \ldots, B_n)$, take any $\Lambda\mathcal{I}$-theory Σ that contains $\Box(B_1, \ldots, B_{i-1}, \Gamma, B_{i+1}, \ldots, B_n)$, and put

$$\Delta = \{B : \Box(B_1, \ldots, B_{i-1}, B, B_{i+1}, \ldots, B_n) \in \Sigma\}.$$

We want to show $C \in \Sigma$, i.e. $A \in \Delta$, so as $\Gamma \vdash^+_{\Lambda\mathcal{I}} A$ it suffices to show that Δ is a $\Lambda\mathcal{I}$-theory containing Γ.

The definitions immediately give $\Gamma \subseteq \Delta$. Since Λ is normal, Necessitation implies that if $B \in \Lambda$ then $C_A[B] \in \Lambda$, so $C_A[B] \in \Sigma$, hence $B \in \Delta$. Thus $\Lambda \subseteq \Delta$.

Also from normality, Σ contains all instances of the schema

$$K: \quad C_A[B \to D] \to (C_A[B] \to C_A[D]),$$

so as Σ is closed under Detachment, this shows that if $B \to D, B \in \Delta$, then $C_A[D] \in \Sigma$, whence $D \in \Delta$. Thus Δ is closed under Detachment.

Finally, to show that Δ is $\mathcal{I}_\omega$-closed, suppose $(\Theta, D) \in \mathcal{I}_\omega$ and $\Theta \subseteq \Delta$. Then $C_A[\Theta] \subseteq \Sigma$, where

$$C_A[\Theta] = \{C_A[B] : B \in \Theta\}.$$

But $(C_A[\Theta], C_A[D]) \in \mathcal{I}_\omega$, since $\mathcal{I}_\omega$ satisfies BR, and Σ is $\mathcal{I}_\omega$-closed, so $C_A[D] \in \Sigma$, hence $D \in \Delta$.

This completes the proof that $\vdash^+_{\Lambda\mathcal{I}}$ satisfies BR, and hence the proof of the Theorem.

$\square$

Corollary 9.3.4 $\Gamma \vdash_{\Lambda\mathcal{I}} A$ implies $\Gamma \vdash^+_{\Lambda\mathcal{I}} A$.

Proof. $\vdash_{\Lambda\mathcal{I}}$ was defined to be the *smallest* relation satisfying the properties just proven for $\vdash^+_{\Lambda\mathcal{I}}$ in the Theorem. $\square$

Consistency

A set Γ is $\vdash_{\Lambda\mathcal{I}}$ *-consistent* if $\Gamma \nvdash_{\Lambda\mathcal{I}} \bot$, and *finitely* $\vdash_{\Lambda\mathcal{I}}$ *-consistent* if each finite subset of Γ is $\vdash_{\Lambda\mathcal{I}}$-consistent. Γ is *negation complete* if

for all $A \in \Phi$, either $A \in \Gamma$ or $\neg A \in \Gamma$.

Lemma 9.3.5

(1) *If Γ is finitely $\vdash_{\Lambda\mathcal{I}}$ -consistent, then so is one of $\Gamma \cup \{A\}$ and $\Gamma \cup \{\neg A\}$ for any A.*

(2) *If Γ is negation complete and finitely $\vdash_{\Lambda\mathcal{I}}$ -consistent, then Γ is closed under Detachment and contains Λ.*

(3) *If Γ is finitely $\vdash_{\Lambda\mathcal{I}}$ -consistent, then Γ is maximally finitely $\vdash_{\Lambda\mathcal{I}}$ - consistent if, and only if, it is negation complete.*

Proof. Most of these are familiar results from standard (finitary) propositional logic, but we go over the arguments to see just what properties of the deducibility relation $\vdash_{\Lambda\mathcal{I}}$ are involved.

(1) If the conclusion of (1) fails, $\Gamma_0 \cup \{A\} \vdash_{\Lambda\mathcal{I}} \bot$ and $\Gamma_1 \cup \{\neg A\} \vdash_{\Lambda\mathcal{I}} \bot$ for some A and some finite subsets Γ_0, Γ_1 of Γ. Then by DT and Monotonicity (9.2.2(2)), we get

$$\Gamma_0 \cup \Gamma_1 \vdash_{\Lambda\mathcal{I}} A \to \bot, \quad \text{and} \quad \Gamma_0 \cup \Gamma_1 \vdash_{\Lambda\mathcal{I}} \neg A \to \bot.$$

But as

$$\{A \to \bot, \neg A \to \bot\} \vdash_\Lambda \bot$$

by tautological consequence, applying CT leads to the conclusion that $\Gamma_0 \cup \Gamma_1 \vdash_{\Lambda\mathcal{I}} \bot$, so Γ is not finitely $\vdash_{\Lambda\mathcal{I}}$-consistent.

(2) Let Γ be negation complete and finitely $\vdash_{\Lambda\mathcal{I}}$-consistent. Then Γ is closed under Detachment, for if $A, A \to B \in \Gamma$ but $B \notin \Gamma$, then $\neg B \in \Gamma$, so Γ contains the finite $\vdash_{\Lambda\mathcal{I}}$-inconsistent set $\{A, A \to B, \neg B\}$.

To show $\Lambda \subseteq \Gamma$, assume $A \in \Lambda$. Then $(\neg A \to \bot) \in \Lambda$, so $\{\neg A\}$ is a finite $\vdash_{\Lambda\mathcal{I}}$-inconsistent set. Hence $\neg A \notin \Gamma$, and so $A \in \Gamma$.

(3) Let Γ be maximally finitely $\vdash_{\Lambda\mathcal{I}}$-consistent. Then for any formula A, (1) implies that one of $\Gamma \cup \{A\}$ and $\Gamma \cup \{\neg A\}$ is equal to Γ, hence one of A and $\neg A$ is in Γ. Thus Γ is negation complete.

Conversely, if Γ is finitely $\vdash_{\Lambda\mathcal{I}}$-consistent and negation complete, let $\Gamma \subsetneq \Delta$. Then there exists $A \in \Delta$ with $\neg A \in \Gamma$, so Δ contains the $\vdash_{\Lambda\mathcal{I}}$-inconsistent set $\{A, \neg A\}$. Hence Γ has no finitely $\vdash_{\Lambda\mathcal{I}}$-consistent proper extensions, i.e. is maximally finitely $\vdash_{\Lambda\mathcal{I}}$-consistent.

$$\square$$

Corollary 9.3.6 *If Γ is maximally finitely $\vdash_{\Lambda\mathcal{I}}$-consistent and $\mathcal{I}_\omega$-closed, then Γ is a negation complete $\Lambda\mathcal{I}$-theory that is $\vdash_{\Lambda\mathcal{I}}$-closed and (maximally) $\vdash_{\Lambda\mathcal{I}}$-consistent.*

Proof. If Γ is maximally finitely $\vdash_{\Lambda\mathcal{I}}$-consistent and $\mathcal{I}_\omega$-closed, then by (3) and (2) of Lemma 9.3.5 it is also negation complete, contains Λ and is closed under Detachment, so is a $\Lambda\mathcal{I}$-theory.

Thus if $\Gamma \vdash_{\Lambda\mathcal{I}} A$, we have $\Gamma \vdash^+_{\Lambda\mathcal{I}} A$ by Corollary 9.3.4, so $A \in \Gamma$ because $\Lambda\mathcal{I}$-*theories are* $\vdash^+_{\Lambda\mathcal{I}}$-*closed*. This shows that Γ is $\vdash_{\Lambda\mathcal{I}}$-closed.

Consequently, if Γ were $\vdash_{\Lambda\mathcal{I}}$-inconsistent, i.e. $\Gamma \vdash_{\Lambda\mathcal{I}} \bot$, then $\{\bot\} \subseteq \Gamma$, contrary to the fact that Γ is *finitely* $\vdash_{\Lambda\mathcal{I}}$-consistent. Finally, Γ has no proper $\vdash_{\Lambda\mathcal{I}}$-consistent extensions, since it has no proper finitely $\vdash_{\Lambda\mathcal{I}}$-consistent extensions. $\square$

Maximal Theories

A *maximal $\Lambda\mathcal{I}$-theory* is a $\Lambda\mathcal{I}$-theory that is $\vdash_{\Lambda\mathcal{I}}$-consistent and negation complete. Such a theory has no proper $\vdash_{\Lambda\mathcal{I}}$-consistent extensions, and indeed has no proper finitely $\vdash_{\Lambda\mathcal{I}}$-consistent extensions, by Lemma 9.3.5(3). Corollary 9.3.6 shows that a set is a maximal $\Lambda\mathcal{I}$-theory if, and only if, it is maximally finitely $\vdash_{\Lambda\mathcal{I}}$-consistent and $\mathcal{I}_\omega$-closed.

Lemma 9.3.7 (Extension Lemma) *Every $\vdash_{\Lambda\mathcal{I}}$-consistent set can be extended to a maximal $\Lambda\mathcal{I}$-theory.*

Proof. Suppose Γ is $\vdash_{\Lambda\mathcal{I}}$-consistent, i.e. $\Gamma \nvdash_{\Lambda\mathcal{I}} \bot$. Let

$$A_0, \ldots, A_n, \ldots \ldots$$

be an enumeration of the set Φ of all formulae, and

$$(\Sigma_0, B_0), \ldots, (\Sigma_m, B_m), \ldots \ldots$$

an enumeration of $\mathcal{I}_\omega$, which is countable since $\mathcal{I}$ is (Lemma 9.3.1). Let $\Delta_0 = \Gamma$, so that Δ_0 is $\vdash_{\Lambda\mathcal{I}}$-consistent by hypothesis.

Now assume inductively that Δ_n has been defined, and is $\vdash_{\Lambda\mathcal{I}}$-consistent. If $\Delta_n \vdash_{\Lambda\mathcal{I}} A_n$, put

$$\Delta_{n+1} = \Delta_n \cup \{A_n\},$$

so that Δ_{n+1} is $\vdash_{\Lambda\mathcal{I}}$-consistent by Lemma 9.2.2(4).

If however $\Delta_n \nvdash_{\Lambda\mathcal{I}} A_n$, we have possible two cases. If $A_n \neq B_m$ for any m, put

$$\Delta_{n+1} = \Delta_n \cup \{\neg A_n\},$$

which is $\vdash_{\Lambda\mathcal{I}}$-consistent by Lemma 9.2.2(5).

Alternatively, suppose $A_n = B_m$ for some m. Then observe that

$$\Delta_n \cup \{\neg A_n\} \nvdash_{\Lambda\mathcal{I}} B_m$$

i.e.

$$\Delta_n \cup \{\neg A_n\} \nvdash_{\Lambda\mathcal{I}} A_n,$$

or else by DT and the tautology $(\neg A_n \to A_n) \to A_n$ we would contradict the assumption $\Delta_n \nvdash_{\Lambda\mathcal{I}} A_n$.

But $\Sigma_m \vdash_{\Lambda\mathcal{I}} B_m$, as $\vdash_{\Lambda\mathcal{I}}$ extends $\mathcal{I}_\omega$ (since it extends $\mathcal{I}$ and satisfies IR and BR). Hence by CT there must be some formula $C \in \Sigma_m$ with

$$\Delta_n \cup \{\neg A_n\} \nvdash_{\Lambda\mathcal{I}} C.$$

Put

$$\Delta_{n+1} = \Delta_n \cup \{\neg A_n, \neg C\},$$

which again is $\vdash_{\Lambda\mathcal{I}}$-consistent by Lemma 9.2.2(5).

This completes the inductive definition of the sets Δ_n, each of which is $\vdash_{\Lambda\mathcal{I}}$-consistent. Now put

$$\Delta = \bigcup_{n<\omega} \Delta_n.$$

Then Δ is our desired extension of Γ. To see this, observe that Δ is finitely $\vdash_{\Lambda\mathcal{I}}$-consistent, since any finite subset of Δ is contained in some Δ_n and so must be $\vdash_{\Lambda\mathcal{I}}$-consistent, or else by Monotonicity this would contradict the $\vdash_{\Lambda\mathcal{I}}$-consistency of Δ_n. Also it is clear from the construction that Δ is negation complete. Thus by Lemma 9.3.5(3), Δ is maximally finitely $\vdash_{\Lambda\mathcal{I}}$-consistent.

The construction also guarantees that Δ is $\mathcal{I}_\omega$-closed, for if $B_m \notin \Delta$, where $B_m = A_n$, then $A_n \notin \Delta_{n+1}$, so $\Delta_n \nvdash_{\Lambda\mathcal{I}} A_n$. But we arranged in that case that $\neg C \in \Delta_{n+1}$ for some $C \in \Sigma_m$, so $C \notin \Delta$ or else $\neg C, C \in \Delta_p$ for some p, contrary to $\vdash_{\Lambda\mathcal{I}}$-consistency of Δ_p.

Thus all told, Δ is a maximally finitely $\vdash_{\Lambda\mathcal{I}}$-consistent and $\mathcal{I}_\omega$-closed extension of Γ, and hence by Corollary 9.3.6 is a maximal $\vdash_{\Lambda\mathcal{I}}$-theory.
□

We can now show that $\vdash_{\Lambda\mathcal{I}}$ and $\vdash^+_{\Lambda\mathcal{I}}$ are identical:

Corollary 9.3.8 $\Gamma \vdash^+_{\Lambda\mathcal{I}} A$ *implies* $\Gamma \vdash_{\Lambda\mathcal{I}} A$.

Proof. If $\Gamma \nvdash_{\Lambda\mathcal{I}} A$, then $\Gamma \cup \{\neg A\}$ is $\vdash_{\Lambda\mathcal{I}}$-consistent by Lemma 9.2.2(5), so by the Extension Lemma there is some maximal $\Lambda\mathcal{I}$-theory Δ with $\Gamma \cup \{\neg A\} \subseteq \Delta$. Then $A \notin \Delta$ as Δ is (finitely) $\vdash_{\Lambda\mathcal{I}}$-consistent, so Δ is a $\Lambda\mathcal{I}$-theory which contains Γ but not A, showing that $\Gamma \nvdash^+_{\Lambda\mathcal{I}} A$. □

9.4   The Box Lemma

Our ultimate aim is to show that $\vdash_{\Lambda\mathcal{I}}$ axiomatises the semantic relation $\models_{\Lambda\mathcal{I}}$, and to do that we will build an $\mathcal{I}$-sound Λ-model $\mathcal{M}^{\Lambda\mathcal{I}}$ based on the set

$$W^{\Lambda\mathcal{I}} = \{\Gamma : \Gamma \text{ is a } \Lambda\mathcal{I}\text{-maximal theory}\}.$$

The $n+1$-placed relation $R^{\Lambda\mathcal{I}}$ on $W^{\Lambda\mathcal{I}}$ that interprets the n-ary modality $\Box$ is defined by specifying that

$$R^{\Lambda\mathcal{I}}(\Gamma, \Delta_1, \ldots, \Delta_n)$$

is to hold if, and only if, for any formulae $A_1, \ldots, A_n$,

if $\Box(A_1, \ldots, A_n) \in \Gamma$, then for some $i \leq n$, $A_i \in \Delta_i$.

Using the dual modality $\Diamond$ and negation completeness, this last condition is equivalent to:

if $A_i \in \Delta_i$ for all $i \leq n$, then $\Diamond(A_1, \ldots, A_n) \in \Gamma$.

The satisfaction relation in $\mathcal{M}^{\Lambda\mathcal{I}}$ will have

$$\mathcal{M} \models_\Gamma A \quad \text{iff} \quad A \in \Gamma,$$

so to prove $\mathcal{M}^{\Lambda\mathcal{I}}$ obeys model-condition (m3) we will need the following result.

Lemma 9.4.1 (Box Lemma) *If $\Gamma \in W^{\Lambda\mathcal{I}}$ and $\square(A_1, \ldots, A_n) \notin \Gamma$, then there exist $\Delta_1, \ldots, \Delta_n \in W^{\Lambda\mathcal{I}}$ such that $R^{\Lambda\mathcal{I}}(\Gamma, \Delta_1, \ldots, \Delta_n)$ and $A_i \notin \Delta_i$ for all $i \leq n$.*

In the case $n = 1$ the proof of this Lemma is quite direct: we simply show that the set

$$\{A : \square A \in \Gamma\} \cup \{\neg A\}$$

is $\vdash_{\Lambda\mathcal{I}}$-consistent, and apply the Extension Lemma to get the desired Δ. If the set were not $\vdash_{\Lambda\mathcal{I}}$-consistent, then $\{A : \square A \in \Gamma\} \vdash_{\Lambda\mathcal{I}} A$, so by BR, $\{\square A : \square A \in \Gamma\} \vdash_{\Lambda\mathcal{I}} \square A$, whence $\Gamma \vdash_{\Lambda\mathcal{I}} \square A$ by Monotonicity. But Γ is $\vdash_{\Lambda\mathcal{I}}$-closed, so this contradicts the hypothesis $\square A \notin \Gamma$.

In the general case $n > 1$, the proof is more complicated because we have to construct n maximal theories connected by the relation $R^{\Lambda\mathcal{I}}(\Gamma, \Delta_1, \ldots, \Delta_n)$. For finitary logics it is possible to systematically build $\Delta_1, \ldots, \Delta_n$ one after the other (cf. [28, Theorem 2.2.1] for an algebraic version of this). But here we have to ensure that the Δ_i's are all $\mathcal{I}_\omega$-closed, and it seems that to do this we need to inductively define them all simultaneously. For this we will need the following technical results.

Lemma 9.4.2

(1) *If $\vdash_{\Lambda\mathcal{I}} B_i \to C_i$ for all $i \leq n$, then*

$$\vdash_{\Lambda\mathcal{I}} \square(B_1, \ldots, B_n) \to \square(C_1, \ldots, C_n).$$

(2) *Let $\Diamond(B_1, \ldots, B_{i-1}, A, B_{i+1}, \ldots, B_n) \in \Gamma \in W^{\Lambda\mathcal{I}}$. Then*
   (i) *The formula A is $\vdash_{\Lambda\mathcal{I}}$-consistent.*
   (ii) *For all $B \in \Phi$, one of the formulae*

$$\Diamond(B_1, \ldots, B_{i-1}, A \wedge B, B_{i+1}, \ldots, B_n),$$
$$\Diamond(B_1, \ldots, B_{i-1}, A \wedge \neg B, B_{i+1}, \ldots, B_n)$$

   is in Γ.

(3) *If $(\Sigma, C) \in \mathcal{I}_\omega$ and*

$$\Diamond(B_1, \ldots, B_{i-1}, D \wedge \neg C, B_{i+1}, \ldots, B_n) \in \Gamma,$$

   then

$$\Diamond(B_1, \ldots, B_{i-1}, D \wedge \neg C \wedge \neg E, B_{i+1}, \ldots, B_n) \in \Gamma,$$

   for some $E \in \Sigma$.

Proof.

(1) By BR, $\Lambda\mathcal{I}$ is a normal logic, so satisfies the rule of Necessitation and contains the schema K. (1) can then be obtained by repeated application of these.

(2) Let $C = \Diamond(B_1, \ldots, B_{i-1}, A, B_{i+1}, \ldots, B_n)$, and

$$D = \Box(\neg B_1, \ldots, \neg B_{i-1}, \neg A, \neg B_{i+1}, \ldots, \neg B_n),$$

so that $C = \neg D$.

(i) If A is not $\vdash_{\Lambda\mathcal{I}}$-consistent, then $\vdash_{\Lambda\mathcal{I}} \neg A$, so by Necessitation $\vdash_{\Lambda\mathcal{I}} D$. Then $D \in \Gamma$, as Γ is $\vdash_{\Lambda\mathcal{I}}$-closed, so $C \notin \Gamma$ as Γ is $\vdash_{\Lambda\mathcal{I}}$-consistent.

(ii) Tautologically we have

$$\{\neg(A \wedge B), \neg(A \wedge \neg B)\} \vdash_{\Lambda\mathcal{I}} \neg A,$$

so by BR

$$\{D_{\neg A}[A \wedge B], D_{\neg A}[\neg(A \wedge \neg B)]\} \vdash_{\Lambda\mathcal{I}} D.$$

Therefore if $C \in \Gamma$ then $D \notin \Gamma$, so one of the formulae $D_{\neg A}[A \wedge B]$ and $D_{\neg A}[\neg(A \wedge \neg B)]$ must fail to be in Γ. Hence the negation of one of these formulae is in Γ, which is the desired conclusion.

(3) Let

$$H = \Diamond(B_1, \ldots, B_{i-1}, D \wedge \neg C, B_{i+1}, \ldots, B_n) \in \Gamma,$$

and

$$F = \Box(\neg B_1, \ldots, \neg B_{i-1}, A, \neg B_{i+1}, \ldots, \neg B_n),$$

so that $H = \neg F_A[\neg(D \wedge \neg C)]$. Then $F_A[\neg(D \wedge \neg C)] \notin \Gamma$, so $F_A[(D \wedge \neg C) \to C] \notin \Gamma$ (using an instance of result (1)). But since $(\Sigma, C) \in \mathcal{I}_\omega$ and $\mathcal{I}_\omega$ satisfies IR and BR,

$$(F_A[(D \wedge \neg C) \to \Sigma], F_A[(D \wedge \neg C) \to C]) \in \mathcal{I}_\omega,$$

where

$$F_A[(D \wedge \neg C) \to \Sigma] = \{F_A[(D \wedge \neg C) \to E] : E \in \Sigma\}.$$

It follows that for some $E \in \Sigma$, $F_A[(D \wedge \neg C) \to E] \notin \Gamma$, whence $\neg F_A[\neg(D \wedge \neg C \wedge \neg E)] \in \Gamma$, which is the desired conclusion. $\square$

We are now ready to embark on the proof of the Box Lemma, starting from the assumption

$$\Box(A_1, \ldots, A_n) \notin \Gamma \in W^{\Lambda\mathcal{I}}.$$

Let

$$B_0, \ldots, B_m, \ldots\ldots$$

be an enumeration of the set Φ of all formulae, and

$$(\Sigma_0, C_0), \ldots, (\Sigma_k, C_k), \ldots\ldots$$

an enumeration of $\mathcal{I}_\omega$. The theories $\Delta_1, \dots, \Delta_n \in W^{\Lambda \mathcal{I}}$ are constructed in such a way that each Δ_i is a union

$$\bigcup \{\Delta_i^m : m < \omega\}$$

of finite sets Δ_i^m. Defining D_i^m to be the conjunction of the members of Δ_i^m, we will show

(1_m) $\Diamond(D_1^m, \dots, D_i^m, \dots, D_n^m) \in \Gamma.$

For the base case $m = 0$, put $\Delta_i^0 = \{\neg A_i\}$. Then $D_i^0 = \neg A_i$, so from our initial hypothesis and negation completeness of Γ we have

(1_0) $\Diamond(D_1^0, \dots, D_i^0, \dots, D_n^0) \in \Gamma.$

Now make the inductive assumption on m that Δ_i^m has been defined for all i with $1 \leq i \leq n$, and that (1_m) holds. We proceed by an inner induction on i to define Δ_i^{m+1} for $1 \leq i \leq n$ in such a way that (1_{m+1}) is satisfied.

For $1 \leq j \leq n$, we will prove that

(2_j) $\Diamond(D_1^{m+1}, \dots, D_j^{m+1}, D_{j+1}^m, \dots, D_n^m) \in \Gamma.$

Then (2_n) is exactly (1_{m+1}).

So, fix i with $1 \leq i \leq n$ and assume inductively that Δ_j^{m+1} is defined and (2_j) holds for all $j < i$. Let $E_m = B_m$ if

$$\Diamond(D_1^{m+1}, \dots, D_{i-1}^{m+1}, D_i^m \wedge B_m, D_{i+1}^m, \dots, D_n^m) \in \Gamma,$$

and $E_m = \neg B_m$ otherwise. By (2_{i-1}) when $i > 1$, or (1_m) when $i = 1$, Lemma 9.4.2(2)(ii) ensures that

$(\dagger)$ $\Diamond(D_1^{m+1}, \dots, D_{i-1}^{m+1}, D_i^m \wedge E_m, D_{i+1}^m, \dots, D_n^m) \in \Gamma.$

If $E_m \neq \neg C_k$ for any $k < \omega$, put $\Delta_i^{m+1} = \Delta_i^m \cup \{E_m\}$. Then $D_i^{m+1} = D_i^m \wedge E_m$, so (2_i) holds by $(\dagger)$.

If however $E_m = \neg C_k$ for some k, then from $(\dagger)$ and Lemma 9.4.2(3), there exists $E \in \Sigma_k$ such that

$(\ddagger)$ $\Diamond(D_1^{m+1}, \dots, D_{i-1}^{m+1}, D_i^m \wedge E_m \wedge \neg E, D_{i+1}^m, \dots, D_n^m) \in \Gamma.$

Then put $\Delta_i^{m+1} = \Delta_i^m \cup \{E_m, \neg E\}$, so that $(\ddagger)$ implies (2_i).

This completes the induction on i with m fixed. But that takes care of the induction on m, and completes the definition of Δ_i^m for all $m < \omega$ and $1 \leq i \leq n$.

Applying Lemma 9.4.2(2)(i) to (1_m) shows that each formula D_i^m is $\vdash_{\Lambda \mathcal{I}}$-consistent, which implies that the set Δ_i^m is $\vdash_{\Lambda \mathcal{I}}$-consistent. Defining

$$\Delta_i = \bigcup \{\Delta_i^m : m < \omega\},$$

we then have that Δ_i, as the union of a chain of $\vdash_{\Lambda \mathcal{I}}$-consistent sets, is itself finitely $\vdash_{\Lambda \mathcal{I}}$-consistent. Since the construction placed one of B_m

and $\neg B_m$ in Δ_i^{m+1}, Δ_i is negation complete, and therefore maximally finitely $\vdash_{A\mathcal{I}}$-consistent. But it is also $\mathcal{I}_\omega$-closed, for if $(\Sigma_k, C_k) \in \mathcal{I}_\omega$ with $C_k = B_m$, and $C_k \notin \Delta_i$, then $B_m \notin \Delta_i^{m+1}$, so $E_m = \neg B_m = \neg C_k$, and in that case the construction arranged that for some $E \in \Sigma_k$, $\neg E \in \Delta_i^{m+1}$. Hence $E \notin \Delta_i$.

Since Δ_i is maximally finitely $\vdash_{A\mathcal{I}}$-consistent and $\mathcal{I}_\omega$-closed, it belongs to $W^{A\mathcal{I}}$ (Corollary 9.3.6). Also $A_i \notin \Delta_i$ as $\neg A_i \in \Delta_i^0$.

It remains to show $R^{A\mathcal{I}}(\Gamma, \Delta_1, \ldots, \Delta_n)$ to complete the Box Lemma. So, let $\Box(B_1, \ldots, B_n) \in \Gamma$. We want $B_i \in \Delta_i$ for some i. If not, then $\neg B_i \in \Delta_i$ for all $i \leq n$, so we can choose an m with $\neg B_i \in \Delta_i^m$ for all $i \leq n$. Then $\vdash_{A\mathcal{I}} D_i^m \rightarrow \neg B_i$, hence $\vdash_{A\mathcal{I}} B_i \rightarrow \neg D_i^m$, so

$$\vdash_{A\mathcal{I}} \Box(B_1, \ldots, B_n) \rightarrow \Box(\neg D_1^m, \ldots, \neg D_n^m)$$

by Lemma 9.4.2(1). But then $\Box(\neg D_1^m, \ldots, \neg D_n^m) \in \Gamma$, which contradicts (1_m).

$\square$

9.5   Completeness

Let Λ be any normal logic, and $\mathcal{I}$ any countable subset of $2^\Phi \times \Phi$. The *canonical $\Lambda\mathcal{I}$-model* is the structure

$$\mathcal{M}^{\Lambda\mathcal{I}} = \langle W^{\Lambda\mathcal{I}}, R^{\Lambda\mathcal{I}}, \models \rangle,$$

with $W^{\Lambda\mathcal{I}}$ the set of maximal $\Lambda\mathcal{I}$-theories, $R^{\Lambda\mathcal{I}}$ defined as at the beginning of Section 9.4, and $\models$ defined by

$$\mathcal{M}^{\Lambda\mathcal{I}} \models_\Gamma A \quad \text{iff} \quad A \in \Gamma.$$

That $\mathcal{M}^{\Lambda\mathcal{I}}$ satisfies model conditions (m1) and (m2) follows readily from properties of maximal $\Lambda\mathcal{I}$-theories. (m3) follows from the definition of $R^{\Lambda\mathcal{I}}$ and the Box Lemma 9.4.1.

Theorem 9.5.1 (Soundness) $\mathcal{M}^{\Lambda\mathcal{I}}$ *is an $\mathcal{I}$-sound Λ-model.*

Proof. If $(\Sigma, A) \in \mathcal{I}$ and $\mathcal{M}^{\Lambda\mathcal{I}} \models_\Gamma \Sigma$, then $\Sigma \subseteq \Gamma$, so as Γ is $\mathcal{I}_\omega$-closed, $A \in \Gamma$, whence $\mathcal{M}^{\Lambda\mathcal{I}} \models_\Gamma A$. This shows $\mathcal{I}$-soundness.

Also $\mathcal{M}^{\Lambda\mathcal{I}} \models_\Gamma \Lambda$ for any $\Gamma \in W^{\Lambda\mathcal{I}}$, as $\Lambda \subseteq \Gamma$. Hence $\mathcal{M}^{\Lambda\mathcal{I}}$ is a Λ-model.

$\square$

Theorem 9.5.2 *If $\Gamma \subseteq \Phi$ and $A \in \Phi$, the following are equivalent.*

(1) $\Gamma \vdash_{A\mathcal{I}} A$.

(2) $\Gamma \vdash_{A\mathcal{I}}^+ A$.

(3) $\Gamma \models_{A\mathcal{I}} A$.

(4) $\Gamma \models^{\mathcal{M}^{\Lambda\mathcal{I}}} A$.

Proof. We have already shown that (1) and (2) are equivalent (9.3.4, 9.3.8), and that (2) implies (3) (9.3.2). That (3) implies (4) is immediate because $\mathcal{M}^{\Lambda\mathcal{I}}$ is an $\mathcal{I}$-sound Λ-model.

To show (4) implies (1), suppose $\Gamma \nvdash_{\Lambda\mathcal{I}} A$. Then $\Gamma \cup \{\neg A\}$ is $\vdash_{\Lambda\mathcal{I}}$-consistent (9.2.2(5)), so by the Extension Lemma 9.3.7 there is some $\Delta \in W^{\Lambda\mathcal{I}}$ with $\Gamma \subseteq \Delta$ and $A \notin \Delta$. Thus $\mathcal{M}^{\Lambda\mathcal{I}} \models_\Delta \Gamma$ and $\mathcal{M}^{\Lambda\mathcal{I}} \nvDash_\Delta A$, showing that $\Gamma \nvDash^{\mathcal{M}^{\Lambda\mathcal{I}}} A$.   □

Corollary 9.5.3

(1) $\Lambda\mathcal{I}$ is the smallest $\Lambda\mathcal{I}$-theory.
(2) $\Lambda\mathcal{I}$ is the smallest normal logic extending Λ that is closed under $\mathcal{I}_\omega$.
(3) If $\mathcal{I}$ satisfies the rules IR and BR, then $\Lambda\mathcal{I}$ is the smallest normal logic extending Λ that is closed under $\mathcal{I}$.

Proof.

(1) Since
$$\Lambda\mathcal{I} = \{A \colon \mathcal{M} \models A \text{ for all } \mathcal{I}\text{-sound } \Lambda\text{-models } \mathcal{M}\},$$

and
$$\Gamma \models_{\Lambda\mathcal{I}} A \quad \text{iff} \quad \Gamma \models^{\mathcal{M}} A \text{ for all } \mathcal{I}\text{-sound } \Lambda\text{-models } \mathcal{M},$$

we have
$$A \in \Lambda\mathcal{I} \quad \text{iff} \quad \emptyset \models_{\Lambda\mathcal{I}} A.$$

Hence by the last Theorem,
$$A \in \Lambda\mathcal{I} \quad \text{iff} \quad \emptyset \vdash^+_{\Lambda\mathcal{I}} A.$$

In other words
$$\Lambda\mathcal{I} = \bigcap\{\Delta : \Delta \text{ is a } \Lambda\mathcal{I}\text{-theory}\}.$$

But $\Lambda\mathcal{I}$ is itself a $\Lambda\mathcal{I}$-theory, since it contains Λ and is closed under Detachment and $\mathcal{I}_\omega$.

(2) Any normal logic that extends Λ and is closed under $\mathcal{I}_\omega$ is a $\Lambda\mathcal{I}$-theory, and so contains $\Lambda\mathcal{I}$ by (1).

(3) If $\mathcal{I}$ satisfies IR and BR, then $\mathcal{I}_\omega = \mathcal{I}$, so the result follows from (2).   □

Note

Development of the material of this chapter and Section 8.7 has benefited from discussions with Krister Segerberg.

10

The McKinsey Axiom Is Not Canonical

The logic KM is the smallest normal modal logic that includes the *McKinsey axiom*

$$\Box \Diamond \varphi \rightarrow \Diamond \Box \varphi.$$

It is shown here that this axiom is not valid in the canonical frame for KM, answering a question first posed in the Lemmon-Scott manuscript [59].

The result is not just an esoteric counter-example: apart from interest generated by the long delay in a solution being found, the problem has been of historical importance in the development of our understanding of intensional model theory, and is of some conceptual significance, as will now be explained.

The relational semantics for normal modal logics first appeared in [52], where a number of well known systems were shown to be characterised by simple *first-order* conditions on binary relations (frames). This phenomenon was systematically investigated in [59], which introduced the technique of associating with each logic L a canonical frame $\mathcal{F}_L$ which invalidates every non-theorem of L. If, in addition, each L-theorem is valid in $\mathcal{F}_L$, then L is said to be *canonical*. The problem of showing that L is *determined* by some validating condition C, meaning that the L-theorems are precisely those formulae valid in all frames satisfying C, can be solved by showing that $\mathcal{F}_L$ satisfies C – in which case canonicity is also established. Numerous cases were studied, leading to the definition of a first-order condition C_φ associated with each formula φ of the form

$$\Diamond^{m_1} \Box^{n_1} \varphi_1 \wedge \cdots \wedge \Diamond^{m_k} \Box^{n_k} \varphi_k \rightarrow \Psi(\varphi_1, \ldots, \varphi_k),$$

where Ψ is a positive modal formula.

It was proven in [35] (cf. also Section 1.15) that the condition C_φ

is satisfied by the canonical frame for the logic with axiom φ. This result was also obtained independently by Sahlqvist [79], who broadened the class of formulae to which it applied, essentially by allowing the antecedent of φ to be any implication-free formula in which no variable occurs positively in a subformula of the type $\Diamond\varphi_1$ or $(\varphi_1 \vee \varphi_2)$ that is itself within the scope of a $\Box$ (cf. [80] for a recent discussion of the result). The McKinsey axiom is the simplest formula not (equivalent to one) meeting this criterion, and so the main result of the present paper indicates that there is no natural way to extend Sahlqvist's scheme to obtain a larger class of canonical formulae.

The class of all frames for KM is not *elementary*, i.e. is not characterised by any set of first-order conditions. This was shown in [99] by a Löwenheim-Skolem argument, and in [34], where failure of closure under ultraproducts was demonstrated. The latter work was then extended [37, §17] to prove that *any* class of frames that determines KM must fail to be closed under ultraproducts, and hence fail to be elementary (this material may be found in Section 1.17 of the present volume). This suggests that it would not be easy to establish whether KM was determined by its Kripke frames at all (Lemmon had conjectured in [59] that every logic is thus determined, but this was shown not to be so by Thomason [96, 97] and Fine [13]). That matter was soon resolved, however, by Fine [14], who gave completeness theorems for a general class of formulae by an analysis of normal forms. In particular, he showed that KM is determined by its *finite* frames, and is decidable.

The first general result about the connection between first-order definability and canonicity appeared in [15]: if the class of all L-frames determines L and is closed under first-order equivalence, then L is canonical. An example was also given of a logic for which the converse is false. It was also proved [15, Theorem 3] that if L is determined by *some* elementary class, then L is canonical. This clarified the example just mentioned, since that logic had been shown to be canonical by showing that it was determined by a first-order condition which was satisfied by the canonical frame, but not satisfied by all frames for the logic.

It is plausible to conjecture that the converse of Fine's latter result is true, i.e. that

> *if L is canonical, then L is determined by some elementary*
> *subclass of its frames*

(an approach to this is sketched in [29], where the problem is reduced to showing that if L is canonical, then L is preserved by ultrapowers of $\mathcal{F}_L$ – cf. Theorem 11.5.1 of this volume). Until now KM has been the one potential obstacle to this conjecture, as the only logic that was known not to be determined by any elementary class, but whose canonicity was unresolved. If the conjecture is true, it will undoubtedly hold also for other non-classical logics, including multi-modal systems and extensions of intuitionistic propositional logic, and will provide a most satisfactory explanation for the observed connections between intensional and first-order logic.

Frames and Models

All modal formulae will be constructed from a fixed denumerable set $\{p_n : n < \omega\}$ of propositional *variables* by the Boolean connectives $\wedge$, $\vee$, $\neg$, $\rightarrow$, and the modal $\square$. $\diamond$ may be defined as $\neg\square\neg$.

A *frame* is a structure $\mathcal{F} = (X, R)$, with R a binary relation on non-empty set X. If $x \in X$, put $R^x = \{y \in X : xRy\}$. The members of R^x are the *R-alternatives* of x. $\mathcal{F}$ is *generated* by a point $g \in X$ if $X = \{y : gR^*y\}$, where R^* is the reflexive transitive closure (ancestral) of R.

A *model* $\mathcal{M} = (X, R, V)$ on frame (X, R) is given by a *valuation* V which assigns a set $V(p) \subseteq X$ to every variable p. $V(p)$ is to be thought of as the set of points at which p is true. The satisfaction relation "formula φ is true/satisfied at x in $\mathcal{M}$", written $\mathcal{M} \models_x \varphi$, is defined inductively, with

$$\mathcal{M} \models_x p \quad \text{iff} \quad x \in V(p),$$
$$\mathcal{M} \models_x \square\varphi \quad \text{iff} \quad R^x \subseteq \{y : \mathcal{M} \models_y \varphi\},$$

and the Boolean connectives treated as usual, so that

$$\mathcal{M} \models_x \diamond\varphi \quad \text{iff} \quad R^x \cap \{y : \mathcal{M} \models_y \varphi\} \neq \emptyset.$$

φ is *true in model* $\mathcal{M}$, $\mathcal{M} \models \varphi$, if it is true at all points in $\mathcal{M}$, and *valid in frame* $\mathcal{F}$, $\mathcal{F} \models \varphi$, if it is true in all models on $\mathcal{F}$.

Logics

A *(normal modal) logic* is a set L of formulae that contains all tautologies and all instances of the schema

$(K) \qquad \square(\varphi \rightarrow \psi) \rightarrow (\square\varphi \rightarrow \square\psi),$

and is closed under Detachment (Modus Ponens), and the rule of *Necessitation*, i.e.

$$\varphi \in L \quad \text{implies} \quad \square\varphi \in L.$$

The intersection of any collection of logics is a logic, and so for any set Γ of formulae there is a *smallest* logic containing Γ.

The members of a logic L are the *L-theorems*. An *L-model* is one in which all L-theorems are true, and an *L-frame* is one in which all L-theorems are valid.

A set Γ of formulae is *L-consistent* if $\neg\varphi \notin L$ for any conjunction φ of finitely many members of Γ. For this to hold it suffices that every finite subset Γ_0 of Γ be *L*-consistent. Hence it suffices that any such finite Γ_0 be satisfied at a point in some *L*-model (for the negation of the conjunction of Γ_0 is then false at that point, and so cannot be an *L*-theorem).

There is only one logic that is not self-consistent, viz. the class of all formulae. Setting this case aside, it may be assumed that there exist *L*-consistent sets of formulae. Any such set can be extended to a *maximally L-consistent* set, and the *L*-theorems are precisely those formulae that belong to every maximally *L*-consistent set.

The *canonical L-frame* is $\mathcal{F}_L = (X_L, R_L)$, where X_L is the set of all maximally *L*-consistent sets of formulae, and $\Gamma R_L \Delta$ iff $\{\varphi \colon \Box\varphi \in \Gamma\} \subseteq \Delta$. The *canonical model* $\mathcal{M}_L$ on $\mathcal{F}_L$ is given by the valuation $V_L(p) = \{\Gamma \in X_L \colon p \in \Gamma\}$. An inductive proof shows that $\mathcal{M}_L \models_\Gamma \varphi$ iff $\varphi \in \Gamma$, for all formulae φ. Consequently $\mathcal{M}_L \models \varphi$ iff $\varphi \in L$, and so $\mathcal{F}_L \models \varphi$ implies that φ is an *L*-theorem. If the converse holds ($\varphi \in L$ implies $\mathcal{F}_L \models \varphi$), then *L* is called *canonical*.

Falsifying the McKinsey Axiom

KM is defined to be the smallest logic containing all instances of the schema

(McK) $\Box\Diamond\varphi \rightarrow \Diamond\Box\varphi,$

although we will usually find it more convenient to use the equivalent form

$$\Diamond(\Box\varphi \vee \Box\neg\varphi),$$

which is true at exactly the same points in all models. Since the set $\{\varphi \colon \mathcal{F} \models \varphi\}$ of all formulae valid in a frame $\mathcal{F}$ is a logic, to prove that $\mathcal{F}$ is a KM-frame it suffices to show that $\mathcal{F}$ validates the schema McK.

Now an *end-point* of a frame $\mathcal{F}$ is defined to be a point e for which $R^e = \{e\}$, i.e. eRy iff $e = y$. In any model, a formula of the form $\Box\varphi$ is true at an end-point iff φ is true there. Thus the formula $(\Box\varphi \vee \Box\neg\varphi)$ is always true at an end-point, and so McK cannot be falsified at any point x with the property that there is an end-point in R^x. This property is not however necessary for the truth of McK, as shown by the frame depicted in Figure 10.1, which will be referred to as *The Trellis*.

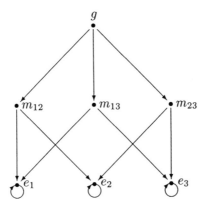

Figure 10.1. The Trellis

Formally, this frame is defined as the structure $T = (X, R)$, where

$$X = \{g, m_{12}, m_{13}, m_{23}, e_1, e_2, e_3\},$$

and R is specified by

$$R^g = \{m_{12}, m_{13}, m_{23}\}, \quad R^{m_{ij}} = \{e_i, e_j\}, \quad R^{e_i} = \{e_i\}.$$

Each point $x \neq g$ has an end-point in R^x, so can never falsify McK. But in any model on T, for each formula φ there must exist $1 \leq i < j \leq 3$ such that φ has constant truth-value on the set $\{e_i, e_j\}$ (i.e. is either true at both points, or false at both). But then $(\Box\varphi \vee \Box\neg\varphi)$ is true at m_{ij}, so $\Diamond(\Box\varphi \vee \Box\neg\varphi)$ is true at g. Hence T is a KM-frame, even though R^g contains no end-points.

This argument will re-emerge in the main model construction below (cf. Theorem 10.4). It could be paraphrased by saying that the alternatives of the generator g do not themselves have enough alternatives to make φ true and false at enough points to falsify McK. A condition under which enough alternatives do exist is given by the following result.

Theorem 10.1. *Let $\mathcal{F}$ be a frame containing a point g with the property that for any $m \in R^g$, R^m is an infinite set that is at least as large in cardinality as R^g itself. Then $\mathcal{F}$ is not a KM-frame.*

Proof. Let κ be the cardinality of R^g, and let $\{m_\lambda : \lambda < \kappa\}$ be an indexing of the members of R^g by the ordinals λ less than κ. For each λ, distinct points $m_{\lambda 0}, m_{\lambda 1} \in R^{m_\lambda}$ will then be defined in such a way that $\{m_{\lambda 0}, m_{\lambda 1}\}$ and $\{m_{\mu 0}, m_{\mu 1}\}$ are disjoint whenever $\lambda \neq \mu < \kappa$. Then putting $V(p) = \{m_{\lambda 1} : \lambda < \kappa\}$ defines a model on $\mathcal{F}$ in which p is false at $m_{\lambda 0}$, and true at $m_{\lambda 1}$, making $(\Box p \vee \Box\neg p)$ false at m_λ. Since this holds for every member m_λ of R^g, $\Diamond(\Box p \vee \Box \neg p)$ fails at g in this model on $\mathcal{F}$.

It remains then to show that the $m_{\lambda i}$ can be defined as claimed. Fix $\lambda < \kappa$, and suppose inductively that $m_{\mu i}$ has been defined for all $\mu < \lambda$ and $i \in \{0, 1\}$, such that $m_{\mu i} \neq m_{\nu j}$ whenever $\mu \neq \nu < \lambda$. Let

$$Y_\lambda = \{m_{\mu 0}, m_{\mu 1} \colon \mu < \lambda\}.$$

Then if λ is a finite ordinal, Y_λ is a finite set, so as R^{m_λ} is infinite, distinct points $m_{\lambda 0}, m_{\lambda 1}$ can be selected from $R^{m_\lambda} - Y_\lambda$. If however λ is infinite, then the cardinality of Y_λ is at most that of λ, and hence is less than κ. But R^{m_λ} has cardinality at least κ, so again the selection of $m_{\lambda 0}, m_{\lambda 1} \in R^{m_\lambda}$ can be made to ensure that $m_{\mu i} \neq m_{\nu j}$ for all $\mu \neq \nu \leq \lambda$. Hence the construction extends to λ, and so goes through by induction.   □

If the cardinal κ in this proof is finite, then R^m need not be infinite: the argument works if R^m is of size at least 2κ. A more important consequence for what follows is

Corollary 10.2. *If a canonical frame $\mathcal{F}_L$ contains a point g with the property that for any $m \in R_L^g$, R_L^m is of cardinality $2^{\aleph_0}$, then $\mathcal{F}_L$ is not a KM-frame.*

Proof. R_L^g has size at most $2^{\aleph_0}$, since X_L itself cannot be bigger than this, there being only countably many formulae. (Actually, it can be shown that X_L has exactly $2^{\aleph_0}$ members.)   □

The main work of this paper will be to show that $\mathcal{F}_{KM}$ contains a point g fulfilling the hypothesis of Corollary 10.2.

Atoms

An *atom of length n* is a formula α of the form

$$\alpha_0 \wedge \cdots \wedge \alpha_{n-1},$$

such that for all $i < n$, α_i is either the variable p_i or its negation $\neg p_i$. Put $|\alpha| = n$, so that $|\alpha|$ denotes the *length* of α. A partial ordering of atoms α, β is defined by letting $\alpha \leq \beta$ iff α is an initial segment of β, i.e. iff $|\alpha| \leq |\beta|$ and $\alpha_i = \beta_i$ for all $i < |\alpha|$.

For each atom α, three *successor* atoms $\alpha^1, \alpha^2, \alpha^3$ are defined as follows:

$$\begin{aligned}
\alpha^1 &= \alpha \wedge p_n \wedge p_{n+1} \\
\alpha^2 &= \alpha \wedge p_n \wedge \neg p_{n+1} \\
\alpha^3 &= \alpha \wedge \neg p_n \wedge p_{n+1}
\end{aligned}$$

(a fourth successor could be defined using $\neg p_n \wedge \neg p_{n+1}$, but this will not be needed). Then the formula α^* is

$$\Box(\Diamond\alpha \rightarrow \bigvee_{1 \leq i < j \leq 3} (\Diamond\alpha^i \wedge \Diamond\alpha^j)).$$

Now let Σ be the closure under successors of the set $\{p_0\}$. Hence $\beta \in \Sigma$ iff there is a finite sequence $\alpha_0 = p_0, \ldots, \alpha_n$, with $\alpha_n = \beta$, such that for all $k < n$, $\alpha_{k+1} = \alpha_k^i$ for some i with $1 \leq i \leq 3$. It follows that if α and β are distinct members of Σ with $\alpha \leq \beta$, then $\alpha^i \leq \beta$ for some i (this fact will be crucial in the proof of Theorem 10.4). Put

$$\Sigma^* = \{\alpha^* : \alpha \in \Sigma\}.$$

Theorem 10.3. *Let g be a point in a canonical frame $\mathcal{F}_L$ such that $\Sigma^* \cup \{\Box\Diamond p_0\} \subseteq g$. Then for every $m \in R_L^g$, R_L^m is of cardinality $2^{\aleph_0}$.*

Proof. Fix an m in R_L^g. Consider the notion of a sequence

$$\sigma = \langle \sigma_0, \ldots, \sigma_n, \ldots \ldots \rangle$$

such that for all n, σ_n is either p_n or $\neg p_n$, and

$$(\text{I}_n) \qquad (\sigma_0 \wedge \cdots \wedge \sigma_{2n}) \in \Sigma \quad \text{and} \quad \Diamond(\sigma_0 \wedge \cdots \wedge \sigma_{2n}) \in m.$$

Given such a σ, for each n there is some $y \in X_L$ with mR_Ly and $(\sigma_0 \wedge \cdots \wedge \sigma_{2n}) \in y$. Hence the set

$$\{\varphi : \Box\varphi \in m\} \cup \{\sigma_0, \ldots, \sigma_{2n}\}$$

is contained in y, and so is L-consistent. It follows that the set

$$\{\varphi : \Box\varphi \in m\} \cup \{\sigma_n : n < \omega\}$$

is L-consistent, and so extends to a point $m_\sigma \in X_L$. Then

$$m_\sigma \in R_L^m, \quad \text{and} \quad \{\sigma_n : n < \omega\} \subseteq m_\sigma.$$

But any two such sequences σ, σ' that are distinct must have σ_n provably equivalent to $\neg\sigma_n'$ for some n, so that $\sigma_n \notin m_{\sigma'}$ and hence $m_\sigma \neq m_{\sigma'}$. Thus if it can be shown that there are $2^{\aleph_0}$ σ's satisfying (I_n), then it will follow that there are $2^{\aleph_0}$ m_σ's in R_L^m.

To construct a σ, observe first that since $\Box\Diamond p_0 \in g$, $\Diamond p_0 \in m$. Hence putting $\sigma_0 = p_0$ gives

$$(\text{I}_0) \qquad \sigma_0 \in \Sigma \quad \text{and} \quad \Diamond\sigma_0 \in m.$$

Next, suppose inductively that $\sigma_0, \ldots, \sigma_{2n}$ have been defined so that (I_n) holds. Let $\alpha = (\sigma_0 \wedge \cdots \wedge \sigma_{2n})$. By (I_n), $\alpha \in \Sigma$. Hence $\alpha^* \in \Sigma^* \subseteq g$, so as gR_Lm,

$$(\Diamond\alpha \to \bigvee_{1\le i<j\le 3} (\Diamond\alpha^i \wedge \Diamond\alpha^j)) \in m.$$

But $\Diamond\alpha \in m$, by (I_n), so there exist $1 \le i < j \le 3$ with

$$\Diamond\alpha^i, \Diamond\alpha^j \in m.$$

Putting $\sigma_{2n+k} = \alpha^i_{2n+k}$ for $k = 1, 2$ then defines $\sigma_{2n+1}, \sigma_{2n+2}$, and gives

(I_{n+1}) $\qquad (\sigma_0 \wedge \cdots \wedge \sigma_{2(n+1)}) \in \Sigma$ and $\Diamond(\sigma_0 \wedge \cdots \wedge \sigma_{2(n+1)}) \in m,$

since Σ is closed under successors, and $(\sigma_0 \wedge \cdots \wedge \sigma_{2(n+1)}) = \alpha^i$.

Alternatively, since also $\Diamond\alpha^j \in m$, σ_{2n+k} could be defined as α^j_{2n+k} and still satisfy (I_{n+1}). But as $i \ne j$, the definition of successors implies that $\alpha^i_{2n+k} \ne \alpha^j_{2n+k}$ for some k, and so this would give a different extended sequence.

This demonstrates that infinite sequences σ of the desired kind can be defined by induction in such a way that at each inductive step there are at least two choices as to how the sequence extends. Hence there are $2^{\aleph_0}$ ways of constructing such sequences σ. $\qquad\square$

Corollary 10.2 and Theorem 10.3 have reduced the problem of showing that $\mathcal{F}_L$ is not a KM-frame to that of showing that $\Sigma^* \cup \{\Box\Diamond p_0\}$ is L-consistent. To deal with that requires a new model construction.

Trellis-Like Frames

A frame $\mathcal{F} = (X, R)$ will be called *trellis-like* if it fulfills the following description.

(1) There is some $g \in X$ such that X is the union of the three sets $\{g\}$, R^g, and E, where

$$E = \bigcup\{R^m : m \in R^g\},$$

and these three sets are mutually disjoint. In particular, g generates $\mathcal{F}$. The members of R^g are called the *middle points* of $\mathcal{F}$.

(2) E is the set of all end-points of $\mathcal{F}$, and for each $m \in R^g$ there is an end-point in R^m.

Condition (2) implies that in a model on a trellis-like frame, McK is true at any middle or end point. Hence by (1), in such a model McK can only be false, if at all, at the generator g.

Returning now to Σ, for each n put

$$\Sigma_n = \{\alpha \in \Sigma : |\alpha| \le n\}, \quad \text{and} \quad \Sigma^*_n = \{\alpha^* : \alpha \in \Sigma_n\}.$$

Theorem 10.4. *For any n, there exists a model $\mathcal{M}_n$, based on a trellis-like KM-frame, such that $\Sigma^*_{2n+1} \cup \{\Box \Diamond p_0\}$ is (simultaneously) satisfied at the generator in $\mathcal{M}_n$.*

Proof. The construction takes place within Σ_{2n+3}. Let

$$E_n = \{\alpha \in \Sigma_{2n+3} : |\alpha| = 2n + 3\}.$$

The members of E_n will be the end-points of $\mathcal{M}_n$. Members of the set $\Sigma_{2n+3} - E_n$ of atoms of length less than $2n + 3$ will be referred to as *interior points*.

If $\alpha \in \Sigma_{2n+3}$, a *binary subtree starting at* α is defined to be any set Θ satisfying

(1) $\alpha \in \Theta \subseteq \{\beta \in \Sigma_{2n+3} : \alpha \leq \beta\}$;

(2) If $\alpha \leq \gamma \leq \beta$, and $\beta \in \Theta$, then $\gamma \in \Theta$;   and

(3) If $\beta \in \Theta$ and β is an interior point, then exactly *two* of the successors of β belong to Θ.

Observe that if α is an end-point, i.e. $\alpha \in E_n$, then $\{\alpha\}$ is, by this definition, a binary subtree starting at α.

Let $\mathcal{M}_n$ be the set of all binary subtrees in Σ_{2n+3} that start at the shortest atom p_0. The members Θ of $\mathcal{M}_n$ will be the middle points of $\mathcal{M}_n$, and any such Θ will be related precisely to the end-points that belong to Θ. Note that there will always exist such end-points: any binary subtree starting at p_0 must contain members of length $2n + 3$. Truth-values are assigned in $\mathcal{M}_n$ by making each $\alpha \in E_n$ act as a valuation of its own variables, so that α becomes true at α.

Formally, let $\mathcal{F}_n = (X_n, R_n)$, and $\mathcal{M}_n = (X_n, R_n, V_n)$, where

$$X_n = \{g_n\} \cup M_n \cup E_n, \quad \text{with } g_n \notin M_n \cup E_n,$$

$$V_n(p_k) = \begin{cases} \{\alpha \in E_n : \alpha_k = p_k\}, & \text{for } k < 2n + 3 \\ \emptyset, & \text{for } k \geq 2n + 3, \end{cases}$$

and the relation R_n is specified by

$$R^{g_n}_n = M_n,$$
$$R^{\Theta}_n = \Theta \cap E_n, \quad \text{for } \Theta \in M_n,$$
$$R^{\alpha}_n = \{\alpha\}, \quad \text{for } \alpha \in E_n.$$

This ensures that $\mathcal{F}_n$ is trellis-like. By definition of V_n, if $\alpha \in E_n$ then

$$\mathcal{M}_n \models_\alpha p_k \quad \text{iff} \quad \alpha_k = p_k,$$

for $k < 2n + 3$. From this it follows readily that

(i) $\mathcal{M}_n \models_\alpha \beta$ iff $\beta \leq \alpha$, for all $\beta \in \Sigma_{2n+3}$. Hence, in particular, p_0 is true at every end-point α in $\mathcal{M}_n$, so as $\mathcal{F}_n$ is trellis-like, $\square \lozenge p_0$ is true at the generator g_n.

Now let $\beta \in \Sigma_{2n+1}$. To show that β^* is true at g_n, suppose $\lozenge \beta$ is true in $\mathcal{M}_n$ at some middle point Θ. Then β is true at some α with $\Theta R_n \alpha$, i.e. $\alpha \in \Theta \cap E_n$. By (i), $\beta \leq \alpha$. But $|\beta| \leq 2n + 1 < |\alpha| = 2n + 3$, so then $\beta^i \leq \alpha$ for some $1 \leq i \leq 3$. By clause (2) of the definition of binary subtree, $\beta^i \in \Theta$, so by (i), β^i is true in $\mathcal{M}_n$ at α, making $\lozenge \beta^i$ true at Θ. Moreover, β is an interior point, and so by (3), there is some $j \neq i$ such that $\alpha^j \in \Theta$. But then by applying (3) repeatedly, a γ may be constructed with $\alpha^j \leq \gamma \in \Theta \cap E_n$. Then α^j is true at γ, whence $\lozenge \alpha^j$ true at Θ, in $\mathcal{M}_n$. This establishes that β^* is true at g_n, and completes the proof that $\Sigma^*_{2n+1} \cup \{\square \lozenge p_0\}$ is satisfied at the generator in $\mathcal{M}_n$.

It remains to prove that $\mathcal{F}_n$ is a KM-frame, and for this it suffices to show that if φ is any formula, and $\mathcal{M}$ is any model based on $\mathcal{F}_n$, then $\lozenge(\square \varphi \vee \square \neg \varphi)$ is true in $\mathcal{M}$. Since $\mathcal{F}_n$ is trellis-like, the only issue is whether the latter formula is true at g_n in $\mathcal{M}$.

Lemma. *For any formula φ and any $\alpha \in \Sigma_{2n+3}$, there is a binary subtree Θ_α starting at α such that φ has constant truth value on $\Theta_\alpha \cap E_n$ in $\mathcal{M}$.*

Proof. This proceeds by reverse induction on the length of atom α. If α is of maximal length in Σ_{2n+3}, i.e. $\alpha \in E_n$, put $\Theta_\alpha = \{\alpha\}$. As noted above, in this case Θ_α is a binary subtree since α is an end-point. Also φ has constant truth value on $\Theta_\alpha \cap E_n = \{\alpha\}$.

Now suppose α is an interior point. Then the induction hypothesis may be made that the Lemma holds for all members of Σ_{2n+3} of length greater than $|\alpha|$. In particular, it holds for the successors of α. Hence for each $1 \leq i \leq 3$, there is a binary subtree Θ_i starting at α^i such that φ has constant truth-value on $\Theta_i \cap E_n$ in $\mathcal{M}$. But then there must exist $1 \leq i < j \leq 3$ such that φ has constant truth-value on $(\Theta_i \cap E_n) \cup (\Theta_j \cap E_n)$. Then

$$\Theta_\alpha = \{\alpha\} \cup \Theta_i \cup \Theta_j$$

is a binary subtree starting at α, with φ having constant truth-value on

$$\Theta_\alpha \cap E_n = (\Theta_i \cap E_n) \cup (\Theta_j \cap E_n).$$

Hence the Lemma holds for α, and so, by induction, it holds in general. $\square$

Applying the Lemma in the case $\alpha = p_0$, it follows that there is a subtree $\Theta \in \mathcal{M}_n$ such that φ has constant truth-value on $\Theta \cap E_n = R^\Theta_n$ in $\mathcal{M}$. Thus in the model $\mathcal{M}$, $(\square \varphi \vee \square \neg \varphi)$ is true at Θ, and so

$\Diamond(\Box\varphi\vee\Box\neg\varphi)$ is true at g_n, as desired to complete the proof of Theorem 10.4. $\qquad\Box$

Corollary 10.5. *KM is not canonical.*

Proof. If Γ is any finite subset of $\Sigma^* \cup \{\Box\Diamond p_0\}$, then Γ is contained in $\Sigma^*_{2n+1} \cup \{\Box\Diamond p_0\}$ for some n, so is satisfied at a point of a KM-model $(\mathcal{M}_n)$, and hence is KM-consistent.

It follows that $\Sigma^* \cup \{\Box\Diamond p_0\}$ is itself KM-consistent, and so extends to a maximally KM-consistent set g. Then by Corollary 10.2 and Theorem 10.3, there is a model on $\mathcal{F}_{KM}$ in which the McKinsey axiom is false at g.

$\qquad\Box$

Readers with a predilection for combinatorics will be interested to note that the essence of the proof, in Theorem 10.4, that $\mathcal{F}_n$ is a KM-frame resides in the following Ramsey-like property of trees:

> *given any 2-colouring of the leaves of a finite ternary tree T, there is a binary subtree having the same height as T, with all its leaves of the same colour.*

In conclusion, I would like to thank Max Cresswell for many helpful conversations, over the years, about this and related topics.

11

Elementary Logics are Canonical and Pseudo-Equational

Let Λ be a normal propositional modal logic that is *elementary*, i.e. is determined by some first-order axiomatisable class of Kripke frames. Then it is known that Λ is *canonical*, i.e. is validated by its canonical frame $\mathcal{F}^\Lambda$ (this was first shown in [15, Theorem 3]).

In this article we will prove a considerable strengthening and refinement of this result. First we show (Theorem 11.3.1) that if Λ is elementary, then it is determined by the *first-order theory* of $\mathcal{F}^\Lambda$, i.e.

> $\vdash_\Lambda A$ *if, and only if, A is valid in all frames elementarily equivalent to $\mathcal{F}^\Lambda$.*

Then we introduce the class of *pseudo-equational* first-order sentences, those which are preserved by the fundamental constructions of subframes, bounded morphisms, and disjoint unions of frames. We prove (Theorem 11.4.2) that if Λ is elementary, then it is determined by the *pseudo-equational theory* of $\mathcal{F}^\Lambda$, i.e.

> $\vdash_\Lambda A$ *if, and only if, A is valid in all frames satisfying the pseudo-equational sentences true of $\mathcal{F}^\Lambda$.*

The most significant question that remains unresolved in this area is whether, conversely, every canonical logic is elementary (cf. the introduction to the previous chapter for some discussion of the ramifications of this question). One natural way to tackle the problem would be to try to prove that if Λ is canonical then it is determined by the class $\mathcal{K}_\Lambda$ of frames elementarily equivalent to $\mathcal{F}^\Lambda$. Our results show that this is indeed an appropriate strategy: if Λ is determined by any elementary class at all, then it is determined by $\mathcal{K}_\Lambda$.

In the later part of this article we examine further connections between elementarity and canonicity, involving closure properties of classes of general frames (i.e. first-order frames in the sense of Section 1.3).

The analysis given here for modal logic can be extended to other kinds of intensional logic, and generalises algebraically to give results about the structure of varieties of Boolean algebras with operators. This is developed in the paper [31].

11.1   Review of Operations on Frames

Recall that a *Kripke frame* $\mathcal{F} = \langle W, R \rangle$ (or simply "frame" for now) comprises a binary relation R on a set W, and that a *model* $\mathcal{M} = \langle \mathcal{F}, V \rangle$ on $\mathcal{F}$ is given by a *valuation* V assigning a subset $V(p)$ of W to each propositional variable p. The relation "*formula A is true at point x in model $\mathcal{M}$*", denoted $\mathcal{M} \models_x A$, is defined in a way that is familiar from other chapters. Then A is *true in* $\mathcal{M}$, $\mathcal{M} \models A$, if it is true at all points of $\mathcal{M}$. A is *valid in frame* $\mathcal{F}$, $\mathcal{F} \models A$, is if A is true in all models on $\mathcal{F}$. If Λ is a set of formulae, then we write $\mathcal{F} \models \Lambda$ if every member of Λ is valid in $\mathcal{F}$.

$\langle W', R' \rangle$ is a *subframe* of $\langle W, R \rangle$ if $W' \subseteq W$, R' is the restriction of R to W', and

if xRy and $x \in W'$, then $y \in W'$.

(Subframes are sometimes referred to as "generated subframes" in the modal logic literature, and are called "inner substructures" in [31].)

A *bounded morphism* $\varphi : \langle W, R \rangle \to \langle W', R' \rangle$ is a function $\varphi : W \to W'$ such that

- xRy implies $\varphi(x)R'\varphi(y)$;
- $\varphi(x)R'z$ implies $\exists y(xRy$ and $\varphi(y) = z)$.

If φ is surjective, then it is called a *bounded epimorphism*, and $\langle W', R' \rangle$ is a *bounded epimorphic image* of $\langle W, R \rangle$. (Bounded morphisms are the "frame homomorphisms" of Chapter 1, and are also known as "p-morphisms" in the literature.)

If $\{\mathcal{F}_z : z \in Z\}$ is a collection of frames, with $\mathcal{F}_z = \langle W_z, R_z \rangle$, then frame $\mathcal{F} = \langle W, R \rangle$ is the *bounded union of the* $\mathcal{F}_z$'s if each $\mathcal{F}_z$ is a subframe of $\mathcal{F}$, and $W = \bigcup\{W_z : z \in Z\}$. $\mathcal{F}'$ is a *disjoint union* of the $\mathcal{F}_z$'s if it is the union of a collection $\{\mathcal{F}'_z : z \in Z\}$ of pairwise disjoint isomorphic copies of the $\mathcal{F}_z$'s, i.e. $\mathcal{F}'_z \cong \mathcal{F}_z$, and $W'_z \cap W'_w = \emptyset$ when $z \neq w \in Z$. Then each $\mathcal{F}'_z$ is a subframe of $\mathcal{F}'$, so $\mathcal{F}'$ is the bounded union of the $\mathcal{F}'_z$'s.

Observe that if $\mathcal{F}$ is the bounded union of the $\mathcal{F}_z$'s, and $\mathcal{F}'$ is their disjoint union, then the isomorphisms $\mathcal{F}'_z \cong \mathcal{F}_z$ combine to give a function $\mathcal{F}' \rightarrow \mathcal{F}$ which is a bounded epimorphism. Thus *a bounded union of structures is a bounded epimorphic image of their disjoint union.*

For any set Λ of formulae, the class

$$Fr(\Lambda) = \{\mathcal{F} : \mathcal{F} \models \Lambda\}$$

of all frames that validate Λ is closed under subframes, bounded epimorphic images (which includes isomorphic images) and disjoint unions. Hence by the last observation of the previous paragraph, $Fr(\Lambda)$ is closed under bounded unions as well.

In order to handle these constructions more conveniently, we introduce some notation for operations on a class $\mathcal{K}$ of frames:

$\mathbb{S}\mathcal{K}$	$=$	the class of isomorphic images of subframes of members of $\mathcal{K}$.
$\mathbb{H}\mathcal{K}$	$=$	the class of bounded epimorphic images of members of $\mathcal{K}$.
$\mathbb{U}\mathrm{d}\,\mathcal{K}$	$=$	the class of disjoint unions of collections of frames isomorphic to members of $\mathcal{K}$.
$\mathbb{U}\mathrm{b}\,\mathcal{K}$	$=$	the class of bounded unions of collections of frames isomorphic to members of $\mathcal{K}$.
$\mathbb{P}\mathrm{u}\,\mathcal{K}$	$=$	the class of isomorphic images of ultraproducts of collections of frames in $\mathcal{K}$.
$\mathbb{P}\mathrm{w}\,\mathcal{K}$	$=$	the class of isomorphic images of ultrapowers of frames in $\mathcal{K}$.
$\mathbb{R}\mathrm{u}\,\mathcal{K}$	$=$	the class of frames $\mathcal{F}$ having some ultrapower $\mathcal{F}^J/U$ isomorphic to a member of $\mathcal{K}$.

$\mathbb{R}\mathrm{u}\,\mathcal{K}$ is the class of *ultraroots* of $\mathcal{K}$.

Lemma 11.1.1 *For any class $\mathcal{K}$ of frames,*

(1) $\mathbb{X}\mathbb{X}\mathcal{K} = \mathbb{X}\mathcal{K}$ for $\mathbb{X} = \mathbb{H}, \mathbb{S}, \mathbb{U}\mathrm{d}, \mathbb{U}\mathrm{b}$.

(2) $\mathbb{U}\mathrm{d}\,\mathcal{K} \subseteq \mathbb{U}\mathrm{b}\,\mathcal{K} \subseteq \mathbb{H}\mathbb{U}\mathrm{d}\,\mathcal{K} = \mathbb{H}\mathbb{U}\mathrm{b}\,\mathcal{K} = \mathbb{U}\mathrm{b}\,\mathbb{H}\mathcal{K}$.

(3) $\mathbb{X}\mathbb{H}\mathcal{K} \subseteq \mathbb{H}\mathbb{X}\mathcal{K}$ for $\mathbb{X} = \mathbb{S}, \mathbb{U}\mathrm{d}, \mathbb{P}\mathrm{u}, \mathbb{P}\mathrm{w}$.

(4) $\mathbb{S}\,\mathbb{U}\mathrm{d}\,\mathcal{K} = \mathbb{U}\mathrm{d}\,\mathbb{S}\,\mathcal{K}$.

(5) $\mathbb{S}\,\mathbb{U}\mathrm{b}\,\mathcal{K} \subseteq \mathbb{U}\mathrm{b}\,\mathbb{S}\,\mathcal{K} \subseteq \mathbb{H}\mathbb{S}\,\mathbb{U}\mathrm{d}\,\mathcal{K} = \mathbb{H}\mathbb{S}\,\mathbb{U}\mathrm{b}\,\mathcal{K}$.

(6) $\mathbb{P}\mathrm{u}\,\mathbb{S}\,\mathcal{K} \subseteq \mathbb{S}\,\mathbb{P}\mathrm{u}\,\mathcal{K}$ and $\mathbb{P}\mathrm{w}\,\mathbb{S}\,\mathcal{K} \subseteq \mathbb{S}\,\mathbb{P}\mathrm{w}\,\mathcal{K}$.

(7) $\mathbb{S}\,\mathbb{R}\mathrm{u}\,\mathcal{K} \subseteq \mathbb{R}\mathrm{u}\,\mathbb{S}\,\mathcal{K}$ and $\mathbb{H}\mathbb{R}\mathrm{u}\,\mathcal{K} \subseteq \mathbb{R}\mathrm{u}\,\mathbb{H}\mathcal{K}$.

(8) If $\mathcal{K} = Fr(\Lambda)$ for some set Λ of formulae, then $\mathbb{X}\mathcal{K} = \mathcal{K}$ for $\mathbb{X} = \mathbb{H}, \mathbb{S}, \mathbb{U}\mathrm{d}, \mathbb{U}\mathrm{b}, \mathbb{R}\mathrm{u}$.

Proof. Some of these results are well-known, and the remainder are left to the reader, who can find further discussion in [31, Theorem 2.1].

Note that (8) expresses the fact that validity of formulae is preserved by bounded epimorphic images, subframes, disjoint and bounded unions, and ultraroots.   □

Lemma 11.1.2 *For any class $\mathcal{K}$ of frames,*

$$\mathbb{P}u\,\mathbb{U}b\,\mathcal{K} \subseteq \mathbb{U}b\,\mathbb{P}u\,\mathcal{K}.$$

Proof. This is given in Theorem 2.4 of [31], the argument being a generalisation of that of the proof of Lemma 3.5 of [29], which itself showed that $\mathbb{P}w\,\mathbb{U}d\,\mathcal{K} \subseteq \mathbb{U}b\,\mathbb{P}u\,\mathcal{K}$.

Let $\mathcal{F}$ be an ultraproduct $\prod_Z \mathcal{F}_z/U$ with each $\mathcal{F}_z$ being a bounded union $\bigcup\{\mathcal{F}_{zj}: j \in J_z\}$ of subframes $\mathcal{F}_{zj}$ isomorphic to members of $\mathcal{K}$.

We want to prove $\mathcal{F} \in \mathbb{U}b\,\mathbb{P}u\,\mathcal{K}$, and for this it suffices to show that for each $t \in \mathcal{F}$ there is a subframe $\mathcal{F}_t$ of $\mathcal{F}$ with $t \in \mathcal{F}_t \in \mathbb{P}u\,\mathcal{K}$.

But if $t \in \mathcal{F}$, then $t = f_t/U$ for some $f_t \in \prod_Z \mathcal{F}_z$. For each $z \in Z$, since $\mathcal{F}_z$ is the union of the $\mathcal{F}_{zj}$'s there exists $j_{zt} \in J_z$ with $f_t(z) \in \mathcal{F}_{zj_{zt}}$. Let

$$\mathcal{G}_t = \prod_Z \mathcal{F}_{zj_{zt}}/U \in \mathbb{P}u\,\mathcal{K}.$$

Then the inclusion $\prod_Z \mathcal{F}_{zj_{zt}} \hookrightarrow \prod_Z \mathcal{F}_z$ induces an injection $\varphi_t: \mathcal{G}_t \rightarrowtail \mathcal{F}$ which proves to be a bounded morphism because each $\mathcal{F}_{zj_{zt}}$ is an inner substructure of $\mathcal{F}_z$ [29, p. 225]. Moreover, t is in the image of φ_t. Hence the desired $\mathcal{F}_t$ is given by this image, which is a subframe of $\mathcal{F}$ and an isomorphic copy of $\mathcal{G}_t \in \mathbb{P}u\,\mathcal{K}$.   □

11.2   The Role of Ultrapowers

Let Λ be a normal propositional modal logic. A model $\mathcal{M}$ *determines* Λ if the Λ-theorems are precisely those formulae that are true in $\mathcal{M}$:

$$\vdash_\Lambda A \quad \text{iff} \quad \mathcal{M} \models A.$$

Similarly, Λ is determined by a frame $\mathcal{F}$, or by a class of frames $\mathcal{K}$, if the Λ-theorems are precisely those formulae that are valid in $\mathcal{F}$, or valid in all members of $\mathcal{K}$, respectively.

The *canonical Λ-frame* is $\mathcal{F}^\Lambda = \langle W^\Lambda, R^\Lambda \rangle$, with W^Λ being the set of all maximally Λ-consistent sets of formulae, and, for $x, y \in W^\Lambda$,

$$xR^\Lambda y \quad \text{iff} \quad \{A : \Box A \in x\} \subseteq y.$$

The *canonical Λ-model* is $\mathcal{M}^\Lambda = \langle \mathcal{F}^\Lambda, V^\Lambda \rangle$, where

$$V^\Lambda(p) = \{x \in W^\Lambda : p \in x\}.$$

This model determines Λ, and consequently

$$\mathcal{F}^\Lambda \models A \quad \text{implies} \quad \vdash_\Lambda A.$$

If the converse holds, i.e. if $\mathcal{F}^\Lambda$ validates all Λ-theorems, then Λ is called a *canonical logic*.

Theorem 11.2.1 *If a model $\mathcal{M} = \langle \mathcal{F}, V \rangle$ determines Λ, then $\mathcal{F}^\Lambda$ is a bounded epimorphic image of an ultrapower $\mathcal{F}^J/U$ of $\mathcal{F}$.*

Proof. This is an "ultrapowers version" of the construction due to Fine [15], which is presented in Section 19 of Chapter 1 (cf. also Lemma 3.1 of [29] or §3.6 of [28]).

Given an ultrapower $\mathcal{F}^J/U = \langle W^J/U, R_U \rangle$ of $\mathcal{F}$, define a model $\mathcal{M}_U = \langle \mathcal{F}^J/U, V_U \rangle$ by declaring

$$V_U(p) = \{ f/U \in W^J/U : \{ j \in J : f(j) \in V(p) \} \in U \}.$$

Then it can be shown that for any formula A,

$$\mathcal{M}_U \models_{f/U} A \quad \text{iff} \quad \{ j \in J : \mathcal{M} \models_{f(j)} A \} \in U.$$

From this it follows that

$$\mathcal{M} \models A \quad \text{implies} \quad \mathcal{M}_U \models A,$$

and hence in particular that $\mathcal{M}_U$ is an Λ-model ($\mathcal{M}_U \models \Lambda$), because $\mathcal{M}$ is a Λ-model. It also follows that for each w in $\mathcal{M}$,

(†) $\mathcal{M} \models_w A$ iff $\mathcal{M}_U \models_{\widetilde{w}} A$,

where $\widetilde{w} \in W^J/U$ is the function on J constantly equal to w.

Thus for each element x of $\mathcal{F}^J/U$, the set of formulas

$$\varphi(x) = \{ A : \mathcal{M}_U \models_x A \}$$

is maximally Λ-consistent, and therefore this construction defines a function $\varphi : \mathcal{F}^J/U \to \mathcal{F}^\Lambda$. It is readily shown that

$$x R_U y \quad \text{implies} \quad \varphi(x) R^\Lambda \varphi(y).$$

To show that φ is a bounded epimorphism we must show that it is surjective and that

$$\varphi(x) R^\Lambda \Gamma \quad \text{implies} \quad \exists y (x R_U y \text{ and } \varphi(y) = \Gamma).$$

For this purpose the ultrafilter U must be chosen appropriately to ensure that the ultrapower $\mathcal{F}^J/U$ has certain *saturation* properties. To describe these, we say that a set Γ of formulae is *satisfiable* in a given model if there is some point in the model at which all members of Γ are true. Putting

$$\Diamond \Gamma = \{ \Diamond(A_1 \wedge \cdots \wedge A_n) : A_1, \ldots, A_n \in \Gamma \},$$

then the properties required of $\mathcal{M}_U$ are:

- if every finite subset of Γ is satisfiable in $\mathcal{M}_U$, then Γ is satisfiable in $\mathcal{M}_U$;

- if $\mathcal{M}_U \models_x \Diamond \Gamma$, then there exists y with $xR_U y$ and $\mathcal{M}_U \models_y \Gamma$.

It is explained in the above references that, by results of first-order model-theory, a suitable U can indeed be found that ensures these properties hold for $\mathcal{M}_U$ (cf. [15, Lemma 8], [28, pp. 211–212]).

Thus if $\Gamma \in W^\Lambda$, then each finite subset of Γ is Λ-consistent, hence is true at some point of $\mathcal{M}$ because $\mathcal{M}$ determines Λ, and so is true at some point of $\mathcal{M}_U$ by (†). This show that every finite subset of Γ is satisfiable in $\mathcal{M}_U$. Hence by saturation there exists $x \in W^J/U$ with $\mathcal{M}_U \models_x \Gamma$. Then $\Gamma \subseteq \varphi(x)$, and so $\varphi(x) = \Gamma$ as Γ is *maximally* Λ-consistent. It follows that φ is surjective.

Finally, suppose that $\varphi(x) R^\Lambda \Gamma$ in $\mathcal{M}^\Lambda$. Then if $A_1, \ldots, A_n \in \Gamma$, the formula $\Diamond(A_1 \wedge \cdots \wedge A_n)$ belongs to $\varphi(x)$, so is true in $\mathcal{M}_U$ at x. Thus $\mathcal{M}_U \models_x \Diamond \Gamma$, so by saturation there exists $y \in W^J/U$ with $xR_U y$ and $\mathcal{M}_U \models_y \Gamma$. Then $\Gamma \subseteq \varphi(y)$, and so $\varphi(y) = \Gamma$. □

Corollary 11.2.2 *Let Λ be a normal propositional modal logic.*

(1) *If Λ is determined by a class of frames $\mathcal{K}$, then $\mathcal{F}^\Lambda \in \mathbb{HPwUd}\mathcal{K}$.*

(2) *If Λ is determined by a class $\mathcal{K}$ that is closed under ultraproducts, then $\mathcal{F}^\Lambda \in \mathbb{HUb}\mathcal{K}$.*

Proof.

(1) If formula A is not a Λ-theorem, then as $\mathcal{K}$ determines Λ there is a Λ-model $\mathcal{M}_A = \langle \mathcal{F}_A, V_A \rangle$ with $\mathcal{M}_A \nvDash A$ and $\mathcal{F}_A \in \mathcal{K}$. Let $\mathcal{M} = \langle \mathcal{F}, V \rangle$ be the disjoint union of the collection

$$\{\mathcal{M}_A : A \text{ is a formula and } \nvdash_\Lambda A\},$$

i.e. $\mathcal{F}$ is the disjoint union of the $\mathcal{F}_A$'s, and $V(p)$ is the disjoint union of the $V_A(p)$'s. Then in general

$$\mathcal{M} \models B \quad \text{iff} \quad \text{for all } A \text{ such that } \nvdash_\Lambda A, \mathcal{M}_A \models B,$$

so $\mathcal{M}$ is a Λ-model, and $\mathcal{M} \nvDash A$ whenever $\nvdash_\Lambda A$. Hence $\mathcal{M}$ determines Λ. Thus by Theorem 11.2.1 there exists an ultrapower $\mathcal{F}^J/U$ of $\mathcal{F}$ and a bounded epimorphism from $\mathcal{F}^J/U$ onto $\mathcal{F}^\Lambda$. In other words, $\mathcal{F}^\Lambda$ is a bounded epimorphic image of an ultrapower of a disjoint union of frames from $\mathcal{K}$.

(2) If $\mathbb{Pu}\mathcal{K} = \mathcal{K}$, then applying Lemma 11.1.2, to result (1),

$$\mathcal{F}^\Lambda \in \mathbb{HPwUd}\mathcal{K} \subseteq \mathbb{HPuUb}\mathcal{K} \subseteq \mathbb{HUbPu}\mathcal{K} = \mathbb{HUb}\mathcal{K}.$$

□

11.3   The First-Order Theory of $\mathcal{F}^\Lambda$

A class $\mathcal{K}$ of frames is *elementary* if it is first-order axiomatisable, i.e. if it is the class of all models of some set of sentences in the first-order language L_2 (with identity) of a binary relation. A logic Λ is *elementary* if it is determined by some elementary class of frames.

Two frames $\mathcal{F}$ and $\mathcal{G}$ are *elementarily equivalent*, $\mathcal{F} \equiv \mathcal{G}$, if they satisfy exactly the same L_2-sentences. For any logic Λ, the class

$$\mathcal{K}_\Lambda = \{\mathcal{F} : \mathcal{F} \equiv \mathcal{F}^\Lambda\}$$

of frames that are elementarily equivalent to $\mathcal{F}^\Lambda$ is elementary: $\mathcal{K}_\Lambda$ is the class of all models of the *first-order theory*

$$\{\sigma : \sigma \text{ is an } \mathsf{L}_2\text{-sentence and } \mathcal{F}^\Lambda \models \sigma\}$$

of the structure $\mathcal{F}^\Lambda$.

There are a number of characterisations of elementary classes. Thus $\mathcal{K}$ is elementary if, and only if, it is closed under ultraproducts and elementary equivalence. Alternatively, $\mathcal{K}$ is elementary if, and only if, it is closed under ultraproducts and ultraroots.

Thus the assumption that $\mathcal{K}$ is $\mathbb{P}$u-closed is weaker than the assumption that $\mathcal{K}$ is elementary. However the assumption that a logic Λ is determined by a $\mathbb{P}$u-closed class is *not* weaker than the assumption that Λ is elementary, since the former implies the latter, as is shown by the next result.

Theorem 11.3.1 *If Λ is determined by some class of frames that is closed under ultraproducts, then Λ is determined by the elementary class $\mathcal{K}_\Lambda$.*

Proof. Suppose $\mathcal{K}$ determines Λ and $\mathbb{P}$u$\mathcal{K} = \mathcal{K}$. Then by Corollary 11.2.2(2), there exists a frame $\mathcal{G} \in \mathbb{U}b\mathcal{K}$ and a bounded epimorphism $\varphi : \mathcal{G} \twoheadrightarrow \mathcal{F}^\Lambda$.

Now let $\mathcal{F} \in \mathcal{K}_\Lambda$. Since $\mathcal{F} \equiv \mathcal{F}^\Lambda$, the Keisler-Shelah Ultrapower Theorem states that there exist isomorphic ultrapowers $\mathcal{F}^J/U \cong (\mathcal{F}^\Lambda)^J/U$ for some ultrafilter U. Then applying Lemma 11.1.2 to the corresponding ultrapower of $\mathcal{G}$,

$$\mathcal{G}^J/U \in \mathbb{P}\text{w}\,\mathbb{U}\text{b}\mathcal{K} \subseteq \mathbb{U}\text{b}\,\mathbb{P}\text{u}\mathcal{K} = \mathbb{U}\text{b}\mathcal{K}.$$

Since $\mathcal{K}$ is a class of Λ-frames, and bounded unions preserve validity, this yields $\mathcal{G}^J/U \models \Lambda$. But the function φ lifts to a bounded epimorphism $\mathcal{G}^J/U \twoheadrightarrow (\mathcal{F}^\Lambda)^J/U$ (taking f/U to $(\varphi \circ f)/U$) so this in turn yields $(\mathcal{F}^\Lambda)^J/U \models \Lambda$. Then $\mathcal{F}^J/U \models \Lambda$, as $\mathcal{F}^J/U \cong (\mathcal{F}^\Lambda)^J/U$, and hence $\mathcal{F} \models \Lambda$ because validity is preserved by ultraroots.

We have now established that every member of $\mathcal{K}_\Lambda$ validates Λ, i.e.

$\mathcal{K}_\Lambda \subseteq Fr(\Lambda)$. Since $\mathcal{F}^\Lambda \in \mathcal{K}_\Lambda$, and the model $\mathcal{M}^\Lambda$ on $\mathcal{F}^\Lambda$ falsifies all non-theorems of Λ, this shows that $\mathcal{K}_\Lambda$ determines Λ. $\square$

It is noteworthy that in general $\mathcal{K}_\Lambda \neq \mathcal{K}$ even when $\mathcal{K}$ is elementary (otherwise Theorem 11.3.1 would be rather uninteresting!). For example, if $\mathcal{K}$ is the class of partial orderings (reflexive, transitive, antisymmetric frames), then $\mathcal{K}$ determines the logic $\Lambda = $ S4. But the canonical S4-frame is not antisymmetric, hence is in $\mathcal{K}_\Lambda$ but not $\mathcal{K}$. Indeed in this example $\mathcal{K}$ and $\mathcal{K}_\Lambda$ are *disjoint* elementary classes that each determine the logic Λ.

11.4 The Pseudo-Equational Theory of $\mathcal{F}^\Lambda$

A formula in the first-order language L_2 of a binary relation is called *essentially atomic* if it is constructed from amongst atomic formulae and the constants $\bot$ and $\top$ using at most $\wedge$ (conjunction), $\vee$ (disjunction), and *bounded* universal and existential quantifiers

$$\forall x(xRy \rightarrow \phi)$$
$$\exists x(xRy \wedge \phi)$$

with $x \neq y$.

An L_2-sentence will be called *pseudo-equational* if it is of the form $\forall x \phi$ with ϕ essentially atomic. Such sentences look nothing like equations of course, but the name is derived from the fact that their models give rise to equational classes of modal algebras (cf. [31, §4]).

Any pseudo-equational sentence is preserved by $\mathbb{H}$, $\mathbb{S}$, and $\mathbb{U}d$ (hence by $\mathbb{U}b$). Conversely, if a set of L_2-sentences is preserved by these three operations, then it is logically equivalent to a set of pseudo-equational sentences. This *preservation theorem* was first shown by van Benthem [100, 103]. An alternative proof for more general relational structures was developed in [28, §4], where the notion of bounded union was first introduced.

For an arbitrary class of frames $\mathcal{K}$, let $\Psi_\mathcal{K}$ be the *pseudo-equational theory of $\mathcal{K}$*, i.e. the set of all pseudo-equational sentences that are true in all members of $\mathcal{K}$, and let

$$Mod\,\Psi_\mathcal{K} = \{\mathcal{F} : \mathcal{F} \models \Psi_\mathcal{K}\}$$

be the class of all models of $\Psi_\mathcal{K}$. Then $Mod\,\Psi_\mathcal{K}$ is an elementary class containing $\mathcal{K}$. By a careful analysis of the version of the preservation theorem given in [28, §4], it can be shown that when $\mathcal{K}$ is closed under ultraproducts, then the members of $Mod\,\Psi_\mathcal{K}$ can be constructed from $\mathcal{K}$ by operations that preserve validity of modal formulae. This analysis is presented in detail in [31, §7], where it is shown that

$$\mathbb{P}u\,\mathcal{K} = \mathcal{K} \quad \text{implies} \quad Mod\,\Psi_{\mathcal{K}} = \mathbb{R}u\,\mathbb{U}b\,\mathbb{R}u\,\mathbb{U}b\,\mathbb{R}u\,\mathbb{H}\mathbb{S}\,\mathcal{K}.$$

Corollary 11.4.1 *If Λ is determined by a class of frames $\mathcal{K}$ that is closed under ultraproducts, then Λ is determined by the elementary class $Mod\,\Psi_{\mathcal{K}}$.*

Proof. Suppose $\mathcal{K}$ determines Λ and $\mathbb{P}u\,\mathcal{K} = \mathcal{K}$. Then by the result just quoted, since modal validity is preserved by $\mathbb{R}u$, $\mathbb{U}b$, $\mathbb{H}$, and $\mathbb{S}$, every member of $Mod\,\Psi_{\mathcal{K}}$ validates Λ.

But by Corollary 11.2.2(2), $\mathcal{F}^{\Lambda} \in \mathbb{H}\mathbb{U}b\mathcal{K}$. Since truth of pseudo-equational sentences is preserved by bounded unions and bounded epimorphic images, this implies that $\mathcal{F}^{\Lambda} \in Mod\,\Psi_{\mathcal{K}}$. Hence $Mod\,\Psi_{\mathcal{K}}$ determines Λ. □

We can now obtain the main result of this article. For a logic Λ, let Ψ_{Λ} be the *pseudo-equational theory of $\mathcal{F}^{\Lambda}$*, i.e. the set of pseudo-equational sentences that are true of the canonical frame of Λ. Since $\mathcal{F}^{\Lambda}$ is elementarily equivalent to all members of $\mathcal{K}_{\Lambda}$, it follows that Ψ_{Λ} is also the pseudo-equational theory of the class $\mathcal{K}_{\Lambda}$. Symbolically: $\Psi_{\Lambda} = \Psi_{\mathcal{K}_{\Lambda}}$.

Theorem 11.4.2 *If Λ is determined by some class of frames that is closed under ultraproducts, then Λ is determined by the elementary class $Mod\,\Psi_{\Lambda}$ of all models of the pseudo-equational theory of $\mathcal{F}^{\Lambda}$.*

Proof. Suppose $\mathcal{K}$ determines Λ and $\mathbb{P}u\,\mathcal{K} = \mathcal{K}$. Then by Corollary 11.4.1 all members of $Mod\,\Psi_{\mathcal{K}}$ validate Λ, i.e.

$$Mod\,\Psi_{\mathcal{K}} \subseteq Fr(\Lambda).$$

Also, by Corollary 11.2.2(2) $\mathcal{F}^{\Lambda} \in \mathbb{H}\mathbb{U}b\mathcal{K}$, so every pseudo-equational sentence true throughout $\mathcal{K}$ will be true in $\mathcal{F}^{\Lambda}$, i.e. $\Psi_{\mathcal{K}} \subseteq \Psi_{\Lambda}$. Hence

$$Mod\,\Psi_{\Lambda} \subseteq Mod\,\Psi_{\mathcal{K}}$$

and thus altogether $Mod\,\Psi_{\Lambda} \subseteq Fr(\Lambda)$. Since $\mathcal{F}^{\Lambda} \in Mod\,\Psi_{\Lambda}$ by definition, it follows that $Mod\,\Psi_{\Lambda}$ determines Λ. □

Counter Examples

Of course Theorem 11.4.2 is only an advance on Corollary 11.4.1 if there are cases of a logic Λ, determined by some $\mathbb{P}u$-closed class $\mathcal{K}$, for which $Mod\,\Psi_{\Lambda} \neq Mod\,\Psi_{\mathcal{K}}$. We saw in the proof of 11.4.2 that for such a logic we have $Mod\,\Psi_{\Lambda} \subseteq Mod\,\Psi_{\mathcal{K}}$, but in fact the converse of this last inclusion does not always hold. This may be seen from the example of KMT, the smallest normal modal logic that contains all formulae of the form

$$\Diamond((\Box A_1 \rightarrow A_1) \wedge \cdots \wedge (\Box A_n \rightarrow A_n)).$$

KMT is studied in [46], where the following are shown.

(1) KMT is determined by the class $\mathcal{K}_H$ of all frames satisfying the pseudo-equational condition

$$1(i) \qquad \forall x \exists y (xRy \wedge yRy).$$

(2) The canonical KMT-frame $\mathcal{F}^{KMT}$ satisfies 1(i) (hence KMT is canonical).

(3) A frame $\langle W, R \rangle$ validates KMT if, and only if, for each $x \in W$ the set $R_x = \{y : xRy\}$ is not finitely colourable.

Here a *colouring* is an assignment of colours to points in such a way that if yRz then y and z are assigned different colours. Notice that $\mathfrak{N} = \langle \omega, < \rangle$ is a KMT-frame by (3), but fails 1(i) and so is not in $\mathcal{K}_H$.

Now let $K_{\mathfrak{N}}$ be the class of all structures elementarily equivalent to $\mathfrak{N}$. Then for $\langle X, R \rangle \in K_{\mathfrak{N}}$, each set R_x will be infinite and linearly ordered by R, so cannot be finitely coloured. Thus by (3), all members of $K_{\mathfrak{N}}$ validate KMT.

Further, let $\mathcal{K} = \mathcal{K}_H \cup K_{\mathfrak{N}}$. Then $\mathcal{K}$ is the union of two elementary classes, so is itself an elementary class, hence is $\mathbb{P}u$-closed. All members of $\mathcal{K}$ validate KMT, and $\mathcal{K}$ contains $\mathcal{F}^{KMT}$, so $\mathcal{K}$ determines KMT.

Since $\mathfrak{N} \in \mathcal{K}$, it is immediate that $\mathfrak{N} \in Mod\Psi_{\mathcal{K}}$, i.e any pseudo-equational sentence true throughout $\mathcal{K}$ is true in $\mathfrak{N}$. But $\mathfrak{N} \notin Mod\Psi_\Lambda$, because Ψ_Λ is the pseudo-equational theory of the canonical frame $\mathcal{F}^{KMT}$, which is in $\mathcal{K}_H$ by (2). Hence the sentence 1(i) is in Ψ_Λ but not in $\Psi_{\mathcal{K}}$ since it fails in $\mathfrak{N}$, and indeed $\mathfrak{N} \nvDash \Psi_\Lambda$. Thus in this example we have

$$\Psi_{\mathcal{K}} \subsetneqq \Psi_\Lambda \quad \text{and} \quad Mod\Psi_\Lambda \subsetneqq Mod\Psi_{\mathcal{K}}.$$

Next we show that the inclusion

$$Mod\Psi_{\mathcal{K}} \subseteq Fr(\Lambda)$$

established in the proof of Corollary 11.4.1 is not in general an equality. For this, let Λ be the smallest normal logic containing the schema

$$\Diamond\Box A \rightarrow (\Diamond\Box(A \wedge B) \vee \Diamond\Box(A \wedge \neg B)),$$

and let $\mathcal{K}$ be the class of all frames satisfying the sentence

$$\forall x \forall y (xRy \rightarrow \exists z (xRz \wedge \forall u \forall v (zRu \wedge zRv \rightarrow u = v \wedge yRv))).$$

Then Fine [15] shows that $\mathcal{F}^\Lambda \in \mathcal{K} \subseteq Fr(\Lambda)$, and so Λ is determined by the elementary class $\mathcal{K}$. Hence $Mod\Psi_{\mathcal{K}} \subseteq Fr(\Lambda)$ by 11.4.1. But it is also shown in [15] that $Fr(\Lambda)$ is not closed under elementary equivalence, and so is not an elementary class. Since $Mod\Psi_{\mathcal{K}}$ is an elementary class by definition, we have $Mod\Psi_{\mathcal{K}} \neq Fr(\Lambda)$.

Finally, consider the the inclusion

$$\mathcal{K}_\Lambda \subseteq Mod\Psi_\Lambda,$$

asserting that any frame elementarily equivalent to $\mathcal{F}^\Lambda$ is a model of the pseudo-equational theory of $\mathcal{F}^\Lambda$, which is true by definition. Now take Λ to be S4 again. As mentioned at the end of Section 11.3, in this case $\mathcal{F}^\Lambda$ is not antisymmetric, and contains distinct points x, y with $xR^\Lambda y R^\Lambda x$. By identifying such pairs we can collapse $\mathcal{F}^\Lambda$ to an anti-symmetric quotient frame $\mathcal{G}$ which is a bounded epimorphic image of $\mathcal{F}^\Lambda$. Then pseudo-equational sentences are preserved in passing from $\mathcal{F}^\Lambda$ to $\mathcal{G}$, so $\mathcal{G}$ is a model of Ψ_Λ, i.e. $\mathcal{G} \in Mod\,\Psi_\Lambda$. But $\mathcal{G} \notin \mathcal{K}_\Lambda$, i.e. $\mathcal{G} \not\equiv \mathcal{F}^\Lambda$, since the two structures are distinguished by the antisymmetry condition.

Summary

The main results of the last two sections can be summarised by the following statement.

> If a logic Λ is determined by some class $\mathcal{K}$ of frames that is closed under ultraproducts, then
>
> $$\mathcal{F}^\Lambda \in \mathcal{K}_\Lambda \subseteq Mod\,\Psi_\Lambda \subseteq Mod\,\Psi_\mathcal{K} \subseteq Fr(\Lambda),$$
>
> with none of these set inclusions being an equality in general.

11.5   Stability Properties for General Frames

It remains an open question as to whether a canonical logic must be elementary. In the absence of a solution it is natural to look for alternative versions of these notions, in the hope of finding some reformulation that might allow their true relationship to be determined. We now consider some such characterisations, based on preservation properties of *general frames*.

By a general frame we will mean a structure $\mathcal{F} = \langle W, R, P \rangle$, of the type introduced in Definition 1.3.4 of Chapter 1, with $\langle W, R \rangle$ being a Kripke frame, and P a non-empty collection of subsets of W closed under the Boolean set operations and the operation

$$l_R(S) = \{x \in W : \forall y(xRy \text{ implies } y \in S)\}.$$

A model $\mathcal{M} = \langle \mathcal{F}, V \rangle$ on such a frame is given by a valuation V that has $V(p) \in P$ for all variables p. The notions of truth in $\mathcal{M}$ and validity in $\mathcal{F}$ are then defined as usual.

A general frame is *full* if $P = 2^W$. A Kripke frame $\langle W, R \rangle$ can be identified with the full frame $\langle W, R, 2^W \rangle$, and we will usually present a general frame in the form $\langle \mathcal{F}, P \rangle$, with $\mathcal{F}$ a Kripke frame.

The *canonical general frame* for a normal logic Λ is $\mathcal{H}^\Lambda = \langle \mathcal{F}^\Lambda, P^\Lambda \rangle$, with $\mathcal{F}^\Lambda$ the canonical Kripke frame for Λ, and

$$P^\Lambda = \{|A|: A \text{ is a formula}\},$$

where $|A| = \{x \in W^\Lambda : A \in x\}$. This frame determines Λ:

$$\vdash_\Lambda A \quad \text{iff} \quad \mathcal{H}^\Lambda \models A.$$

We identify a property of frames with the class Π of all frames having that property. Some more interesting properties are:

- The class of *atomic* frames, those having $\{x\} \in P$ for all $x \in W$;
- The class of *image-closed* frames, for which $\{y : xRy\} \in P$ for all $x \in W$;
- The class of *iterated-image-closed* frames, for which $\{y : xR^ny\} \in P$ for all $x \in W$ and all n;
- The class Π_d of *definably-closed* frames, for which P contains every subset X of W that is L_2-definable in the sense that there is some L_2-formula $\phi(v, v_1, \ldots, v_n)$ and some $w_1, \ldots, w_n \in W$ such that

$$X = \{x \in W : \langle W, R \rangle \models \phi[x, w_1, \ldots, w_n]\}.$$

Note that the L_2-definable sets include all finite and cofinite sets, and all sets of the form $\{y : xR^ny\}$.

All of the properties just listed are preserved by *ultraproducts* of general frames, a notion that was described in Definition 1.7.6 of Chapter 1. Given a collection

$$\{\langle \mathcal{F}_j, P_j \rangle : j \in J\}$$

of general frames, and an ultrafilter U on J, the associated ultraproduct $\mathcal{F}_U$ has the form $\langle \prod_J \mathcal{F}_j/U, P_U \rangle$, where $\prod_J \mathcal{F}_j/U$ is the ultraproduct of the Kripke frames, and P_U is a collection of subsets of $\prod_J \mathcal{F}_j/U$ defined from the ultraproduct of the P_j's. The crucial property of this construction is that for any modal formula A,

$$\mathcal{F}_U \models A \quad \text{iff} \quad \{j : \mathcal{F}_j \models A\} \in U$$

(cf. Corollary 1.7.13), a result that does not hold for ultraproducts of Kripke frames. An immediate consequence of this property is that if each $\mathcal{F}_j$ is a Λ-frame, where Λ is some logic, or set of formulae, then $\mathcal{F}_U$ is a Λ-frame.

In the particular case of an ultra*power*, when $\mathcal{F}_j = \mathcal{F}$ for all $j \in J$, we get

$$(*) \quad \mathcal{F}_U \models \Lambda \quad \text{iff} \quad \mathcal{F} \models \Lambda.$$

If Π is a class of frames, then a set Λ of formulae is Π-*stable* if its validity is preserved by *full expansions* of Π-frames, which means that

> if $\langle \mathcal{F}, P \rangle \models \Lambda$ and $\langle \mathcal{F}, P \rangle \in \Pi$ (where $\mathcal{F}$ is a Kripke frame),
> then $\mathcal{F} \models \Lambda$.

Thus if $\Pi_1 \subseteq \Pi_2$, then Π_2-stable formulae are Π_1-stable.

If a logic Λ is Π-stable for some Π that includes the canonical general frame $\mathcal{H}^\Lambda = \langle \mathcal{F}^\Lambda, P^\Lambda \rangle$, then since in general $\mathcal{H}^\Lambda \models \Lambda$ it follows from Π-stability that $\mathcal{F}^\Lambda \models \Lambda$, i.e. that Λ is canonical. This observation is the basis for the following characterisations of elementarity.

Theorem 11.5.1 *For any normal logic Λ, the following are equivalent.*

(1) Λ is determined by some class of Kripke frames that is closed under ultraproducts.

(2) Λ is determined by the elementary class $\mathcal{K}_\Lambda$ of Kripke frames elementarily equivalent to $\mathcal{F}^\Lambda$.

(3) Λ is determined by the elementary class $Mod\,\Psi_\Lambda$ of all models of the pseudo-equational theory of $\mathcal{F}^\Lambda$.

(4) Λ is Π-stable for some class Π of general frames that includes $\langle \mathcal{F}^\Lambda, P^\Lambda \rangle$ and is closed under ultrapowers.

(5) Λ is valid in all full ultrapowers $(\mathcal{F}^\Lambda)^J/U$ of $\mathcal{F}^\Lambda$.

(6) Λ is canonical and Π-stable for some class Π of general frames that includes the full frame $\mathcal{F}^\Lambda$ and is closed under ultrapowers.

Proof. We have already established the equivalence of (1), (2), and (3).

(2) *implies* (4): Let Π be the class of all ultrapowers of the general frame $\langle \mathcal{F}^\Lambda, P^\Lambda \rangle$. Then if $\langle \mathcal{F}, P \rangle \in \Pi$, $\mathcal{F}$ is an ultrapower of the Kripke frame $\mathcal{F}^\Lambda$, so $\mathcal{F} \equiv \mathcal{F}^\Lambda$, i.e. $\mathcal{F} \in \mathcal{K}_\Lambda$. Assuming $\mathcal{K}_\Lambda$ determines Λ then gives $\mathcal{F} \models \Lambda$. Hence Λ is Π-stable.

(4) *implies* (5): Let Π satisfy (4). Since $\langle \mathcal{F}^\Lambda, P^\Lambda \rangle \models \Lambda$, any ultrapower

$$\langle (\mathcal{F}^\Lambda)^J/U, (P^\Lambda)^J/U \rangle$$

validates Λ by $(*)$ above, and belongs to Π by assumption, so

$$(\mathcal{F}^\Lambda)^J/U \models \Lambda$$

by Π-stability.

(5) *implies* (6): If (5) holds, then in particular Λ is validated by $\mathcal{F}^\Lambda$ (since $\mathcal{F}^\Lambda \cong (\mathcal{F}^\Lambda)^J/U$ when U is principal), hence Λ is canonical.

Then taking Π to be the class of general ultrapowers of the full frame $\mathcal{F}^\Lambda$, every member of Π has the form

$$\langle (\mathcal{F}^\Lambda)^J/U, (2^{W^\Lambda})^J/U \rangle,$$

so Π-stability is immediate from (5).

(6) *implies* (2): Let $\mathcal{G} \in \mathcal{K}_\Lambda$. It is enough to show $\mathcal{G} \models \Lambda$. But since $\mathcal{G} \equiv \mathcal{F}^\Lambda$, by the Keisler-Shelah Ultrapower Theorem there exist

isomorphic ultrapowers $\mathcal{G}^J/U \cong (\mathcal{F}^\Lambda)^J/U$ for some ultrafilter U. Let

$$\mathcal{H} = \langle (\mathcal{F}^\Lambda)^J/U, (2^{W^\Lambda})^J/U \rangle.$$

Then assuming (6), we have $\mathcal{F}^\Lambda \models \Lambda$, hence $\mathcal{H} \models \Lambda$ by (∗), and $\mathcal{H} \in \Pi$ by the assumed property of Π, whence $(\mathcal{F}^\Lambda)^J/U \models \Lambda$ by Π-stability. Then $\mathcal{G}^J/U \models \Lambda$, so as validity is preserved by ultraroots, $\mathcal{G} \models \Lambda$ as desired.   □

An illustration of condition (6) of Theorem 11.5.1 is given by the class Π_d of general frames for which P contains all first-order definable subsets of the underlying Kripke frame. Π_d includes $\mathcal{F}^\Lambda$ (since it includes all full frames), and is closed under ultrapowers (indeed under ultraproducts). Hence the Theorem implies

every canonical Π_d-stable logic Λ is determined by $\mathcal{K}_\Lambda$.

There are however canonical logics that are determined by $\mathcal{K}_\Lambda$ but are not Π_d-stable. An example is $K4M$, the smallest normal logic to contain the schemata

$$\Box A \to \Box\Box A,$$
$$\Box\Diamond A \to \Diamond\Box A.$$

The class $Fr(K4M)$ of Kripke frames for $K4M$ is the class of all transitive frames satisfying the (pseudo-equational) sentence

$$\forall x \exists y (xRy \land \forall z(yRz \to \forall w(yRw \to z = w)))$$

[103, Lemma 7.2]. The canonical $K4M$-frame is in this class (as has been known since [59]), so $K4M$ is canonical and elementary. But if P is the set of finite and cofinite subsets of ω, then the frame $\langle \omega, <, P \rangle$ is in Π_d [103, Lemma 9.16] and validates $K4M$, whereas $\langle \omega, < \rangle$ does not validate the second of the above two schemata. Hence $K4M$ is not Π_d-stable.

Another illustration of 11.5.1(6) is given by considering the *monadic second-order* language of a binary relation, which we will denote L_2^2. This extends L_2 by adding quantifiable variables ranging over subsets of a frame. A given general frame becomes an L_2^2-structure by allowing P to be the range of quantification of the set variables. It can be shown that ultraproducts of general frames preserve truth of L_2^2-sentences [103, Theorem 4.12], in the same way that they preserve validity of modal formulae. Thus if a class Π of general frames is defined by some property expressible by a set of L_2^2-sentences, then Π must be closed under ultraproducts. We conclude from 11.5.1(4) that

if a logic Λ is Π-stable for some L_2^2-definable property Π possessed by $\langle \mathcal{F}^\Lambda, P^\Lambda \rangle$, then Λ is determined by $\mathcal{K}_\Lambda$;

and from 11.5.1(6) that

if a canonical logic Λ is Π-stable for some L_2^2-definable property Π possessed by $\mathcal{F}^\Lambda$, then Λ is determined by $\mathcal{K}_\Lambda$.

The notion of Π-stability an also be used to discuss the question as to when the class $Fr(\Lambda)$ of all Kripke frames validating Λ is an elementary class. In general $Fr(\Lambda)$ is closed under ultraroots (i.e. its complement is closed under ultrapowers) so it will be an elementary class if, and only if, it is closed under ultraproducts (cf. Corollary 1.16.3(ii) in Chapter 1). This implies

Theorem 11.5.2 *Let Π be any property of general frames that is possessed by all full frames and is preserved by ultraproducts. Then if Λ is Π-stable, it follows that $Fr(\Lambda)$ is elementary.* $\square$

As special cases we can conclude that $Fr(\Lambda)$ is elementary if Λ is Π_d-stable, or if it is Π-stable for some L_2^2-definable property that is possessed by all full frames [103, Theorem 13.4].

Also it follows that if Λ is Π-stable, where Π is as in the first sentence of 11.5.2, and Λ is *complete*, i.e. is determined by *some* class of Kripke frames, then Λ is canonical. This is because $Fr(\Lambda)$ is then elementary by 11.5.2, and for a complete logic to be canonical it suffices that $Fr(\Lambda)$ be closed under elementary equivalence ([15, Theorem 2], cf. also Corollary 1.20.15).

The most telling conclusion we can draw from Theorem 11.5.1 is that in order to prove that a particular canonical logic Λ is elementary, it is both necessary and sufficient to prove that Λ is valid in all ultrapowers $(\mathcal{F}^\Lambda)^J/U$ of its canonical frame. This offers us both a method for proving that Λ is elementary, by showing $(\mathcal{F}^\Lambda)^J/U \models \Lambda$, and a way of showing that it is not, by finding a counter example – an ultrapower of $\mathcal{F}^\Lambda$ that falsifies Λ. Neither approach seems easy, and it is perhaps a fitting way to end this article, and this volume, by placing that challenge in front of the reader.

Bibliography

[1] ALAGIC, S., AND M. A. ARBIB. *The Design of Well-Structured and Correct Programs*. Springer-Verlag, 1978.

[2] AMEMIYA, I., AND H. ARAKI. A Remark on Piron's Paper. *Publications of the Research Institute of Mathematical Sciences, Kyoto University, Series A2, 2* (1966), 423–427.

[3] BARWISE, JON. *Admissible Sets and Structures*. Springer-Verlag, Berlin, Heidelberg, 1975.

[4] BELL, J. L., AND A. B. SLOMSON. *Models and Ultraproducts*. North-Holland, Amsterdam, 1969.

[5] BERBERIAN, S. K. *Introduction to Hilbert Space*. Oxford University Press, London, 1961.

[6] BIRKHOFF, G. *Lattice Theory*, third ed. American Mathematical Society Colloquium Publications. American Mathematical Society, Providence, R. I., 1966. no. 25.

[7] BULL, R. A. MIPC as the Formalisation of an Intuitionist Concept of Modality. *Journal of Symbolic Logic 31* (1966), 609–616.

[8] CHELLAS, B. F. *Modal Logic: An Introduction*. Cambridge University Press, 1980.

[9] DIEGO, A. Sur les Algèbras de Hilbert. *Collection de Logique Mathématiques, Série A, 21* (1966). published in Paris.

[10] DOWKER, C. H., AND D. PAPERT. Quotient Frames and Subspaces. *Proceedings of the London Mathematical Society 16* (1966), 275–296.

[11] EILENBERG, S., AND N. STEENROD. *Foundations of Algebraic Topology*. Princeton University Press, 1952.

[12]   FINE, KIT. Propositional Quantifiers in Modal Logic. *Theoria 36* (1970), 336–346.

[13]   FINE, K. An Incomplete Logic Containing S4. *Theoria 40* (1974), 23–29.

[14]   FINE, K. Normal Forms in Modal Logic. *Notre Dame Journal of Formal Logic 16* (1975), 229–234.

[15]   FINE, K. Some Connections between Elementary and Modal Logic. In *Proceedings of the Third Scandinavian Logic Symposium*, ed. Stig Kanger. North-Holland, 1975, pp. 15–31.

[16]   FISCHER, M. J., AND R. E. LADNER. Propositional Dynamic Logic of Regular Programs. *Journal of Computer and Systems Sciences 18* (1979), 194–211.

[17]   FITTING, M. C. *Intuitionistic Logic, Model Theory & Forcing.* North-Holland, 1969.

[18]   FOULIS, D. J., AND C. H. RANDALL. Lexicographic Orthogonality. *Journal of Combinatorial Theory 11* (1971), 157–162.

[19]   FREYD, PETER. Aspects of Topoi. *Bulletin of the Australian Mathematical Society 7* (1972), 1–76.

[20]   GÖDEL, K. Eine Interpretation des Intuitionistischen Aussagenkalküls. *Ergebnisse eines mathematischen Kolloquiums 4* (1933). English translation in Kurt Gödel, Collected Works, Volume 1, Solomon Feferman et. al. (eds.), Oxford University Press, 1986, 296–303.

[21]   GOLDBLATT, ROB. Arithmetical Necessity, Provability, and Intuitionistic Logic. *Theoria 44* (1978), 38–46.

[22]   GOLDBLATT, ROBERT. Diodorean Modality in Minkowski Spacetime. *Studia Logica 39* (1980), 219–236.

[23]   GOLDBLATT, ROBERT. *Axiomatising the Logic of Computer Programming*, vol. 130 of *Lecture Notes in Computer Science*. Springer-Verlag, 1982.

[24]   GOLDBLATT, ROBERT. An Abstract Setting for Henkin Proofs. *Topoi 3* (1984), 37–41.

[25]   GOLDBLATT, ROBERT. Orthomodularity is Not Elementary. *The Journal of Symbolic Logic 49* (1984), 401–404.

[26]   GOLDBLATT, ROBERT. *Topoi*, second ed., vol. 98 of *Studies in Logic*. North-Holland, Amsterdam, 1984.

[27] GOLDBLATT, ROBERT. On the Role of the Baire Category Theorem and Dependent Choice in the Foundations of Logic. *The Journal of Symbolic Logic 50* (1985), 412–422.

[28] GOLDBLATT, ROBERT. Varieties of Complex Algebras. *Annals of Pure and Applied Logic 44* (1989), 173–242.

[29] GOLDBLATT, ROBERT. On Closure Under Canonical Embedding Algebras. In *Algebraic Logic*, ed. H. Andréka, J.D. Monk, and I. Németi. North-Holland, 1991, pp. 217–229. Colloquia Mathematica Societatis János Bolyai **54**.

[30] GOLDBLATT, ROBERT. The McKinsey Axiom is Not Canonical. *The Journal of Symbolic Logic 56* (1991), 554–562.

[31] GOLDBLATT, ROBERT. Elementary Generation and Canonicity for Varieties of Boolean Algebras with Operators. Research Report 92–101, Mathematics Department, Victoria University of Wellington, October 1992.

[32] GOLDBLATT, ROBERT. *Logics of Time and Computation*, second ed., vol. 7 of *CSLI Lecture Notes*. Center for the Study of Language and Information, Stanford University, 1992. Distributed by University of Chicago Press.

[33] GOLDBLATT, R. I. Semantic Analysis of Orthologic. *Journal of Philosophical Logic 3* (1974), 19–35.

[34] GOLDBLATT, R. I. First-Order Definability in Modal Logic. *Journal of Symbolic Logic 40* (1975), 35–40.

[35] GOLDBLATT, R. I. Solution to a Completeness Problem of Lemmon and Scott. *Notre Dame Journal of Formal Logic 16* (1975), 405–408.

[36] GOLDBLATT, R. I. Metamathematics of Modal Logic, Part I. *Reports on Mathematical Logic 6* (1976), 41–78.

[37] GOLDBLATT, R. I. Metamathematics of Modal Logic, Part II. *Reports on Mathematical Logic 7* (1976), 21–52.

[38] GOLDBLATT, R. I. On the Incompleteness of Hoare's Rule for While-commands. *Notices of the American Mathematical Society 26*, A–524 (1979).

[39] GRÄTZER, G. *Universal Algebra*. D. Van Nostrand, 1968.

[40] GRZEGORCZYK, A. Some Relational Systems and the Associated Topological Spaces. *Fundamenta Mathematicae 60* (1967), 223–231.

[41] HALMOS, P. R. *Introduction to Hilbert Space and the Theory of Spectral Multiplicity.* Chelsea, New York, 1975.

[42] HENKIN, H., J. D. MONK, AND A. TARSKI. *Cylindric Algebras I.* North-Holland, 1971.

[43] HENKIN, LEON. The Completeness of the First-Order Functional Calculus. *Journal of Symbolic Logic 14* (1949), 159–166.

[44] HENKIN, LEON. A Generalisation of the Concept of ω-Completeness. *Journal of Symbolic Logic 22* (1957), 1–14.

[45] HOARE, C. A. R. An Axiomatic Basis for Computer Programming. *Communications of the Association for Computing Machinery 12* (1969), 576–580, 583.

[46] HUGHES, G. E. Every World Can See a Reflexive World. *Studia Logica 49* (1990), 175–181.

[47] HUGHES, G. E., AND M. J. CRESSWELL. *An Introduction to Modal Logic.* Methuen, 1968.

[48] JÓNSSON, B., AND A. TARSKI. Boolean Algebras with Operators, Part I. *American Journal of Mathematics 73* (1951), 891–939.

[49] KOCK, A., AND G. B. WRAITH. *Elementary Toposes*, vol. 30 of *Aarhus Lecture Notes.* Aarhus, 1971.

[50] KOPPERMAN, R. *Model Theory and its Applications.* Allyn and Bacon, 1972.

[51] KRIPKE, S. A Completeness Theorem in Modal Logic. *Journal of Symbolic Logic 24* (1959), 1–14.

[52] KRIPKE, S. Semantic Analysis of Modal Logic I. *Zeitschrift für Mathematische Logik und Grundlagen der Mathematik 9* (1963), 67–96.

[53] KRIPKE, S. Review of "Algebraic Semantics for Modal Logics, Parts I and II". *Mathematical Reviews 34* (1967), 5660–5661.

[54] LACHLAN, A. H. A Note on Thomason's Refined Structures for Tense Logics. *Theoria 40* (1974), 114–120.

[55] LAWVERE, F. W. Quantifiers and Sheaves. *Actes des Congrès International des Mathématiques 1* (1970), 329–334.

[56] LAWVERE, F. W. *Toposes, Algebraic Geometry, and Logic*, vol. 274 of *Lecture Notes in Mathematics.* Springer-Verlag, 1972.

[57] LEMMON, E. J. Algebraic Semantics for Modal Logics I. *Journal of Symbolic Logic 31* (1966), 46–65.

[58] LEMMON, E. J. Algebraic Semantics for Modal Logics II. *Journal of Symbolic Logic 31* (1966), 191–218.

[59] LEMMON, E. J., AND D. SCOTT. *Intensional Logic.* Preliminary draft of initial chapters by E. J. Lemmon, Stanford University (finally published as *An Introduction to Modal Logic*, American Philosophical Quarterly Monograph Series, no. 11 (ed. by Krister Segerberg), Basil Blackwell, Oxford, 1977), July 1966.

[60] LÖB, M. H. Solution of a Problem of Leon Henkin. *The Journal of Symbolic Logic 20* (1955), 115–118.

[61] MAC LANE, SAUNDERS. Sets, Topoi, and Internal Logic in Categories. In *Logic Colloquium 1973*, ed. H. E. Rose and J. C. Shepherdson. North-Holland, Amsterdam, 1975, pp. 119–134.

[62] MACLAREN, M. D. Atomic Orthocomplemented Lattices. *Pacific Journal of Mathematics 14* (1964), 597–612.

[63] MACNAB, D. S. *An Algebraic Study of Modal Operators on Heyting Algebras, with Applications to Topology and Sheafification.* PhD thesis, Aberdeen, 1976.

[64] MAEDA, F., AND S. MAEDA. *Theory of Symmetric Lattices.* Springer-Verlag, Berlin, 1970.

[65] MAKINSON, D. C. A Generalisation of the Concept of Relational Model for Modal Logic. *Theoria 36* (1970), 331–335.

[66] MALCOLM, W. G. *Ultraproducts and Higher Order Models.* PhD thesis, Victoria University of Wellington, 1972.

[67] MCKINSEY, J. C. C. A Solution of the Decision Problem for the Lewis Systems S2 and S4 with an Application to Topology. *Journal of Symbolic Logic 6* (1941), 117–134.

[68] MCKINSEY, J. C. C., AND A. TARSKI. Some Theorems about the Sentential Calculi of Lewis and Heyting. *Journal of Symbolic Logic 13* (1948), 1–15.

[69] PAREIGIS, B. *Categories and Functors.* Academic Press, 1970.

[70] PARIKH, ROHIT. The Completeness of Propositional Dynamic Logic. In *Mathematical Foundations of Computer Science 1978*. Springer Verlag, 1978, pp. 403–415. Lecture Notes in Computer Science 64.

[71] PRATT, V. R. Semantical Considerations on Floyd-Hoare Logic. In *Proceedings of the 17th Annual IEEE Symposium on Foundations of Computer Science* (October 1976), pp. 109–121.

[72] PRIOR, A. *Past, Present, and Future*. Clarendon Press, Oxford, 1967.

[73] RANDALL, C. H., AND D. J. FOULIS. An Approach to Empirical Logic. *American Mathematic Monthly 77* (1970), 363–374.

[74] RASIOWA, H., AND R. SIKORSKI. *The Mathematics of Metamathematics*. PWN–Polish Scientific Publishers, Warsaw, 1963.

[75] ROBB, A. A. *A Theory of Time and Space*. Cambridge, 1914.

[76] ROBINSON, A. *Non-Standard Analysis*. North-Holland, 1966.

[77] ROBINSON, A. Germs. In *Applications of Model Theory to Algebra, Analysis, and Probability*, ed. W. A. Luxembourg. Holt, Rinehart, and Winston, 1969, pp. 138–149.

[78] SACKS, GERALD E. *Saturated Model Theory*. W. A. Bejamin, Reading, Massachusetts, 1972.

[79] SAHLQVIST, H. Completeness and Correspondence in First and Second Order Semantics for Modal Logic. In *Proceedings of the Third Scandinavian Logic Symposium*, ed. S. Kanger. North-Holland, 1975, pp. 110–143.

[80] SAMBIN, G., AND V. VACCARO. A New Proof of Sahlqvist's Theorem on Modal Definability and Completeness. *The Journal of Symbolic Logic 54* (1989), 992–999.

[81] SCOTT, DANA. Continuous Lattices. In *Toposes, Algebraic Geometry, and Logic*, ed. F.W. Lawvere, vol. 274. Springer-Verlag, 1972. Lecture Notes in Mathematics, vol. 274.

[82] SCOTT, DANA. Lambda Calculus and Recursion Theory. In *Proceedings of the Third Scandinavian Logic Symposium*, ed. Stig Kanger. North-Holland, 1975, pp. 154–193.

[83] SEGERBERG, K. Decidability of S4.1. *Theoria 34* (1968), 7–20.

[84] SEGERBERG, KRISTER. Propositional Logics Related to Heyting's and Johansson's. *Theoria 34* (1968), 28–61.

[85] SEGERBERG, KRISTER. Modal Logics with Linear Alternative Relations. *Theoria 36* (1970), 301–322.

[86] SEGERBERG, KRISTER. *An Essay in Classical Modal Logic*, vol. 13 of *Filosofiska Studier*. Uppsala Universitet, 1971.

[87] SEGERBERG, KRISTER. A Completeness Theorem in the Modal Logic of Programs. *Notices of the American Mathematical Society 24* (1977), A–552.

[88] SEGERBERG, KRISTER. A Completeness Theorem in the Modal Logic of Programs. In *Universal Algebra and Applications*, ed. T. Traczyck. PWN–Polish Scientific Publishers, Warsaw, 1982, pp. 31–46. Banach Center Publications, vol. 9.

[89] SHEHTMAN, V. B. Modal Logics of Domains on the Real Plane. *Studia Logica 42* (1983), 63–80.

[90] SIKORSKI, R. *Boolean Algebras*, second ed. Springer-Verlag, Berlin, 1964.

[91] SOLOVAY, R. Provability Interpretations of Modal Logic. *Israel Journal of Mathematics 25* (1976), 287–304.

[92] STONE, M. H. Topological Representation of Distributive Lattices and Brouwerian Logics. *Casopis pro Pestovani Matematiky a Fysiky* (1937), 1–25.

[93] TARSKI, A. Contributions to the Theory of Models III. *Nederlandse Aka. van Wetenschappen Proc. Ser. A 58* (1955), 56–64.

[94] THOMASON, R. H. Some Completeness Results for Modal Predicate Calculi. In *Philosophical Problems in Logic*, ed. K. Lambert. D. Reidel, 1970, pp. 56–76.

[95] THOMASON, S. K. Noncompactness in Propositional Modal Logic. *Journal of Symbolic Logic 37* (1972), 716–720.

[96] THOMASON, S. K. Semantic Analysis of Tense Logic. *Journal of Symbolic Logic 37* (1972), 150–158.

[97] THOMASON, S. K. An Incompleteness Theorem in Modal Logic. *Theoria 40* (1974), 30–34.

[98] THOMASON, S. K. Categories of Frames for Modal Logic. *Journal of Symbolic Logic 40* (1975), 439–442.

[99] VAN BENTHEM, J. A. F. K. A Note on Modal Formulas and Relational Properties. *Journal of Symbolic Logic 40* (1975), 55–58.

[100] VAN BENTHEM, J. A. F. K. *Modal Correspondence Theory*. PhD thesis, University of Amsterdam, 1976.

[101] VAN BENTHEM, J. A. F. K. Modal Formulas are Either Elementary or not $\Sigma\Delta$-Elementary. *Journal of Symbolic Logic 41* (1976), 436–438.

[102] VAN BENTHEM, J. A. F. K. Some Kinds of Modal Completeness. *Studia Logica 39* (1980), 125–141.

[103] VAN BENTHEM, J. A. F. K. *Modal Logic and Classical Logic.* Bibliopolis, Naples, 1983.

[104] VARADARAJAN, V. S. *Geometry of Quantum Theory*, vol. 1. Van Nostrand, Princeton, 1968.

[105] WAND, MITCHELL. A New Incompleteness Result for Hoare's System. *Journal of the Association for Computing Machinery 25* (1978), 168–175.

Index

CSLI Publications

Reports

The following titles have been published in the CSLI Reports series. These reports may be obtained from CSLI Publications, Ventura Hall, Stanford, CA 94305-4115.

An Architecture for Tuning Rules
and Cases Yoshio Nakatami and
David Israel CSLI–92–173

The Linguistic Information in
Dynamic Discourse Megumi
Kameyama CSLI–92–174

Epsilon Substitution Method for
Elementary Analysis Grigori Mints
and Sergei Tupailo CSLI–93–175

Information Spreading and Levels of
Representation in LFG Avery D.
Andrews and Christopher D. Manning
CSLI–93–176

1992 Annual Report CSLI–93–177

Lecture Notes

The titles in this series are distributed by
the University of Chicago Press and may
be purchased in academic or university
bookstores or ordered directly from the
distributor at 11030 South Langley
Avenue, Chicago, IL 60628 (USA) or by
phone 1-800-621-2736, (312)568-1550.

A Manual of Intensional Logic. Johan van
Benthem, second edition, revised and
expanded. Lecture Notes No. 1.
0-937073-29-6 (paper), 0-937073-30-X
(cloth)

Emotion and Focus. Helen Fay
Nissenbaum. Lecture Notes No. 2.
0-937073-20-2 (paper)

*Lectures on Contemporary Syntactic
Theories.* Peter Sells. Lecture Notes
No. 3. 0-937073-14-8 (paper),
0-937073-13-X (cloth)

*An Introduction to Unification-Based
Approaches to Grammar.* Stuart M.
Shieber. Lecture Notes No. 4.
0-937073-00-8 (paper), 0-937073-01-6
(cloth)

The Semantics of Destructive Lisp. Ian
A. Mason. Lecture Notes No. 5.
0-937073-06-7 (paper), 0-937073-05-9
(cloth)

An Essay on Facts. Ken Olson. Lecture
Notes No. 6. 0-937073-08-3 (paper),
0-937073-05-9 (cloth)

Logics of Time and Computation. Robert
Goldblatt, second edition, revised and
expanded. Lecture Notes No. 7.
0-937073-94-6 (paper), 0-937073-93-8
(cloth)

*Word Order and Constituent Structure in
German.* Hans Uszkoreit. Lecture
Notes No. 8. 0-937073-10-5 (paper),
0-937073-09-1 (cloth)

*Color and Color Perception: A Study in
Anthropocentric Realism.* David
Russel Hilbert. Lecture Notes No. 9.
0-937073-16-4 (paper), 0-937073-15-6
(cloth)

Prolog and Natural-Language Analysis.
Fernando C. N. Pereira and Stuart M.
Shieber. Lecture Notes No. 10.
0-937073-18-0 (paper), 0-937073-17-2
(cloth)

*Working Papers in Grammatical Theory
and Discourse Structure: Interactions
of Morphology, Syntax, and Discourse.*
M. Iida, S. Wechsler, and D. Zec
(Eds.) with an Introduction by Joan
Bresnan. Lecture Notes No. 11.
0-937073-04-0 (paper), 0-937073-25-3
(cloth)

*Natural Language Processing in the 1980s:
A Bibliography.* Gerald Gazdar, Alex
Franz, Karen Osborne, and Roger
Evans. Lecture Notes No. 12.
0-937073-28-8 (paper), 0-937073-26-1
(cloth)

Information-Based Syntax and Semantics.
Carl Pollard and Ivan Sag. Lecture
Notes No. 13. 0-937073-24-5 (paper),
0-937073-23-7 (cloth)

Non-Well-Founded Sets. Peter Aczel.
Lecture Notes No. 14. 0-937073-22-9
(paper), 0-937073-21-0 (cloth)

Partiality, Truth and Persistence. Tore
Langholm. Lecture Notes No. 15.
0-937073-34-2 (paper), 0-937073-35-0
(cloth)

*Attribute-Value Logic and the Theory of
Grammar.* Mark Johnson. Lecture
Notes No. 16. 0-937073-36-9 (paper),
0-937073-37-7 (cloth)

The Situation in Logic. Jon Barwise.
Lecture Notes No. 17. 0-937073-32-6
(paper), 0-937073-33-4 (cloth)

The Linguistics of Punctuation. Geoff
Nunberg. Lecture Notes No. 18.
0-937073-46-6 (paper), 0-937073-47-4
(cloth)

Anaphora and Quantification in Situation Semantics. Jean Mark Gawron and Stanley Peters. Lecture Notes No. 19. 0-937073-48-4 (paper), 0-937073-49-0 (cloth)

Propositional Attitudes: The Role of Content in Logic, Language, and Mind. C. Anthony Anderson and Joseph Owens. Lecture Notes No. 20. 0-937073-50-4 (paper), 0-937073-51-2 (cloth)

Literature and Cognition. Jerry R. Hobbs. Lecture Notes No. 21. 0-937073-52-0 (paper), 0-937073-53-9 (cloth)

Situation Theory and Its Applications, Vol. 1. Robin Cooper, Kuniaki Mukai, and John Perry (Eds.). Lecture Notes No. 22. 0-937073-54-7 (paper), 0-937073-55-5 (cloth)

The Language of First-Order Logic (including the Macintosh program, Tarski's World). Jon Barwise and John Etchemendy, second edition, revised and expanded. Lecture Notes No. 23. 0-937073-74-1 (paper)

Lexical Matters. Ivan A. Sag and Anna Szabolcsi, editors. Lecture Notes No. 24. 0-937073-66-0 (paper), 0-937073-65-2 (cloth)

Tarski's World. Jon Barwise and John Etchemendy. Lecture Notes No. 25. 0-937073-67-9 (paper)

Situation Theory and Its Applications, Vol. 2. Jon Barwise, J. Mark Gawron, Gordon Plotkin, Syun Tutiya, editors. Lecture Notes No. 26. 0-937073-70-9 (paper), 0-937073-71-7 (cloth)

Literate Programming. Donald E. Knuth. Lecture Notes No. 27. 0-937073-80-6 (paper), 0-937073-81-4 (cloth)

Normalization, Cut-Elimination and the Theory of Proofs. A. M. Ungar. Lecture Notes No. 28. 0-937073-82-2 (paper), 0-937073-83-0 (cloth)

Lectures on Linear Logic. A. S. Troelstra. Lecture Notes No. 29. 0-937073-77-6 (paper), 0-937073-78-4 (cloth)

A Short Introduction to Modal Logic. Grigori Mints. Lecture Notes No. 30. 0-937073-75-X (paper), 0-937073-76-8 (cloth)

Linguistic Individuals. Almerindo E. Ojeda. Lecture Notes No. 31. 0-937073-84-9 (paper), 0-937073-85-7 (cloth)

Computer Models of American Speech. M. Margaret Withgott and Francine R. Chen. Lecture Notes No. 32. 0-937073-98-9 (paper), 0-937073-97-0 (cloth)

Verbmobil: A Translation System for Face-to-Face Dialog. Martin Kay, Mark Gawron, and Peter Norvig. Lecture Notes No. 33. 0-937073-95-4 (paper), 0-937073-96-2 (cloth)

The Language of First-Order Logic (including the Windows program, Tarski's World). Jon Barwise and John Etchemendy, third edition, revised and expanded. Lecture Notes No. 34. 0-937073-90-3 (paper)

Turing's World. Jon Barwise and John Etchemendy. Lecture Notes No. 35. 1-881526-10-0 (paper)

Syntactic Constraints on Anaphoric Binding. Mary Dalrymple. Lecture Notes No. 36. 1-881526-06-2 (paper), 1-881526-07-0 (cloth)

Situation Theory and Its Applications, Vol. 3. Peter Aczel, David Israel, Yasuhiro Katagiri, and Stanley Peters, editors. Lecture Notes No. 37. 1-881526-08-9 (paper), 1-881526-09-7 (cloth)

Theoretical Aspects of Bantu Grammar. Mchombo, editor. Lecture Notes No. 38. 0-937073-72-5 (paper), 0-937073-73-3 (cloth)

Logic and Representation. Robert C. Moore. Lecture Notes No. 39. 1-881526-15-1 (paper), 1-881526-16-X (cloth)

Meanings of Words and Contextual Determination of Interpretation. Paul Kay. Lecture Notes No. 40. 1-881526-17-8 (paper), 1-881526-18-6 (cloth)

Language and Learning for Robots. Colleen Crangle and Patrick Suppes. Lecture Notes No. 41. 1-881526-19-4 (paper), 1-881526-20-8 (cloth)

Hyperproof. Jon Barwise and John Etchemendy. Lecture Notes No. 42. 1-881526-11-9 (paper)

Mathematics of Modality. Robert Goldblatt. Lecture Notes No. 43. 1-881526-23-2 (paper), 1-881526-24-0 (cloth)

Feature Logics, Infinitary Descriptions, and Grammar. Bill Keller. Lecture Notes No. 44. 1-881526-25-9 (paper), 1-881526-26-7 (cloth)

Other CSLI Titles Distributed by UCP

Agreement in Natural Language: Approaches, Theories, Descriptions. Michael Barlow and Charles A. Ferguson, editors. 0-937073-02-4 (cloth)

Papers from the Second International Workshop on Japanese Syntax. William J. Poser, editor. 0-937073-38-5 (paper), 0-937073-39-3 (cloth)

The Proceedings of the Seventh West Coast Conference on Formal Linguistics (WCCFL 7). 0-937073-40-7 (paper)

The Proceedings of the Eighth West Coast Conference on Formal Linguistics (WCCFL 8). 0-937073-45-8 (paper)

The Phonology-Syntax Connection. Sharon Inkelas and Draga Zec (Eds.) (co-published with The University of Chicago Press). 0-226-38100-5 (paper), 0-226-38101-3 (cloth)

The Proceedings of the Ninth West Coast Conference on Formal Linguistics (WCCFL 9). 0-937073-64-4 (paper)

Japanese/Korean Linguistics. Hajime Hoji, editor. 0-937073-57-1 (paper), 0-937073-56-3 (cloth)

Experiencer Subjects in South Asian Languages. Manindra K. Verma and K. P. Mohanan, editors. 0-937073-60-1 (paper), 0-937073-61-X (cloth)

Grammatical Relations: A Cross-Theoretical Perspective. Katarzyna Dziwirek, Patrick Farrell, Errapel Mejías Bikandi, editors. 0-937073-63-6 (paper), 0-937073-62-8 (cloth)

The Proceedings of the Tenth West Coast Conference on Formal Linguistics (WCCFL 10). 0-937073-79-2 (paper)

On What We Know We Don't Know. Sylvain Bromberger. 0-226-075400 (paper), (cloth)

The Proceedings of the Twenty-fourth Annual Child Language Research Forum. Eve V. Clark, editor. 1-881526-05-4 (paper), 1-881526-04-6 (cloth)

Japanese/Korean Linguistics, Vol. 2. Patricia M. Clancy, editor. 1-881526-13-5 (paper), 1-881526-14-3 (cloth)

Arenas of Language Use. Herbert H. Clark. 0-226-10782-5 (paper), (cloth)

Japanese/Korean Linguistics, Vol. 3. Sonja Choi, editor. 1-881526-21-6 (paper), 1-881526-22-4 (cloth)

The Proceedings of the Eleventh West Coast Conference on Formal Linguistics (WCCFL 11). 1-881526-12-7 (paper)

Books Distributed by CSLI

The Proceedings of the Third West Coast Conference on Formal Linguistics (WCCFL 3). 0-937073-44-X (paper)

The Proceedings of the Fourth West Coast Conference on Formal Linguistics (WCCFL 4). 0-937073-43-1 (paper)

The Proceedings of the Fifth West Coast Conference on Formal Linguistics (WCCFL 5). 0-937073-42-3 (paper)

The Proceedings of the Sixth West Coast Conference on Formal Linguistics (WCCFL 6). 0-937073-31-8 (paper)

Hausar Yau Da Kullum: Intermediate and Advanced Lessons in Hausa Language and Culture. William R. Leben, Ahmadu Bello Zaria, Shekarau B. Maikafi, and Lawan Danladi Yalwa. 0-937073-68-7 (paper)

Hausar Yau Da Kullum Workbook. William R. Leben, Ahmadu Bello Zaria, Shekarau B. Maikafi, and Lawan Danladi Yalwa. 0-93703-69-5 (paper)

Ordering Titles Distributed by CSLI

Titles distributed by CSLI may be ordered directly from CSLI Publications, Ventura Hall, Stanford, CA 94305-4115 or by phone (415)723-1712, (415)723-1839. Orders can also be placed by FAX (415)723-0758 or e-mail (pubs@csli.stanford.edu).

All orders must be prepaid by check or Visa or MasterCard (include card name, number, and expiration date). California residents add 8.25% sales tax. For shipping and handling, add $2.50 for first book and $0.75 for each additional book; $1.75 for first report and $0.25 for each additional report.

For overseas shipping, add $4.50 for first book and $2.25 for each additional book; $2.25 for first report and $0.75 for each additional report. All payments must be made in U.S. currency.

Overseas Orders

The University of Chicago Press has offices worldwide which serve the international community.

Canada: David Simpson, 164 Hillsdale Avenue East, Toronto, Ontario M4S 1T5, Canada. Telephone: (416) 484-8296.

Mexico, Central America, South America, and the Caribbean (including Puerto Rico): EDIREP, 5500 Ridge Oak Drive, Austin, Texas 78731 U. S. A. Telephone: (512)451-4464.

United Kingdom, Europe, Middle East, and Africa (except South Africa): International Book Distributors, Ltd., 66 Wood Lane End, Hemel Hempstead HP4 3RG, England. Telephone: 0442 231555. FAX: 0442 55618.

Australia, New Zealand, South Pacific, Eastern Europe, South Africa, and India: The University of Chicago Press, Foreign Sales Manager, 5801 South Ellis Avenue, Chicago, Illinois 60637 U.S.A. Telephone: (312)702-0289. FAX: (312)702-9756.

Japan: Libraries and individuals should place their orders with local booksellers. Booksellers should place orders with our agent: United Publishers Services, Ltd., Kenkyu-sha Building, 9 Kanda Surugadai 2-chome, Chiyoda-ku, Tokyo, Japan. Telephone: (03)291-4541.

China (PRC), Hong Kong, and Southeast Asia: Peter Ho Hing Leung, The America University Press Group, P.O. Box 24279, Aberdeen Post Office, Hong Kong.

Korea and Taiwan (ROC): The American University Press Group, 3-21-18-206 Higashi-Shinagawa, Shinagawa-ku, Tokyo, 140 Japan. Telephone: (813)450-2857. FAX: (813)472-9706.